192
Advances in Biochemical Engineering/Biotechnology

Aims and Scope

This book series reviews current trends in modern biotechnology and biochemical engineering. Its aim is to cover all aspects of these interdisciplinary disciplines, where knowledge, methods and expertise are required from chemistry, biochemistry, microbiology, molecular biology, chemical engineering and computer science. This series supports and advances the UN's sustainable development goals, mainly SDG7 (Affordable and Clean Energy), SDG3 (Good Health and Well Being), and SDG13 (Climate Change).

Volumes are organized topically and provide a comprehensive discussion of developments in the field over the past 3–5 years. The series also discusses new discoveries and applications. Special volumes are dedicated to selected topics which focus on new biotechnological products and new processes for their synthesis and purification.

In general, volumes are edited by well-known guest editors. The series editor and publisher will, however, always be pleased to receive suggestions and supplementary information. Manuscripts are accepted in English.

In references, Advances in Biochemical Engineering/Biotechnology is abbreviated as *Adv. Biochem. Engin./Biotechnol.* and cited as a journal.

Dirk Holtmann
Editor

Unconventional Organisms in Biotechnology

With contributions by

T. Baur · B. Blombach · P. Bolay · B. Bühler · D. B. Diaz · D. Casey · D. Cholewa · L. Diaz-Garcia · A. Diepold · J. Gescher · A. Grünberger · M. Hädrich · J. Hoff · D. Holtmann · L. Kneuer · J. E. Kufs · S. Lang · B. Molitor · L. Mühling · K. Rabe · L. Regestein · C. Reimer · E. Schepp · C. Schulze · K. E. Schulz · A. Sivanesapillai · J. Stiefelmaier · C. Stolle · K. L. Tee · J. Toepel · R. Ulber · T. S. Wong · T. Wormuth · R. Wurst · M. Yu

Springer

Editor
Dirk Holtmann
Institute of Bio- and Food Technology
Karlsruhe Institute of Technology
Karlsruhe, Germany

ISSN 0724-6145 ISSN 1616-8542 (electronic)
Advances in Biochemical Engineering/Biotechnology
ISBN 978-3-032-08736-2 ISBN 978-3-032-08737-9 (eBook)
https://doi.org/10.1007/978-3-032-08737-9

This Springer imprint is published by the registered company Springer Nature Switzerland AG
The registered company address is: Gewerbestrasse 11, 6330 Cham, Switzerland

Preface

Biotechnological production processes currently rely on a small set of well-characterised microorganisms, such as *Escherichia coli*, *Corynebacterium glutamicum* and *Saccharomyces cerevisiae*, for the production of enzymes, pharmaceuticals, fuels and chemicals. While these model organisms have enabled significant progress, they represent only a tiny fraction of Earth's microbial diversity. In fact, the vast majority of microorganisms remain uncultured, unexplored and unused for industrial purposes.

The book *Unconventional Organisms in Biotechnology* is based on the idea of broadening the scope of bioprocess engineering by unlocking the potential of these microorganisms. From extremophiles that can survive in hot springs to anaerobic fungi that break down lignocellulosic biomass in the guts of ruminants, nature provides a plethora of biological solutions for a sustainable bioeconomy. Exploring this hidden diversity could lead to processes based on novel substrates and metabolic pathways, as well as more robust processes under extreme conditions. Ultimately, this would reduce our reliance on fossil resources and help us to achieve more sustainable processes.

The following chapters introduce promising alternative organisms and cultivation strategies, highlight recent breakthroughs in genetic engineering and bioprocess design and discuss the challenges that must be overcome to implement these new systems in technical applications. Ultimately, the book aims to inspire researchers, engineers and policymakers to look beyond traditional model species and embrace biodiversity as a driver for innovation.

Advancing the bioeconomy requires interdisciplinary collaboration, bringing together microbiologists, biotechnologists, process engineers and environmental scientists. Only by combining fundamental understanding with engineering expertise can we develop scalable, efficient and sustainable processes. The aim of this book is to contribute to this effort and promote greater awareness of the largely unexplored biological diversity.

Finally, I would like to thank all the authors and the publisher for their excellent cooperation.

Karlsruhe, Germany Dirk Holtmann

Contents

Adv Biochem Eng Biotechnol (2025) 192: 1–26
https://doi.org/10.1007/10_2025_288

Published online: 9 September 2025

Exploitation of Biodiversity in Bioeconomy: Examples, Opportunities, and Challenges

Alexander Grünberger, Emily Schepp, Selina Lang, Kevin Edward Schulz, Daniel Baron Diaz, Arabi Sivanesapillai, Andreas Diepold, Camilla Stolle, Kersten Rabe, Roland Ulber, and Dirk Holtmann

Contents

A. Grünberger
Institute of Process Engineering in Life Sciences, Microsystems in Bioprocess Engineering, Karlsruhe Institute of Technology, Karlsruhe, Germany

E. Schepp, S. Lang, K. E. Schulz, D. B. Diaz, A. Sivanesapillai, and D. Holtmann (✉)
Institute of Process Engineering in Life Sciences, Electrobiotechnology, Karlsruhe Institute of Technology, Karlsruhe, Germany
e-mail: dirk.holtmann@kit.edu

A. Diepold
Institute of Applied Biosciences, Applied Biology, Karlsruhe Institute of Technology, Karlsruhe, Germany

C. Stolle and K. Rabe
Institute for Biological Interfaces 1, Biomolecular Micro- and Nanostructures, Karlsruhe Institute of Technology, Karlsruhe, Germany

R. Ulber
Institute of Bioprocess Engineering, University of Kaiserslautern-Landau, Kaiserslautern, Germany

Abstract While bioprocesses using *Escherichia coli*, *Corynebacterium glutamicum*, various species of *Bacillus*, lactic acid bacteria, *Clostridia*, the yeasts *Saccharomyces cerevisiae* and *Pichia pastoris*, fungi such as *Aspergillus niger*, and Chinese hamster ovary cells are well established, the high level of microbial diversity has not yet been exploited industrially. However, the use of alternative organisms has the potential to significantly expand the process window of bioprocesses. These extensions include the use of alternative substrates (e.g., CO_2, syngas, or methane), unusual process conditions (e.g., temperatures, media conditions), or use of microbial diversity itself (e.g., biofilm growth, alternative electron transport systems, product secretion, surface presentation of enzymes). Bringing these hidden champions into application will enable more efficient processes with environmental and ecological benefits. Some of the most promising alternative microorganisms are described and discussed in detail in this book. The aim of this introductory chapter is to provide an overview of additional organisms and cultivation strategies. Finally, it addresses future challenges to bring the organism into technical applications.

Graphical Abstract

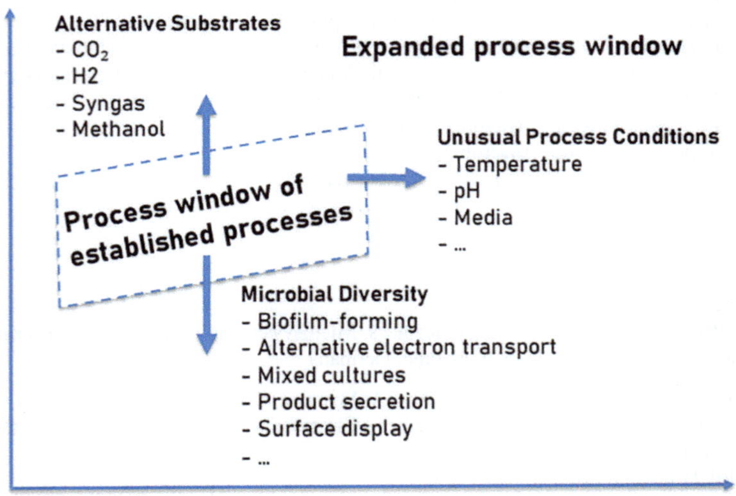

Keywords Alternative organisms, Alternative substrates, Biofilms, C1-substrates, Electro-biotechnology, Process windows, Unusual process conditions

1 Introduction

Microbes are the most diverse and abundant group of organisms on Earth, accounting for up to 60% of the total biomass. On a global scale, soils and oceans are composed of approximately 4–5 × 10^{30} and 3.6 × 10^{29} microbial cells, respectively [1]. It is estimated that the Earth is home to more than 1 trillion (10^{12}) different species of microbes [2]. The abundance and diversity of microbes lead to some amazing observations: for example, there are 100 million times as many bacteria in the oceans as there are stars in the known universe, the number of microorganisms in a single teaspoon of soil (10^9) is the same as the number of people currently living in Africa, and dental plaque is so densely packed that one gram contains about 10^{11} bacteria, roughly equal to the total number of people who have ever lived [3]. The exact figures of overall biomass and different microbial species are not crucial, but it is clear that we are often unable to describe the functions of even those microbes we already know from this vast number of microbes, most of which we do not even know about. Many of the organisms that are known cannot be cultivated using the methods available today. Environmental microbiologists estimate that less than 2% of bacterial species are suitable for cultivation under standard laboratory conditions [4]. For instance, most of the microorganisms in the colonic microflora are thought to be unculturable in the standard systems [4]. Therefore, on the basis of numerical evidence alone, it is highly likely that unculturable and therefore poorly understood organisms play a role in various human diseases. A well-documented example is syphilis, a disease caused by the bacterium *Treponema pallidum. Treponema pallidum* was identified as the cause of syphilis over 120 years ago [5]. It continued to resist in vitro cultivation for decades until long-term in vitro cultivation of *T. pallidum* was first achieved in 2017 [6]. A deeper understanding of the biology of *T. pallidum*, in particular the elucidation of its microaerophilic nature, and its improved survival in the presence of mammalian cells, has led to breakthroughs in culturing these cells [7].

The fields of medical or environmental microbiology and biotechnological applications share common ground, but differ significantly in their focus and requirements. Microbiology studies microorganisms to understand their structure, function, and ecological role. Biotechnology, on the other hand, focuses on the exploitation of microorganisms and their biological processes for technical applications. While microbiologists aim to generate knowledge about microorganisms, biotechnologists apply this knowledge to solve practical problems and create innovative solutions. Together, these fields are driving forces in the advancement in science, industry, and health, demonstrating the immense potential of microorganisms to address global challenges. Biotechnologists require microorganisms to have specific properties to ensure their suitability for industrial and technical applications. First and foremost, microorganisms must have high metabolic efficiency to convert raw materials into desired products, such as enzymes, biofuels, or pharmaceuticals, with minimal waste. Genetic stability is another critical factor, ensuring consistent performance over long production cycles without genetic drift or loss of desired traits. In addition, microorganisms should be adaptable to industrial conditions. Ease of genetic modification is also a key requirement. Microorganisms must be responsive to genetic engineering techniques, allowing scientists to enhance or introduce metabolic pathways for optimized production. Scalability is also important, as microorganisms need to grow efficiently in large bioreactors while maintaining product yield and quality. Finally, safety and environmental sustainability play an important role: Microorganisms used in biotechnology should be safe for humans or ecosystems and should be manageable under controlled conditions. Established organisms in industrial biotechnology such as *Escherichia coli*, *Corynebacterium glutamicum*, *Bacillus subtilis*, *Clostridium acetobutylicum*, *Lactobacillus delbrueckii* ssp. *bulgaricus*, *Saccharomyces cerevisiae*, *Pichia pastoris*, *Aspergillus niger*, and *Penicillium chrysogenum* as well as Chinese hamster ovary cells fulfill these requirements (in some cases with slight restrictions). In most cases, these systems result in relatively similar process designs: aerobic or anaerobic processes in stirred tanks or bubble columns as reactors, using carbohydrates as energy and carbon sources in liquid cultures.

Biotechnology, as part of the bioeconomy, plays a central role in addressing some of society's most pressing challenges, including climate change, resource scarcity, energy supply, and food security. By using renewable biological resources such as microbes and renewable raw materials, bioeconomy offers sustainable alternatives to fossil-based industries. For example, bioplastics made from plant materials can reduce dependence on oil, thereby reducing greenhouse gas emissions and minimizing plastic pollution [8]. In addition, biofuels produced from algae or waste biomass provide cleaner energy solutions, supporting the transition to a low-carbon economy [9]. New substrates are increasingly the focus of industrial biotechnology to avoid competition with carbohydrates that can be used in food and feed, e.g., CO_2, formate, acetate, syngas, methanol, and also waste streams [10].

To achieve this goal, the potential of microbial biodiversity must be further exploited, by exploring organisms that have been rarely or never used in biotechnology. The industrial application of unconventional biological production systems is gaining traction due to their potential for sustainable and efficient production of biofuels, chemicals, and other valuable products. Examples include extremophiles, gas-utilizing organisms, photo- and electro-phototrophic organisms, new class of cable bacteria, higher eukaryotic cells (e.g., amoeba, mixotrophic algae), innovative enzyme cascades, and mixed-culture biofilms. While these systems offer unique advantages, they also present specific challenges. In this book, some of these microbes, their unique metabolisms, and specific challenges associated with their use are described in detail:

- *Methanothermobacter thermautotrophicus* – an industrially relevant methanogen, which converts CO_2 into methane [11]
- *Cupriavidus necator* – an organism able to convert a wide range of substrates, including carbohydrates, waste streams, and autotrophically CO_2 in combination with H_2 and O_2 [12]
- *Vibrio natriegens* – an extremely fast-growing halotolerant bacterium [13]

- *Shewanella oneidensis* – a metal-reducing bacterium that uses solid metals or electrodes as electron acceptors [14]
- *Synechocystis* and *Synechococcus* strains – phototropic cyanobacteria with biotechnological applications [15]
- Cable bacteria and their biotechnological application [16]
- *Dictyostelium discoideum* – an amoeba for the bioproduction of complex molecules [17]
- *Euglena gracilis* – an algae that grows not only as photoautotroph, but also as mixotroph and heterotroph

Many other organisms meet the general requirements but have not been widely used in biotechnology and are increasingly attracting interest. Some of these organisms are briefly described in the following section.

2 Examples of Metabolic Diversity and Biotechnological Processes

2.1 Acetobacterium woodii*: An Acetogenic Model Organism*

Acetobacterium woodii is a strictly anaerobic, Gram-positive, nonspore-forming bacterium belonging to the family *Eubacteriaceae*, which grows optimally at temperatures around 30°C [18]. As a model organism among acetogenic bacteria, its genome has been fully sequenced, and its energy conservation pathway is well understood [19]. A defining characteristic of *A. woodii* is its acetyl-CoA pathway (Wood-Ljungdahl pathway), which enables the conversion of CO_2 and H_2 into the platform chemical acetate. While the Wood-Ljungdahl pathway, likely the oldest CO_2-fixing pathway in nature, has been studied for decades, the exact mechanism of energy conservation in *A. woodii* during growth on $H_2 + CO_2$ has been much later revealed [20–22]. The Wood-Ljungdahl pathway itself does not conserve energy but functions to re-oxidize electron carriers and supply acetyl-CoA for biomass synthesis [23]. For energy conservation, *A. woodii* employs a sodium ion-translocating ferredoxin NAD^+ oxidoreductase (Rnf) instead of a proton motive force, operating without cytochromes and quinones [24]. The Rnf complex consists of six subunits containing flavins and iron-sulfur centers [25]. The electron transfer from reduced ferredoxin to NAD^+ is exergonic, allowing the Rnf complex to transport sodium ions from the cytoplasm. The resulting sodium ion gradient drives ATP synthesis via a membrane-integral Na^+-F_1F_0 ATP synthase [21, 25]. This modular separation of energy conservation and carbon fixation presents interesting possibilities for biotechnological applications. Theoretically, any reductive pathway could be coupled to this energy-conserving module, and indeed, acetogens have been found to reduce nitrate [26, 27], phenylacrylates [28], fumarate [29], and even acetyl-CoA to ethanol [30] or butyrate [31]. *A. woodii* has demonstrated great potential for biotechnological applications, particularly for acetate production during gas fermentations. Studies have shown that volumetric productivity in batch processes with continuous $H_2 + CO_2$ gas supply increases with rising hydrogen partial pressure [32]. Under controlled pH, acetate concentrations of up to 44 g L^{-1} were achieved after 11 days, with volumetric productivities of 7.4 g L^{-1} day^{-1} [32]. In another study, a submerged membrane bioreactor was established for continuous gas fermentation, reaching a volumetric productivity of 148 g L^{-1} day^{-1} at the highest dilution rate with full cell retention [33]. Further metabolic engineering efforts have boosted acetate production to over 50 g L^{-1} in less than 4 days under autotrophic conditions [34]. Moreover, genetically modified *A. woodii* strains also produce solvents such as isopropanol (14.5 mM) and acetone (7.6 mM), highlighting their potential for sustainable C1-based chemical synthesis [35]. Beyond gas fermentation, *A. woodii* has emerged as a promising host for bioproduction from formate, a cheap and renewable one-carbon substrate. Continuous cultivations on formate alone exhibited high specific substrate uptake rates (47 mmol g^{-1} h^{-1}), while co-utilization with H_2, CO, CO_2, or fructose proceeded without catabolite repression and resulted in acetate as the sole metabolic product. These insights emphasize the potential of *A. woodii* for flexible and simultaneous bioconversion of multiple substrates, supporting the development of efficient, sustainable bioprocesses for bulk chemicals and fuels [36].

2.2 **Thermoanaerobacter kivui***: A Thermophilic Acetogen*

Thermoanaerobacter kivui is a thermophilic, strictly anaerobic bacterium that utilizes the Wood-Ljungdahl pathway, to produce acetate from H_2 and CO_2. It thrives at temperatures around 66°C and has a temperature tolerance of 55–70°C [37]. Besides $H_2 + CO_2$, *T. kivui* has been described to only use pyruvate, formate, and the sugars glucose, fructose, and mannose to mainly produce acetate [37]. It can be adapted to grow on CO, broadening its potential applications, especially in converting industrial waste gases (syngas) [38]. It was shown that the organism needs sufficient time to adapt to the CO metabolism; however, it was ultimately able to grow within a 100% CO atmosphere. With a comparably short doubling time of approximately 2 h when growing on H_2 and CO_2, and a minimal vitamin requirement [37], which is a great advantage for biotechnological applications, *T. kivui* emerges as a strong candidate for bioengineering approaches in syngas fermentation. Genome analysis has revealed key differences in energy conservation compared to mesophilic acetogens like *A. woodii* and *Clostridium ljungdahlii*. While *T. kivui* contains genes for the Wood-Ljungdahl pathway, it lacks genes encoding the Rnf complex. Instead, energy conservation relies on one of the two energy-conserving hydrogenase (Ech) complexes [39]. Unlike many acetogens, *T. kivui* utilizes protons for chemiosmotic ATP synthesis rather than Na^+, employing an Ech to catalyze the oxidation of Fd_{red} coupled to the buildup of a proton gradient, which subsequently drives ATP production through an H^+-dependent F_1F_0 ATP synthase [39].

Recently, a novel group of enzymes that catalyze the direct hydrogenation of CO_2 to formic acid at high rates, the hydrogen-dependent CO_2 reductases (HDCRs), was discovered in the acetogenic bacteria *A. woodii* and *T. kivui* [40]. Cell suspensions reached specific formate production rates of 234 mmol $g_{protein}^{-1}$ h^{-1} (152 mmol g_{CDW}^{-1} h^{-1}), while the volumetric formate production rate was 270 mmol L^{-1} h^{-1} [41]. Therefore, *T. kivui* can be regarded as a promising candidate for hydrogen storage, CO_2 capture, and syngas conversion to formate. An alternative way to supply *A. kivui* with electron donors beyond gaseous H_2 for CO_2 reduction is their integration into a bioelectrochemical system. In this setup, also known as microbial electrosynthesis (MES), acetogens grow in the cathode compartment, where they utilize electrons either directly or indirectly through in situ H_2 production. A recent study reported high concentrations of acetic acid of 29 g L^{-1} in a continuous bioelectrochemical system [42]. Furthermore, the existing genetic system for *T. kivui* could be leveraged to expand its product spectrum beyond acetate, enabling the production of compounds such as ethanol through CO_2 reduction [43].

2.3 Methylorubrum extorquens: *A Methylotrophic Production Host*

Methylorubrum extorquens is a Gram-negative, strictly aerobic methylotroph meaning that it uses single-carbon compounds such as methanol enabling cellular growth [44, 45]. To achieve this, carbon is assimilated via the serine cycle. Nevertheless, the organism is not solely limited to C1 compounds but is also able to grow on a diverse range of multicarbon substrates. In addition to methanol and formate, it is capable of growth on ethanol, acetate, succinate, ethylamine, and pyruvate [46, 47]. A unique feature of the organism is its distinctive color. Wild-type strains of *M. extorquens* produce pinkish carotenoids and an abundance of cytochrome c that are responsible for colorizing the cells scarlet red. High-density cultivations were achieved with a cell concentration of 233 g L^{-1} under controlled carbon-nitrogen ratio [48]. *M. extorquens* forms intracellularly poly-3-hydroxybutyrate (PHB), which has been exploited for decades [44]. PHB yields of the wild type reach 30% with the potential to optimize the production up to 70% of the dry weight [49]. Whole-genome sequencing of *M. extorquens* AM1 in 2009 paved the way for more profound investigations with the organism [50]. The ability to naturally produce polyhydroxybutyrate (PHB [49]) paves the way for an environmentally friendly production of biopolymers. The involved genes and their partitioning in the PHB cycle in *M. extorquens* were studied to maximize plastic yields [46, 51]. To improve the mechanical properties of the rather stiff PHB, a copolymerization into poly (3-hydroxybutyrate-*co*-3-hydroxyvalerate) (PHBV) was investigated. In the copolymer, over 70% hydroxyvalerate could be achieved by the addition of short-chain fatty acids with yields of 14% polyhydroxyalkanoates (PHAs) relative to cell dry weight [52]. Besides being used as a production host for enzymes and high-value acids, the organism itself is also used as single-cell protein resource for aquafeeds. It was shown that up to 10% of soybean meal could be replaced in a rainbow trout diet without any negative effects on the growth and health of the fish [53]. In various biotechnological applications, *M. extorquens* has already served as a production organism. Exemplarily, itaconic acid was produced by a genetically modified strain with a titer of 31.6 mg L^{-1} [47]. Further engineered strains are producing dicarboxylic acids [54], terpenes [55], terpene-derived aroma components [56], or vitamins [57]. Furthermore, polymer precursors can be synthesized by coupling electrochemical CO_2 reduction to formate with a genetically modified *M. extorquens* strain [58].

Genetic modifications of the organism are distinctly facilitated by the growing molecular toolbox. Genome deletions in *M. extorquens* have been described for the *cre-lox* system [59, 60] as well as with an unmarked allelic exchange with the suicide vector pCM433 [47, 61]. Inducible expression systems using cumate and tetracycline [62], investigations of the impact of using transposons in the genome [60], and operon-based brick vectors and promoters [63] are only some of the molecular tools newly developed in *M. extorquens*.

2.4 Parageobacillus thermoglucosidasius*: A Versatile Thermophilic Carboxydotroph*

Parageobacillus thermoglucosidasius, formerly known as *Geobacillus thermoglucosidasius*, an organism belonging to the *Bacillaceae* family, is a thermophilic, facultative anaerobic, and Gram-positive bacterium [64, 65]. First isolated from a soil sample in 1983, it thrives optimally at temperatures between 61 and 63°C [64]. Nonetheless, members of the *Geobacillus* genus have been identified in vastly diverse environments, from high temperature oil fields [66] to the height of the Andes mountains [67]. This remarkable ubiquity despite the strictly thermophilic nature of the genus has been studied in the past, and it is likely linked to their spore-forming capacity [68]. A notable characteristic of distinct strains of *P. thermoglucosidasius* is the ability to utilize CO as energy source, facilitated by the Ni-Fe group carbon monoxide dehydrogenase/energy-converting hydrogenase (CODH/ECH) enzyme complex [69, 70]. Unlike other biological hydrogenogenic processes, such as dark fermentation, which are often restricted to obligate anaerobes [71], *P. thermoglucosidasius* is able to grow both with or without oxygen. Additionally, it can degrade complex substrates such as hemicellulose [72, 73] and metabolize both pentose (C5) and hexose (C6) [74, 75] sugars to produce ethanol, acetate, glyoxylate, lactate, succinate, and formate [76–78]. Remarkably, *P. thermoglucosidasius* possesses a relaxed catabolite repression system [74], which is beneficial for complex substrate degradation [79]. The genome of several *P. thermoglucosidasius* strains has been sequenced, and there are numerous tools available to modify the genome for tailored applications [80–83].

One of the most relevant potential uses of *P. thermoglucosidasius* is lignocellulose degradation [79]. It has been shown that the metabolic capability of *P. thermoglucosidasius* makes it possible to reduce the number of enzymes needed when converting palm kernel cake to ethanol, due to its hemicellulose-degrading capability. Through this approach, the strain TM242 was able to produce about five times more ethanol from PKC than a conventional yeast [84]. Another approach tried to obtain ethanol via fermentation of corn stover with *P. thermoglucosidasius*, achieving a final concentration of 9.04 g L^{-1} [85]. Additionally, the strain NBCB1 naturally produces a cellulase, which showed promising activity on a cellulose analogue under optimized conditions [86]. In an industrial process, the thermophilic nature of this organism could accelerate enzymatic activity, minimize contamination risk from mesophilic organisms, and lower cooling costs [87]. Furthermore, due to its thermophilic nature, this organism could be a valuable source of thermostable enzymes for diverse industrial applications, such as proteases, lipases, and amylases [88–90]. Although current ethanol production primarily relies on mesophilic microorganisms, *P. thermoglucosidasius* stands out for its potential in bioethanol production. Engineered strains, such as TM242, exhibit high ethanol yields due to redirected metabolic pathways and could become competitive through gas stripping that reduces toxicity and processing costs to purify ethanol [91]. The concept of hot bubble stripping for continuous production of ethanol has been tested, resulting in increased productivities due to lower product inhibition through continuous removal of ethanol. Additionally, the stripped ethanol-water vapor mixture could be cheaper to purify on an industrial scale compared to the yeast process, which starts at lower temperatures [92]. Similarly, *P. thermoglucosidasius* has shown potential for the production of butanediol and lactic acid. The strain DSM 2542 has been engineered to produce (2*R*,3*R*)-2,3-butanediol (BDO) with an enantiomeric purity exceeding 97% at high temperatures [93, 94]. Titers of both isomers of lactic acid using engineered strains have reached over 150 g L^{-1}, offering an advantage in the downstream purification over chemical synthesis methods that yield racemic mixtures [95]. In general, its ability to degrade diverse biomass, such as potato waste, bread, and palm kernel cake into biofuels highlights its possible role to realize second-generation bioprocesses [73, 85, 96]. Beyond commodity chemicals, engineered *P. thermoglucosidasius* strains have shown promise in producing riboflavin and terpenes from waste feedstocks, achieving significant yields [97, 98]. In addition, the organisms' ability to carry out the water-gas shift (WGS) reaction ($CO + H_2O \rightarrow CO_2 + H_2$) aligns with current efforts to increase the renewable production of hydrogen. Its ability to function under high CO concentrations in the gas phase and tolerate oxygen simplifies bioprocess design and reduces costs [77, 99, 100]. In comparison to biohydrogen production in mesophilic systems, the production of high-purity hydrogen at elevated temperatures could offer advantages through enhanced kinetics of the WGS reaction [101]. Using syngas as a feedstock, the hydrogen content in syngas streams could be increased further via the WGS reaction, converting toxic CO emissions into less harmful CO_2 [79]. This capability, combined with the previously highlighted characteristics, suggests the possibility of an integrated bioprocess where thermostable enzymes or bulk chemicals are produced using energy supplied by the WGS reaction. While *P. thermoglucosidasius* exhibits exceptional metabolic versatility, challenges remain in optimizing genetic modifications and fermentation conditions to fully exploit its potential. Currently, the mechanism of the CODH/ECH complex and its relationship with the central carbon metabolism are not completely understood. Scalability and economic feasibility require further evaluation to ensure successful industrial implementation. Nonetheless, continuous advancements in genetic engineering and process integration might be able to position this bacterium as a cornerstone of second-generation biotechnology.

2.5 Thermus thermophilus*: A Thermophilic Model for Extremophile Biotechnology*

Thermus spp. is a worldwide distributed genus of thermophilic and extremely thermophilic, obligate heterotrophic, Gram-negative bacteria, which are found in neutral to alkaline environments and can grow over a large range of temperatures with an optimum around 80°C [102]. The genus *Thermus* represents a notable source of thermophilic enzymes, known as thermozymes. Among these, the DNA polymerase from *Thermus aquaticus* stands out as the most prominent example [103]. Thermozymes have interesting properties that make them valuable for various applications, such as serving as biocatalysts in industrial processes [104, 105]. Thermozymes remain stable and active at very high operational temperatures (from >60 to 120°C), common in enzymatic processes, as well as resistant to denaturation by chemical agents [106]. In addition, the application of thermozymes reduces the risk of microbial contamination, increases the solubility of substrates and products, while reducing cooling costs [102]. Thermozymes can often be produced in mesophilic hosts, with their thermal stability allowing for easy purification via heat denaturation. However, over 40% cannot be expressed in an active form in these hosts, highlighting the need for thermophilic expression systems [107–111].

Among *Thermus* spp., *Thermus thermophilus* can be considered a suitable laboratory model and a promising “hot” expression system [107, 108]. The halotolerant *T. thermophilus* was isolated in 1968 from water of a natural hot spring in Japan, containing high concentrations of NaCl [102, 112, 113]. *T. thermophilus* exhibits high growth rates and cell densities at an optimal growth temperature ranging from 62 to 75°C [107, 113]. The genomes of the two closely related strains HB8 and HB27 have been sequenced, and the complete sequences, containing a high G+C content (69%) and a high coding density (95%), are available [114–117]. Both strains are naturally competent [107]. The HB8 strain can grow anaerobically, due to a membrane-bound nitrate reductase [102, 118], while the HB27 strain is restricted to aerobic growth. HB27, which contains no extra-chromosomal plasmids and exhibits very high natural competence efficiency, is a preferable host for transformation [107, 119–121]. The most frequently used expression plasmid for

T. thermophilus HB27 is the shuttle vector pMK18, which can be transformed with equal efficiency into *E. coli* and *T. thermophilus* using two distinct replication origins [119]. Several constitutive and inducible promoters have been utilized to produce enzymes such as Mn-dependent catalase and DNA polymerase in *T. thermophilus* [107–111, 122]; however, considerable potential for the identification of novel promoters for the efficient overexpression in *T. thermophilus* HB27 remains [123]. Apart from applications as an expression system for thermozymes, *T. thermophilus* also produces interesting small molecule compounds such as rare polyamines and carotenoids [102, 107] and can be utilized to produce the compatible solute mannosylglycerate, which has a stabilizing effect on biomaterials such as proteins [124]. Furthermore, *T. thermophilus* ferment extracts are already being used commercially in skin cosmetics, and research recently also explored co-fermentations for such applications using for example *B. subtilis* [125] and *S. cerevisiae* [126]. Such efforts to utilize *T. thermophilus* as a whole-cell catalyst will benefit greatly from recently established novel genome editing protocols based on CRISPR-Cas systems enabling more detailed and precise genetic engineering of *T. thermophilus* [127, 128]. Interestingly, it has also been suggested that environmental stress such as high salt concentrations leads to increased expression of endogenous CRISPR-associated proteins in *T. thermophilus* as a way to adapt to such stresses [129], which will enable more rational strain engineering approaches exploring additional applications for this interesting thermophile.

2.6 Yersinia enterocolitica*: Harnessing the Type III Secretion System*

Yersinia enterocolitica is a Gram-negative gamma-proteobacterium primarily known as a human pathogen [130]. Its virulence is largely dependent on the type III secretion system (T3SS), a large protein complex with a hollow extracellular needle, which injects effector proteins into eukaryotic cells to manipulate host cellular processes. This secretion mechanism has been repurposed for biotechnological applications, such as antigen delivery for vaccination, genome editing, and cellular reprogramming. Furthermore, the T3SS offers advantages for protein production, including purification and enhanced secretion of toxic proteins.

The T3SS is a highly specialized protein export system that translocates effector proteins directly into host cells [131]. Effector proteins contain N-terminal secretion signals that direct them to the T3SS machinery. This property has been leveraged to secrete heterologous proteins by fusing them with the secretion signals of native effector proteins, which allows for efficient secretion of these through the T3SS [132, 133]. Although the T3SS has been mainly studied in the context of pathogenesis, it is also a promising tool for biotechnology and protein production. Engineered strains of *Yersinia* and related bacteria have been used to deliver antigens for vaccination, introduce nucleases for genome editing, and influence cell fate via transcription factor injection (reviewed in [134]). For example, injection into epithelial cells of the T cell antigen p60 of *Listeria monocytogenes* as a heterologous T3SS cargo led to a p60 peptide-specific MHC class I-restricted antigen presentation [135]. Similarly, T3SS-dependent translocation of the proapoptotic protein tBID (truncated BH3 interacting domain) induced rapid and efficient tBID-induced apoptosis in HeLa cells [136]. Monobodies targeting the BCR::ABL1 tyrosine kinase were more efficiently delivered into the cytosol of eukaryotic cells than with other methods and retained their functionality, inhibiting oncogenic signaling and specifically inducing apoptosis in BCR::ABL1-dependent cells [137]. Beyond its potential for targeted delivery of proteins into eukaryotic cells, the T3SS can facilitate protein production by continuously secreting proteins into the culture medium, reducing intracellular accumulation and associated toxicity. For instance, spider silk proteins, which form fibrils when produced intracellularly, have been successfully secreted using the *Salmonella* SPI-1 T3SS [138]. *Yersinia enterocolitica* has one of the most efficient and easily controlled known T3SSs. While the T3SS usually only is activated by host cell contact, the *Yersinia* T3SS can also be activated by low calcium concentrations in the medium or in strains that lack a so-called gatekeeper protein [139]. In such conditions, the T3SS efficiently secretes its cargo into the surrounding medium, with over 90% of total protein in the supernatant being secreted by the T3SS [140]. Quantitative studies indicate that approximately 10 mg L^{-1} of a native T3SS effector is secreted per hour under standard conditions [140]. To optimize T3SS-mediated secretion, the native effector proteins can be deleted, which not only renders the strains nonpathogenic, but also increases the efficiency of heterologous protein secretion [141, 142]. Further genetic modifications, such as enhancing cellular resource allocation towards T3SS function [143], could further improve the performance of the system. While the T3SS effectively secretes a variety of proteins, its substrates need to be mechanically unfolded for the transport through the hollow needle, and proteins with highly stable folds may obstruct the secretion pathway [144]. In these cases, engineering approaches have successfully modified such proteins for improved mechanical unfolding (which does not necessarily compromise thermodynamic stability), facilitating their secretion [137, 145]. The type III secretion signals are not removed during secretion. In cases where this may impede the use of the secreted protein, protease cleavage sites can be used and the protease can even be co-secreted for post-secretion processing [136]. Although long-term secretion has been demonstrated, the duration and robustness of T3SS activity, an important aspect for biotechnological application, remain uncertain [146, 147]. Secreting bacteria exhibit reduced or halted growth, possibly due to limiting factors such as pH or nutrient availability. Approaches like optogenetically controlled T3SS activation [148] may enable fine-tuned modulation of secretion to sustain prolonged activity.

Taken together, the *Yersinia* T3SS represents a powerful tool for biotechnological applications, enabling efficient secretion and direct injection into eukaryotic cells of a wide range of proteins. Dedicated engineering efforts could enhance its robustness and further enhance its productivity, making it a valuable system for industrial protein production and therapeutic delivery.

2.7 Zymomonas mobilis: *Microbial Biofilm Former*

Zymomonas mobilis is a Gram-negative, facultative anaerobic bacterium commonly used for ethanol production from sugars with high efficiency, resulting in high product yields. Compared to yeasts, *Z. mobilis* utilizes the Entner-Doudoroff pathway to rapidly convert sugars, such as glucose, fructose, or lactose, into ethanol while generating less biomass. It is also known for its ability to form biofilms on various surfaces under specific conditions, which enhances its stress tolerance and the metabolic activity [149]. Due to the production of extracellular polymeric substances (EPSs), these biofilms have higher tolerance to harsh environmental conditions [149, 150]. In general, (microbial) biofilms are defined as structured communities embedded in a self-produced extracellular matrix, adhering to surfaces, interfaces, and/or each other. The formation of microbial biofilms can provide numerous advantages, such as enhanced tolerance to environmental stress, increased toxic tolerance, and improved survival under chemical treatment [151]. *Z. mobilis* is a biofilm-forming microorganism with various biotechnological applications. These include the production of bioethanol, where application of the biofilm improves ethanol yield, overall operational stability, and bioreactor efficiency [152], enhancing stress resistance in fermentation for ethanol and acid production [153] and utilizing waste biomass, such as fermenting lignocellulosic biomass hydrolysate into valuable second-generation biofuels. *Z. mobilis* is also employed in bioremediation processes for the degradation of toxic compounds [154, 155]. Furthermore, Li et al. reported the improvement of the electroactivity of the strain by the formation of a biofilm when applied in bioelectrical systems [156]. Biofilm formation by *Z. mobilis* has been successfully applied in ethanol production, in example, from rice bran hydrolysate [157] and to enhance benzaldehyde tolerance [149]. Different research groups have demonstrated high ethanol yields using continuous packed bed biofilm reactors with yields up to 92% [155, 158, 159]. Biofilm formation allows *Z. mobilis* to maintain higher survival rates and metabolic activity compared to planktonic cells, leading to improved ethanol production. Using *Z. mobilis* in biofilm bioreactors could reduce overall costs by simplifying the complexity of the upstream process of lignocellulosic materials. Due to higher toxic tolerance of biofilm-associated cells, a separation of toxic compounds is no longer required, making the process more efficient and economically viable [160, 161].

2.8 Neocallimastigales: *Anaerobic Fungi*

Discovered by Liebetanz in 1905 but initially misclassified as protozoa, anaerobic fungi, also known as anaerobic gut fungi (AGF), were identified as *Callimastix*, *Oikomonas*, and *Sphaeromonas* [162]. Only in 1977 did Orpin provide evidence of chitin, proving these organisms to be fungi and classifying them as *Chytridiomycetes* [163]. Later genetic analysis published in 2007 proved the unique phylogenetic position of AGF as a separated basal phylum [164]. The newly created phylum *Neocallimastigomycota* with the order *Neocallimastigales* contains so far all cultivated obligate anaerobic fungi grouped in 4 families and 22 genera [165–167]. In silico analysis shows up to 56 genera of AGF that still need to be isolated [168]. Although recent discoveries in ecological necessities for AGF hosts like specific gut retention proved nonmammalian AGF genera like *Astrotestudinimyces* and *Testudinimyces*, the biological importance of AGF is best understood in ruminants [166]. AGF evolved as primary settlers of lignocellulosic biomass in the rumen of ruminants through co-evolutionary phylosymbiosis and post-evolutionary environmental filtering [168]. This unique ecological niche differentiates AGF from the intracellular parasitic *Cryptomycota* and *Microsporidia*, positioning them as the first known symbiotic phylogenetic evolutionary branch in the fungal tree of life [169]. In addition to the previously mentioned evolutionary adaptations, further genes from rumen bacteria were acquired through horizontal gene transfer, resulting in the greatest variety and quantity of carbohydrate-active enzymes (CAZymes) in the entire fungal kingdom [170–172]. The reorganization of CAZymes into multienzyme complexes called cellulosomes leads to a 12-fold increase in efficiency in breaking down various hemicellulosic biomasses compared to pure enzymatic treatments [173]. Further improvement using fungal thalli penetration via appressoria-like structures results in an unrivalled efficiency in degradation [174]. Hexoses and pentoses produced through enzymatic degradation are metabolized using cytosolic and hydrogenosomal pathways to hydrogen, formic acid, acetate, ethanol, lactate, and succinate [172, 175]. The hydrogenosome, an evolutionarily adapted mitochondrion, serves not only as the primary energy provider of the cell but also as the organelle containing ferredoxin hydrogenases and possibly bifurcating hydrogenases, playing a crucial role in methanogenic growth in the rumen [172, 176, 177]. As key species with high efficiency in degrading hemicellulosic materials, multiple suitable biotechnological applications arise. Potential use cases include waste stream treatment from paper industries, biomass utilization in biogas plants, and production of secondary metabolites [178–180]. Further approaches focus on harnessing the enzymatic diversity of AGF in fast-growing cultures like *E. coli* or yeast through genetic engineering [181]. Recent investigations mainly explore the probiotic addition to ruminant feedstocks to enhance biomass production, leading to better food source utilization [182–184]. Nevertheless, the successful establishment of industrial processes still faces multiple problems. As a relatively new discovery with few scientists working with AGF, due to difficult cultivation strategies, lack of long-term storage methods, unavailability in ATCC and DSMZ, and mostly unknown growth conditions, many questions remain. Neither the full life cycle, especially the spread from host to host in an oxygen atmosphere, nor the key role in the rumen biome is understood. A postulated oxygen-resistant form has not yet been found. Although all fungi are believed to be capable of both asexual and sexual reproduction, AGF have only been observed reproducing asexually. The medical implications of AGF on ruminants, their resistance to plant protectants, different feeding strategies, and their effects on methanogens are not well understood or are only rudimentarily understood.

3 Conclusion and Future Needs

The exploitation of microbial biodiversity presents a vast, untapped potential for advancing the bioeconomy. While traditional bioprocesses have relied on a limited number of well-characterized microorganisms, the exploration of alternative microorganisms offers significant opportunities to expand the capabilities and efficiency of biotechnological applications. These innovations help tackle economic and societal changes and contribute directly to many Sustainable Development Goals (in particular SDGs 2, 7, 9, 12, and 13) [185]. The use of alternative microorganisms can extend the process window of bioprocesses, enabling the utilization of unconventional substrates, extending the portfolio of products, and enabling the establishment of robust bioprocesses under unusual process conditions. Several promising "hidden champions" among these unusual organisms have already been identified and studied, offering great potential for more efficient and environmentally friendly processes.

However, the future of the exploitation of biodiversity has further and many facets that exceed the identification, cultivation, and domestication of novel production hosts. While most established production hosts are used in axenic suspension cultures, biodiversity offers many more biological concepts that can be used to produce value-added products exceeding traditional routes [186]. Emerging concepts that significantly differ from current concepts include the use of mixed cultures [187], productive biofilms [188, 189], and electroactive microorganisms [190]. Using mixed-culture bioprocesses, often referred to as microbial consortia, complex biotransformations can be performed by sharing metabolic tasks within consortia members [191], leading to more efficient and robust bioprocesses. Biofilm-forming microorganisms offer an alternative concept, by facilitating reactions that are difficult to perform in suspension cultures [189]. Meanwhile, the field of electrobiotechnology is emerging fast. Here, bioelectrochemical systems are developed that utilize microorganisms capable of extracellular electron transfer for sustainable energy production and waste treatment [190, 192, 193]. Many successful proof-of-concept studies have already demonstrated the high potential of these biology-inspired production processes.

As with any emerging technology, significant challenges still hinder the path to full realization of these concepts. Often, detailed understanding of the microbial physiology, especially under technical conditions such as in bioreactors, is missing. Here, employing advanced -omics technologies is essential for gaining a comprehensive understanding of microbial physiology and metabolism, which is the key for further progress towards metabolic tailoring and novel application of these systems for bioproduction [194]. Furthermore, all of these fields currently lack adequate development of novel bioreactor concepts that enable efficient and large-scale bioproduction [195, 196]. This includes innovative reactor designs, process control strategies, downstream processing techniques, and efficient transfer from laboratory-scale innovations to industrial-scale production [197, 198].

These new opportunities will undoubtedly raise the question of whether they can truly make our lives easier and better towards the development of novel bioprocesses. All the demonstrated examples show that it may not ultimately be a question of whether something is technically feasible, but rather if it can be achieved in an economical and sustainable way. Let us start to foster the collaboration between biotechnologists and engineers to work towards the realization of these goals.

References

1. Singh BK (2010) Exploring microbial diversity for biotechnology: the way forward. Trends Biotechnol 28(3):111–116
2. Locey KJ, Lennon JT (2016) Scaling laws predict global microbial diversity. Proc Natl Acad Sci 113(21):5970–5975
3. Microbiology by numbers (2011). Nat Rev Microbiol 9(9):628
4. Wade W (2002) Unculturable bacteria – the uncharacterized organisms that cause oral infections. J R Soc Med 95(2):81–83
5. Schaudinn FR, Hoffmann E (1905) Vorläufiger Bericht über das Vorkommen von Spirochaeten in syphilitischen Krankheitsprodukten und bei Papillomen. Springer
6. Edmondson Diane G, Hu B, Norris Steven J (2018) Long-term in vitro culture of the syphilis spirochete *Treponema pallidum* subsp. *pallidum*. mBio 9(3). https://doi.org/10.1128/mbio.01153-18
7. Edmondson DG, Norris SJ (2021) In vitro cultivation of the syphilis spirochete *Treponema pallidum*. Curr Protoc 1(2):e44
8. Yadav K, Nikalje GC (2024) Comprehensive analysis of bioplastics: life cycle assessment, waste management, biodiversity impact, and sustainable mitigation strategies. PeerJ 12:e18013
9. Hannon M et al (2010) Biofuels from algae: challenges and potential. Biofuels 1(5):763–784
10. Minden S et al (2024) Producing food from CO2 using microorganisms: lots to do, little to lose! Trends Food Sci Technol 154:104778
11. Mühling L, Baur T, Molitor B (2024) Methanothermobacter thermautotrophicus and alternative methanogens: archaea-based production. In: Advances in biochemical engineering/biotechnology. Springer, Berlin. https://doi.org/10.1007/10_2024_270
12. Casey D et al (2024) From Knallgas bacterium to promising biomanufacturing host: the evolution of Cupriavidus necator. In: Advances in biochemical engineering/biotechnology. Springer, Berlin. https://doi.org/10.1007/10_2024_269
13. Hädrich M et al (2024) *Vibrio natriegens*: application of a fast-growing halophilic bacterium. In: Advances in biochemical engineering/biotechnology. Springer, Berlin, pp 1–32
14. Kneuer L, Wurst R, Gescher J (2024) Shewanella oneidensis: biotechnological application of metal-reducing bacteria. In: Advances in biochemical engineering/biotechnology. Springer, Berlin. https://doi.org/10.1007/10_2024_272
15. Bolay P, Toepel J, Bühler B (2025) Biotechnological applications of cyanobacteria: Synechocystis and Synechococcus strains. In: Advances in biochemical engineering/biotechnology. Springer, Berlin, pp 1–37
16. Stiefelmaier J (2025) Cable bacteria and their biotechnological application. In: Advances in biochemical engineering/biotechnology. Springer, Berlin, pp 1–21
17. Kufs JE, Reimer C, Regestein L (2025) The amoeba Dictyostelium discoideum as novel production host for complex substances. In: Advances in biochemical engineering/biotechnology. Springer, Berlin, pp 1–14

18. Balch WE et al (1977) Acetobacterium, a new genus of hydrogen-oxidizing, carbon dioxide-reducing, anaerobic bacteria. Int J Syst Evol Microbiol 27(4):355–361
19. Poehlein A et al (2012) An ancient pathway combining carbon dioxide fixation with the generation and utilization of a sodium ion gradient for ATP synthesis. PLoS One 7(3):e33439
20. Biegel E et al (2011) Biochemistry, evolution and physiological function of the Rnf complex, a novel ion-motive electron transport complex in prokaryotes. Cell Mol Life Sci 68(4):613–634
21. Biegel E, Müller V (2010) Bacterial Na+-translocating ferredoxin:NAD+ oxidoreductase. Proc Natl Acad Sci 107(42):18138–18142
22. Heise R et al (1991) A sodium-stimulated ATP synthase in the acetogenic bacterium Acetobacterium woodii. FEBS Lett 295(1–3):119–122
23. Schuchmann K, Müller V (2014) Autotrophy at the thermodynamic limit of life: a model for energy conservation in acetogenic bacteria. Nat Rev Microbiol 12(12):809–821
24. Imkamp F et al (2007) Dissection of the caffeate respiratory chain in the acetogen *Acetobacterium woodii*: identification of an Rnf-type NADH dehydrogenase as a potential coupling site. J Bacteriol 189(22):8145–8153
25. Reidlinger J, Müller V (1994) Purification of ATP synthase from Acetobacterium woodii and identification as a Na+-translocating F1FO-type enzyme. Eur J Biochem 223(1):275–283
26. Seifritz C et al (1993) Nitrate as a preferred electron sink for the acetogen Clostridium thermoaceticum. J Bacteriol 175(24):8008–8013
27. Fröstl JM, Seifritz C, Drake HL (1996) Effect of nitrate on the autotrophic metabolism of the acetogens Clostridium thermoautotrophicum and Clostridium thermoaceticum. J Bacteriol 178(15):4597–4603
28. Bache R, Pfennig N (1981) Selective isolation of Acetobacterium woodii on methoxylated aromatic acids and determination of growth yields. Arch Microbiol 130(3):255–261
29. Matthies C, Freiberger A, Drake HL (1993) Fumarate dissimilation and differential reductant flow by Clostridium formicoaceticum and Clostridium aceticum. Arch Microbiol 160(4): 273–278
30. Tanner RS, Miller LM, Yang D (1993) Clostridium ljungdahlii sp. nov., an acetogenic species in clostridial rRNA homology group I. Int J Syst Evol Microbiol 43(2):232–236
31. Liou JS-C et al (2005) Clostridium carboxidivorans sp. nov., a solvent-producing clostridium isolated from an agricultural settling lagoon, and reclassification of the acetogen Clostridium scatologenes strain SL1 as Clostridium drakei sp. nov. Int J Syst Evol Microbiol 55(5): 2085–2091
32. Demler M, Weuster-Botz D (2011) Reaction engineering analysis of hydrogenotrophic production of acetic acid by Acetobacterium woodii. Biotechnol Bioeng 108(2):470–474
33. Kantzow C, Mayer A, Weuster-Botz D (2015) Continuous gas fermentation by Acetobacterium woodii in a submerged membrane reactor with full cell retention. J Biotechnol 212:11–18
34. Straub M et al (2014) Selective enhancement of autotrophic acetate production with genetically modified Acetobacterium woodii. J Biotechnol 178:67–72
35. Arslan K et al (2022) Engineering Acetobacterium woodii for the production of isopropanol and acetone from carbon dioxide and hydrogen. Biotechnol J 17(5):2100515
36. Neuendorf CS et al (2021) A quantitative metabolic analysis reveals Acetobacterium woodii as a flexible and robust host for formate-based bioproduction. Metab Eng 68:68–85
37. Leigh JA, Mayer F, Wolfe RS (1981) Acetogenium kivui, a new thermophilic hydrogen-oxidizing acetogenic bacterium. Arch Microbiol 129(4):275–280
38. Weghoff Marie C, Müller V (2016) CO metabolism in the thermophilic acetogen Thermoanaerobacter kivui. Appl Environ Microbiol 82(8):2312–2319
39. Hess V et al (2014) A genome-guided analysis of energy conservation in the thermophilic, cytochrome-free acetogenic bacterium Thermoanaerobacter kivui. BMC Genomics 15(1): 1139
40. Schwarz FM, Schuchmann K, Müller V (2018) Hydrogenation of CO2 at ambient pressure catalyzed by a highly active thermostable biocatalyst. Biotechnol Biofuels 11(1):237

41. Schwarz FM, Müller V (2020) Whole-cell biocatalysis for hydrogen storage and syngas conversion to formate using a thermophilic acetogen. Biotechnol Biofuels 13(1):32
42. Deutzmann JS, Spormann AM (2024) High acetate titer obtained from CO2 by thermophilic microbial electrosynthesis with Thermoanaerobacter kivui. Bioresour Technol Rep 25:101740
43. Basen M et al (2018) A genetic system for the thermophilic acetogenic bacterium Thermoanaerobacter kivui. Appl Environ Microbiol 84(3):e02210–e02217
44. Anthony C (1982) The biochemistry of methylotrophs. Academic Press, London
45. Schrader J et al (2009) Methanol-based industrial biotechnology: current status and future perspectives of methylotrophic bacteria. Trends Biotechnol 27(2):107–115
46. Korotkova N, Lidstrom Mary E (2001) Connection between poly-β-hydroxybutyrate biosynthesis and growth on C1 and C2 compounds in the methylotroph Methylobacterium extorquens AM1. J Bacteriol 183(3):1038–1046
47. Lim CK et al (2019) Designing and engineering Methylorubrum extorquens AM1 for itaconic acid production. Front Microbiol 10:1027
48. Suzuki T, Yamane T, Shimizu S (1986) Mass production of poly-β-hydroxybutyric acid by fed-batch culture with controlled carbon/nitrogen feeding. Appl Microbiol Biotechnol 24(5): 370–374
49. Bourque D et al (1992) Production of poly-β-hydroxybutyrate from methanol: characterization of a new isolate of Methylobacterium extorquens. Appl Microbiol Biotechnol 37(1):7–12
50. Vuilleumier S et al (2009) Methylobacterium genome sequences: a reference blueprint to investigate microbial metabolism of C1 compounds from natural and industrial sources. PLoS One 4(5):e5584
51. Korotkova N, Chistoserdova L, Lidstrom Mary E (2002) Poly-β-hydroxybutyrate biosynthesis in the facultative methylotroph Methylobacterium extorquens AM1: identification and mutation of gap11, gap20, and phaR. J Bacteriol 184(22):6174–6181
52. Yoon J et al (2021) Metabolic engineering of Methylorubrum extorquens AM1 for poly (3-hydroxybutyrate-co-3-hydroxyvalerate) production using formate. Int J Biol Macromol 177:284–293
53. Hardy RW et al (2018) Partial replacement of soybean meal with Methylobacterium extorquens single-cell protein in feeds for rainbow trout (Oncorhynchus mykiss Walbaum). Aquac Res 49(6):2218–2224
54. Sonntag F et al (2015) High-level production of ethylmalonyl-CoA pathway-derived dicarboxylic acids by Methylobacterium extorquens under cobalt-deficient conditions and by polyhydroxybutyrate negative strains. Appl Microbiol Biotechnol 99(8):3407–3419
55. Sonntag F et al (2015) Engineering Methylobacterium extorquens for de novo synthesis of the sesquiterpenoid α-humulene from methanol. Metab Eng 32:82–94
56. Beekwilder J et al (2014) Polycistronic expression of a β-carotene biosynthetic pathway in Saccharomyces cerevisiae coupled to β-ionone production. J Biotechnol 192(Pt B):383–392
57. Dudko D, Holtmann D, Buchhaupt M (2023) Methylotrophic bacteria with cobalamin-dependent mutases in primary metabolism as potential strains for vitamin B12 production. Antonie Van Leeuwenhoek 116(3):207–220
58. Hegner R et al (2020) Coupled electrochemical and microbial catalysis for the production of polymer bricks. ChemSusChem 13(19):5295–5300
59. Marx CJ, Lidstrom ME (2002) Broad-host-range cre-lox system for antibiotic marker recycling in gram-negative bacteria. Biotechniques 33(5):1062–1067
60. Zhu L et al (2024) Heterologous production of 3-hydroxypropionic acid in Methylorubrum extorquens by introducing the mcr gene via a multi-round chromosomal integration system based on cre-lox71/lox66 and transposon. Microb Cell Factories 23(1):5
61. Marx CJ (2008) Development of a broad-host-range sacB-based vector for unmarked allelic exchange. BMC Res Notes 1(1):1
62. Chubiz LM et al (2013) A novel pair of inducible expression vectors for use in Methylobacterium extorquens. BMC Res Notes 6(1):183

63. Schada von Borzyskowski L et al (2015) A set of versatile brick vectors and promoters for the assembly, expression, and integration of synthetic operons in Methylobacterium extorquens AM1 and other alphaproteobacteria. ACS Synth Biol 4(4):430–443
64. Suzuki Y et al (1983) Bacillus thermoglucosidasius sp. nov., a new species of obligately thermophilic bacilli. Syst Appl Microbiol 4(4):487–495
65. Aliyu H et al (2016) Phylogenomic re-assessment of the thermophilic genus Geobacillus. Syst Appl Microbiol 39(8):527–533
66. Nazina TN et al (2001) Taxonomic study of aerobic thermophilic bacilli: descriptions of Geobacillus subterraneus gen. nov., sp. nov. and Geobacillus uzenensis sp. nov. from petroleum reservoirs and transfer of Bacillus stearothermophilus, Bacillus thermocatenulatus, Bacillus thermoleovorans, Bacillus kaustophilus, Bacillus thermodenitrificans to Geobacillus as the new combinations G. stearothermophilus, G. th. Int J Syst Evol Microbiol 51(Pt 2): 433–446
67. Marchant R et al (2008) Thermophilic bacteria in cool temperate soils: are they metabolically active or continually added by global atmospheric transport? Appl Microbiol Biotechnol 78(5):841–852
68. Zeigler DR (2014) The Geobacillus paradox: why is a thermophilic bacterial genus so prevalent on a mesophilic planet? Microbiology (Reading) 160(Pt 1):1–11
69. Mohr T et al (2018) Comparative genomic analysis of Parageobacillus thermoglucosidasius strains with distinct hydrogenogenic capacities. BMC Genomics 19(1):880
70. Adachi Y et al (2020) Genetic engineering of carbon monoxide-dependent hydrogen-producing machinery in Parageobacillus thermoglucosidasius. Microbes Environ 35(4):ME20101
71. Manish S, Banerjee R (2008) Comparison of biohydrogen production processes. Int J Hydrog Energy 33(1):279–286
72. De Maayer P et al (2014) Comparative analysis of the Geobacillus hemicellulose utilization locus reveals a highly variable target for improved hemicellulolysis. BMC Genomics 15(1): 836
73. Ibenegbu CC, Leak DJ (2022) Simultaneous saccharification and ethanologenic fermentation (SSF) of waste bread by an amylolytic Parageobacillus thermoglucosidasius strain TM333. Microb Cell Factories 21(1):251
74. Liang J et al (2022) Removing carbon catabolite repression in Parageobacillus thermoglucosidasius DSM 2542. Front Microbiol 13:985465
75. Bidart GN, Gharabli H, Welner DH (2023) Functional characterization of the phosphotransferase system in Parageobacillus thermoglucosidasius. Sci Rep 13(1):7131
76. Ardila MS et al (2024) Effect of different partial pressures on H2 production with Parageobacillus thermoglucosidasius DSM 6285. Fermentation 10(11):592
77. Aliyu H et al (2021) Carbon monoxide induced metabolic shift in the carboxydotrophic Parageobacillus thermoglucosidasius DSM 6285. Microorganisms 9(5):1090
78. Mol V et al (2021) Genome-scale metabolic modeling of P. thermoglucosidasius NCIMB 11955 reveals metabolic bottlenecks in anaerobic metabolism. Metab Eng 65:123–134
79. Paredes-Barrada M et al (2024) Parageobacillus thermoglucosidasius as an emerging thermophilic cell factory. Metab Eng 83:39–51
80. Brumm PJ, Land ML, Mead DA (2015) Complete genome sequence of Geobacillus thermoglucosidasius C56-YS93, a novel biomass degrader isolated from obsidian hot spring in Yellowstone National Park. Stand Genomic Sci 10(1):73
81. Inoue M et al (2019) Draft genome sequence of Parageobacillus thermoglucosidasius strain TG4, a hydrogenogenic carboxydotrophic bacterium isolated from a marine sediment. Microbiol Resour Announc 8(5):e01666-18
82. Lau MSH et al (2021) Development of a suite of tools for genome editing in Parageobacillus thermoglucosidasius and their use to identify the potential of a native plasmid in the generation of stable engineered strains. ACS Synth Biol 10(7):1739–1749

83. Sheng L, Zhang Y, Minton NP (2016) Complete genome sequence of Geobacillus thermoglucosidasius NCIMB 11955, the progenitor of a bioethanol production strain. Genome Announc 4(5):e01065-16
84. Raita M et al (2016) Production of ethanol by thermophilic oligosaccharide utilising Geobacillus thermoglucosidasius TM242 using palm kernel cake as a renewable feedstock. Biomass Bioenergy 95:45–54
85. Madhuvanthi S et al (2022) Optimization of consolidated bioprocessing by response surface methodology in the conversion of corn stover to bioethanol by thermophilic Geobacillus thermoglucosidasius. Chemosphere 304:135242
86. Basak A et al (2023) Optimization and biochemical characterization of a thermotolerant processive cellulase, PtCel1, of Parageobacillus thermoglucosidasius NBCB1. J Basic Microbiol 63(3–4):326–339
87. Turner P, Mamo G, Karlsson EN (2007) Potential and utilization of thermophiles and thermostable enzymes in biorefining. Microb Cell Factories 6(1):9
88. Hussein AH, Lisowska BK, Leak DJ (2015) The genus Geobacillus and their biotechnological potential. Adv Appl Microbiol 92:1–48
89. Saeed AM, El-Shatoury EH, Sayed HAE (2021) Statistical factorial designs for optimum production of thermostable α-amylase by the degradative bacterium Parageobacillus thermoglucosidasius Pharon1 isolated from Sinai, Egypt. J Genet Eng Biotechnol 19(1):24
90. Suleiman AD et al (2020) Effect of cultural conditions on protease production by a thermophilic Geobacillus thermoglucosidasius SKF4 isolated from Sungai Klah Hot Spring Park, Malaysia. Molecules 25(11):2609
91. Cripps RE et al (2009) Metabolic engineering of Geobacillus thermoglucosidasius for high yield ethanol production. Metab Eng 11(6):398–408
92. Calverley J et al (2024) Ethanologenic fermentation by Parageobacillus thermoglucosidasius with continuous hot microbubble gas-stripping. Microb Cell Factories 23(1):165
93. Zhou J, Lian J, Rao CV (2020) Metabolic engineering of Parageobacillus thermoglucosidasius for the efficient production of (2R, 3R)-butanediol. Appl Microbiol Biotechnol 104(10): 4303–4311
94. Sheng L et al (2023) Metabolic engineering for the production of acetoin and 2,3-butanediol at elevated temperature in Parageobacillus thermoglucosidasius NCIMB 11955. Front Bioeng Biotechnol 11:1191079
95. Liu J et al (2024) Metabolic engineering of Geobacillus thermoglucosidasius for polymer-grade lactic acid production at high temperature. Bioresour Technol 393:130164
96. Singhvi M et al (2021) Nanoparticle-associated single step hydrogen fermentation for the conversion of starch potato waste biomass by thermophilic Parageobacillus thermoglucosidasius. Bioresour Technol 337:125490
97. Styles MQ et al (2021) The heterologous production of terpenes by the thermophile Parageobacillus thermoglucosidasius in a consolidated bioprocess using waste bread. Metab Eng 65:146–155
98. Yang Z et al (2021) Engineering thermophilic Geobacillus thermoglucosidasius for riboflavin production. Microb Biotechnol 14(2):363–373
99. Díaz DB, Neumann A, Aliyu H (2022) Thermophilic water gas shift reaction at high carbon monoxide and hydrogen partial pressures in Parageobacillus thermoglucosidasius KP1013. Fermentation 8. https://doi.org/10.3390/fermentation8110596
100. Mohr T et al (2019) Effects of different operating parameters on hydrogen production by Parageobacillus thermoglucosidasius DSM 6285. AMB Express 9(1):207
101. Alfano M, Cavazza C (2018) The biologically mediated water–gas shift reaction: structure, function and biosynthesis of monofunctional [NiFe]-carbon monoxide dehydrogenases. Sustain Energy Fuels 2(8):1653–1670
102. Saghatelyan A, Panosyan H, Birkeland N-K (2021) The genus Thermus: a brief history of cosmopolitan extreme thermophiles: diversity, distribution, biotechnological potential and

applications. In: Egamberdieva D et al (eds) Microbial communities and their interactions in the extreme environment. Springer, Singapore, pp 141–175
103. Saiki RK et al (1988) Primer-directed enzymatic amplification of DNA with a thermostable DNA polymerase. Science 239(4839):487–491
104. Lasa I, Berenguer J (1993) Thermophilic enzymes and their biotechnological potential. Microbiologia 9(2):77–89
105. Pantazaki A, Pritsa A, Kyriakidis D (2002) Biotechnologically relevant enzymes from Thermus thermophilus. Appl Microbiol Biotechnol 58(1):1–12
106. Zeikus JG, Vieille C, Savchenko A (1998) Thermozymes: biotechnology and structure–function relationships. Extremophiles 2(3):179–183
107. Cava F, Hidalgo A, Berenguer J (2009) Thermus thermophilus as biological model. Extremophiles 13(2):213–231
108. Fujino Y et al (2020) Development of a new gene expression vector for Thermus thermophilus using a silica-inducible promoter. Microb Cell Factories 19(1):126
109. Hidalgo A et al (2004) Thermus thermophilus as a cell factory for the production of a thermophilic Mn-dependent catalase which fails to be synthesized in an active form in Escherichia coli. Appl Environ Microbiol 70(7):3839–3844
110. Moreno R et al (2003) Development of a gene expression vector for Thermus thermophilus based on the promoter of the respiratory nitrate reductase. Plasmid 49(1):2–8
111. Park H-S et al (2004) Heterologous gene expression in Thermus thermophilus: β-galactosidase, dibenzothiophene monooxygenase, PNB carboxy esterase, 2-aminobiphenyl-2,3-diol dioxygenase, and chloramphenicol acetyl transferase. J Ind Microbiol Biotechnol 31(4):189–197
112. Oshima T, Imahori K (1971) Isolation of an extreme thermophile and thermostability of its transfer ribonucleic acid and ribosomes. J Gen Appl Microbiol 17(6):513–517
113. Oshima T, Imahori K (1974) Description of Thermus thermophilus (Yoshida and Oshima) comb. nov., a nonsporulating thermophilic bacterium from a Japanese thermal spa. Int J Syst Evol Microbiol 24(1):102–112
114. Brüggemann H, Chen C (2006) Comparative genomics of Thermus thermophilus: plasticity of the megaplasmid and its contribution to a thermophilic lifestyle. J Biotechnol 124(4):654–661
115. Henne A et al (2004) The genome sequence of the extreme thermophile Thermus thermophilus. Nat Biotechnol 22(5):547–553
116. Liebl W (2004) Genomics taken to the extreme. Nat Biotechnol 22(5):524–525
117. Lioliou EE, Pantazaki AA, Kyriakidis DA (2004) Thermus thermophilus genome analysis: benefits and implications. Microb Cell Factories 3(1):5
118. Ramírez-Arcos S et al (1998) Anaerobic growth, a property horizontally transferred by an Hfr-like mechanism among extreme thermophiles. J Bacteriol 180(12):3137–3143
119. de Grado M, Castán P, Berenguer J (1999) A high-transformation-efficiency cloning vector for Thermus thermophilus. Plasmid 42(3):241–245
120. Koyama Y et al (1986) Genetic transformation of the extreme thermophile Thermus thermophilus and of other Thermus spp. J Bacteriol 166(1):338–340
121. Schwarzenlander C, Averhoff B (2006) Characterization of DNA transport in the thermophilic bacterium Thermus thermophilus HB27. FEBS J 273(18):4210–4218
122. Moreno R et al (2005) High-level overproduction of His-tagged Tth DNA polymerase in Thermus thermophilus. Appl Environ Microbiol 71(1):591–593
123. Kirchner L, Müller V, Averhoff B (2023) A temperature dependent pilin promoter for production of thermostable enzymes in Thermus thermophilus. Microb Cell Factories 22(1): 187
124. Egorova K et al (2007) High yield of mannosylglycerate production by upshock fermentation and bacterial milking of trehalose-deficient mutant Thermus thermophilus RQ-1. Appl Microbiol Biotechnol 75(5):1039–1045

125. Wang J et al (2024) Novel Thermus thermophilus and Bacillus subtilis mixed-culture ferment extract provides potent skin benefits in vitro and protects skin from aging. J Cosmet Dermatol 23(12):4334–4342
126. Wei Y-F et al (2024) Isolation, purification and characterization of antioxidant peptides from Spirulina platensis extracts co-fermentation with Thermus thermophilus HB27 and Saccharomyces cerevisiae CH006. LWT 205:116497
127. Wang J et al (2022) High-efficiency genome editing of an extreme thermophile Thermus thermophilus using endogenous type I and type III CRISPR-Cas systems. mLife 1(4):412–427
128. Adalsteinsson BT et al (2021) Efficient genome editing of an extreme thermophile, Thermus thermophilus, using a thermostable Cas9 variant. Sci Rep 11(1):9586
129. Karimi-Fard A et al (2025) CRISPR-Cas-mediated adaptation of Thermus thermophilus HB8 to environmental stress conditions. Arch Microbiol 207(2):41
130. Fredriksson-Ahomaa M, Stolle A, Korkeala H (2006) Molecular epidemiology of Yersinia enterocolitica infections. FEMS Immunol Med Microbiol 47(3):315–329
131. Cornelis GR (2006) The type III secretion injectisome. Nat Rev Microbiol 4(11):811–825
132. Michiels T, Cornelis GR (1991) Secretion of hybrid proteins by the Yersinia Yop export system. J Bacteriol 173(5):1677–1685
133. Sory M-P, Cornelis GR (1994) Translocation of a hybrid YopE-adenylate cyclase from Yersinia enterocolitica into HeLa cells. Mol Microbiol 14(3):583–594
134. Bai F et al (2018) Bacterial type III secretion system as a protein delivery tool for a broad range of biomedical applications. Biotechnol Adv 36(2):482–493
135. Rüssmann H et al (2000) Yersinia enterocolitica-mediated translocation of defined fusion proteins to the cytosol of mammalian cells results in peptide-specific MHC class I-restricted antigen presentation. Eur J Immunol 30(5):1375–1384
136. Ittig SJ et al (2015) A bacterial type III secretion-based protein delivery tool for broad applications in cell biology. J Cell Biol 211(4):913–931
137. Lebon C et al (2024) Cytosolic delivery of monobodies using the bacterial type III secretion system inhibits oncogenic BCR: ABL1 signaling. Cell Commun Signal 22(1):500
138. Widmaier DM et al (2009) Engineering the *Salmonella* type III secretion system to export spider silk monomers. Mol Syst Biol 5(1):309
139. Forsberg Å et al (1991) The surface-located YopN protein is involved in calcium signal transduction in Yersinia pseudotuberculosis. Mol Microbiol 5(4):977–986
140. Wimmi S et al (2024) Cytosolic sorting platform complexes shuttle type III secretion system effectors to the injectisome in Yersinia enterocolitica. Nat Microbiol 9(1):185–199
141. Feldman MF et al (2002) SycE allows secretion of YopE-DHFR hybrids by the Yersinia enterocolitica type III Ysc system. Mol Microbiol 46(4):1183–1197
142. Kudryashev M et al (2013) In situ structural analysis of the Yersinia enterocolitica injectisome. eLife 2:e00792
143. Liang JM et al (2023) Construction of a constitutively active type III secretion system for heterologous protein secretion. Appl Microbiol Biotechnol 107(5):1785–1800
144. Radics J, Königsmaier L, Marlovits TC (2014) Structure of a pathogenic type 3 secretion system in action. Nat Struct Mol Biol 21(1):82–87
145. LeBlanc M-A et al (2021) Type III secretion system effector proteins are mechanically labile. Proc Natl Acad Sci 118(12):e2019566118
146. Milne-Davies B et al (2019) Life after secretion-Yersinia enterocolitica rapidly toggles effector secretion and can resume cell division in response to changing external conditions. Front Microbiol 10:2128
147. Milne-Davies B, Wimmi S, Diepold A (2021) Adaptivity and dynamics in type III secretion systems. Mol Microbiol 115(3):395–411
148. Lindner F et al (2020) LITESEC-T3SS – light-controlled protein delivery into eukaryotic cells with high spatial and temporal resolution. Nat Commun 11(1):2381

149. Li Xuan Z et al (2006) Enhanced benzaldehyde tolerance in Zymomonas mobilis biofilms and the potential of biofilm applications in fine-chemical production. Appl Environ Microbiol 72(2):1639–1644
150. O'Toole G, Kaplan HB, Kolter R (2000) Biofilm formation as microbial development. Annu Rev Microbiol 54:49–79
151. Li XZ, Hauer B, Rosche B (2007) Single-species microbial biofilm screening for industrial applications. Appl Microbiol Biotechnol 76(6):1255–1262
152. Kunduru MR, Pometto III AL (1996) Continuous ethanol production by Zymomonas mobilis and Saccharomyces cerevisiae in biofilm reactors. J Ind Microbiol Biotechnol 16(4):249–256
153. Braga A et al (2021) Zymomonas mobilis as an emerging biotechnological chassis for the production of industrially relevant compounds. Bioresour Bioprocess 8(1):128
154. Cao LY et al (2023) Regulation of biofilm formation in Zymomonas mobilis to enhance stress tolerance by heterologous expression of pfs and luxS. Front Bioeng Biotechnol 11:1130405
155. Todhanakasem T et al (2019) Zymomonas mobilis biofilm reactor for ethanol production using rice straw hydrolysate under continuous and repeated batch processes. Front Microbiol 10: 1777
156. Li K et al (2023) Electrochemical control of cell metabolism improves ethanol production of Zymomonas mobilis in an electro-fermentation system. ACS Sustain Chem Eng 11(6): 2364–2374
157. Todhanakasem T et al (2014) Biofilm production by Zymomonas mobilis enhances ethanol production and tolerance to toxic inhibitors from rice bran hydrolysate. New Biotechnol 31(5): 451–459
158. Todhanakasem T et al (2019) Expression and extracellular secretion of endo-glucanase and xylanase by Zymomonas mobilis. Appl Biochem Biotechnol 187(1):239–252
159. Kunduru MR, Pometto AL (1996) Continuous ethanol production by Zymomonas mobilis and Saccharomyces cerevisiae in biofilm reactors. J Ind Microbiol 16(4):249–256
160. Wirawan F et al (2012) Cellulosic ethanol production performance with SSF and SHF processes using immobilized Zymomonas mobilis. Appl Energy 100:19–26
161. Todhanakasem T et al (2015) Fermentation of rice bran hydrolysate to ethanol using Zymomonas mobilis biofilm immobilization on DEAE-cellulose. Electron J Biotechnol 18(3):196–201
162. Liebetanz E (1905) Die Parasitischen Protozoen Des Wiederkäuermagens. Berliner Tieraerztliche Wochenschrift 18:313
163. Orpin CG (1977) The occurrence of chitin in the cell walls of the rumen organisms Neocallimastix frontalis, Piromonas communis and Sphaeromonas communis. J Gen Microbiol 99(1):215–218
164. Hibbett DS et al (2007) A higher-level phylogenetic classification of the fungi. Mycol Res 111(5):509–547
165. Hanafy RA et al (2022) Taxonomy of the anaerobic gut fungi (Neocallimastigomycota): a review of classification criteria and description of current taxa. Int J Syst Evol Microbiol 72(7). https://doi.org/10.1099/ijsem.0.005322
166. Pratt CJ et al (2023) Testudinimyces gracilis gen. nov, sp. nov. and Astrotestudinimyces divisus gen. nov, sp. nov., two novel, deep-branching anaerobic gut fungal genera from tortoise faeces. Int J Syst Evol Microbiol 73(5). https://doi.org/10.1099/ijsem.0.005921
167. Hanafy RA et al (2023) Phylogenomic analysis of the Neocallimastigomycota: proposal of Caecomycetaceae fam. nov., Piromycetaceae fam. nov., and emended description of the families Neocallimastigaceae and Anaeromycetaceae. Int J Syst Evol Microbiol 73(2). https://doi.org/10.1099/ijsem.0.005735
168. Meili CH et al (2023) Patterns and determinants of the global herbivorous mycobiome. Nat Commun 14(1):3798
169. Grigoriev IV et al (2014) MycoCosm portal: gearing up for 1000 fungal genomes. Nucleic Acids Res 42(D1):D699–D704

170. Ciach MA et al (2024) The interkingdom horizontal gene transfer in 44 early diverging fungi boosted their metabolic, adaptive, and immune capabilities. Evol Lett 8(4):526–538
171. Murphy Chelsea L et al (2019) Horizontal gene transfer as an indispensable driver for evolution of neocallimastigomycota into a distinct gut-dwelling fungal lineage. Appl Environ Microbiol 85(15):e00988–e00919
172. Wilken SE et al (2021) Experimentally validated reconstruction and analysis of a genome-scale metabolic model of an anaerobic Neocallimastigomycota fungus. mSystems 6(1). https://doi.org/10.1128/msystems.00002-21
173. Krauss J, Zverlov VV, Schwarz WH (2012) In vitro reconstitution of the complete Clostridium thermocellum cellulosome and synergistic activity on crystalline cellulose. Appl Environ Microbiol 78(12):4301–4307
174. Ho YW, Abdullah N, Jalaludin S (1988) Penetrating structures of anaerobic rumen fungi in cattle and swamp buffalo. Microbiology 134(1):177–181
175. Boxma B et al (2004) The anaerobic chytridiomycete fungus Piromyces sp. E2 produces ethanol via pyruvate:formate lyase and an alcohol dehydrogenase E. Mol Microbiol 51(5): 1389–1399
176. Boxma B et al (2005) An anaerobic mitochondrion that produces hydrogen. Nature 434(7029): 74–79
177. Li Y et al (2021) Interactions between anaerobic fungi and methanogens in the rumen and their biotechnological potential in biogas production from lignocellulosic materials. Microorganisms 9. https://doi.org/10.3390/microorganisms9010190
178. Gruninger RJ et al (2014) Anaerobic fungi (phylum Neocallimastigomycota): advances in understanding their taxonomy, life cycle, ecology, role and biotechnological potential. FEMS Microbiol Ecol 90(1):1–17
179. Dollhofer V et al (2015) Anaerobic fungi and their potential for biogas production. In: Guebitz GM et al (eds) Biogas science and technology. Springer International Publishing, Cham, pp 41–61
180. Swift CL et al (2021) Anaerobic gut fungi are an untapped reservoir of natural products. Proc Natl Acad Sci 118(18):e2019855118
181. Seppälä S et al (2017) The importance of sourcing enzymes from non-conventional fungi for metabolic engineering and biomass breakdown. Metab Eng 44:45–59
182. Lee SS, Ha JK, Cheng KJ (2000) Influence of an anaerobic fungal culture administration on in vivo ruminal fermentation and nutrient digestion. Anim Feed Sci Technol 88(3):201–217
183. Hartinger T, Zebeli Q (2021) The present role and new potentials of anaerobic fungi in ruminant nutrition. J Fungi 7. https://doi.org/10.3390/jof7030200
184. Agustina S et al (2025) The addition of anaerobic fungi isolates from buffalo rumen to increase fiber digestibility, fermentation, and microbial population in ruminants. Biodivers J Biol Divers 25(1):107
185. Timmis K et al (2017) The contribution of microbial biotechnology to sustainable development goals. Microb Biotechnol 10(5):984–987
186. Volmer J, Schmid A, Bühler B (2015) Guiding bioprocess design by microbial ecology. Curr Opin Microbiol 25:25–32
187. Hakalehto E (2025) Mixed cultures in industrial bioprocesses. In: Ulber R (ed) Advances in biochemical engineering/biotechnology, vol 189. Springer, Cham
188. Schmeckebier A, Zayed A, Ulber R (2022) Productive biofilms: from prokaryotic to eukaryotic systems. J Chem Technol Biotechnol 97(11):3049–3064
189. Philipp L-A et al (2024) Beneficial applications of biofilms. Nat Rev Microbiol 22(5):276–290
190. Stöckl M, Gemünde A, Holtmann D (2024) Microbial electrotechnology – intensification of bioprocesses through the combination of electrochemistry and biotechnology. Phys Sci Rev 9(8):2775–2793
191. Diender M, Parera Olm I, Sousa DZ (2021) Synthetic co-cultures: novel avenues for bio-based processes. Curr Opin Biotechnol 67:72–79

192. Sydow A et al (2014) Electroactive bacteria – molecular mechanisms and genetic tools. Appl Microbiol Biotechnol 98(20):8481–8495
193. Koch C, Harnisch F (2016) Is there a specific ecological niche for electroactive microorganisms? ChemElectroChem 3(9):1282–1295
194. Nielsen J, Keasling JD (2016) Engineering cellular metabolism. Cell 164(6):1185–1197
195. Dharmaraja J et al (2023) Chapter 2: Design and scale-up of photobioreactors. In: Sirohi R et al (eds) Current developments in biotechnology and bioengineering. Elsevier, pp 11–32
196. Kretschmer M et al (2022) A rotating bioreactor for the production of biofilms at the solid–air interface. Biotechnol Bioeng 119(3):895–906
197. Paladino O, Neviani M (2020) Scale-up of photo-bioreactors for microalgae cultivation by π-theorem. Biochem Eng J 153:107398
198. Enzmann F et al (2019) Same but different-Scale up and numbering up in electrobiotechnology and photobiotechnology. Eng Life Sci 19(2):121–132

Adv Biochem Eng Biotechnol (2025) 192: 27–58
https://doi.org/10.1007/10_2024_270

Published online: 4 October 2024

Methanothermobacter thermautotrophicus and Alternative Methanogens: Archaea-Based Production

Lucas Mühling, Tina Baur, and Bastian Molitor

Contents

Abstract Methanogenic archaea convert bacterial fermentation intermediates from the decomposition of organic material into methane. This process has relevance in the global carbon cycle and finds application in anthropogenic processes, such as wastewater treatment and anaerobic digestion. Furthermore, methanogenic archaea that utilize hydrogen and carbon dioxide as substrates are being employed as

---

L. Mühling
Environmental Biotechnology Group, Department of Geosciences, University of Tübingen, Tübingen, Germany

T. Baur and B. Molitor (✉)
Environmental Biotechnology Group, Department of Geosciences, University of Tübingen, Tübingen, Germany

Cluster of Excellence – Controlling Microbes to Fight Infections, University of Tübingen, Tübingen, Germany
e-mail: bastian.molitor@uni-tuebingen.de

biocatalysts for the biomethanation step of power-to-gas technology. This technology converts hydrogen from water electrolysis and carbon dioxide into renewable natural gas (i.e., methane). The application of methanogenic archaea in bioproduction beyond methane has been demonstrated in only a few instances and is limited to mesophilic species for which genetic engineering tools are available. In this chapter, we discuss recent developments for those existing genetically tractable systems and the inclusion of novel genetic tools for thermophilic methanogenic species. We then give an overview of recombinant bioproduction with mesophilic methanogenic archaea and thermophilic non-methanogenic microbes. This is the basis for discussing putative products with thermophilic methanogenic archaea, specifically the species *Methanothermobacter thermautotrophicus*. We give estimates of potential conversion efficiencies for those putative products based on a genome-scale metabolic model for *M. thermautotrophicus*.

Keywords Genetic tools, Metabolic modeling, Methanogens, Methanothermobacter, Recombinant products

1 Introduction

Methanogenic archaea (methanogens) produce methane in a process called methanogenesis [1]. This process allows methanogens to conserve energy and derive carbon for biomass formation. Different methanogenic pathways are known, most relevant are as follows: (1) hydrogenotrophic methanogenesis, in which carbon dioxide is reduced to methane with electrons from hydrogen, formate, and certain alcohols; (2) aceticlastic methanogenesis, in which the methyl carbon from acetate is reduced to methane with electrons from the oxidation of the carbonyl carbon from acetate to carbon dioxide; and (3) methylotrophic methanogenesis, in which methylated compounds, such as methanol, and methylated amines and sulfides, are disproportionated to methane and carbon dioxide [2]. Certain methanogens utilize other substrates, such as carbon monoxide (i.e., carboxydotrophic methanogenesis), methylated aromatics (i.e., methoxydotrophic methanogenesis), and a hydrogen-dependent variation of methylotrophic methanogenesis [3].

Methanogenesis core genes have been identified in metagenome-assembled genomes from diverse environments and suggest that methanogenesis, with potential variations to the described schemes, might be even more widespread in the microbial world than currently acknowledged [4]. Moreover, while methanogens are considered obligate methane producers, this view has been challenged recently by converting an aceticlastic and carboxydotrophic methanogen of the genus *Methanosarcina* into an acetate producer [5]. Importantly, methanogenesis allows methanogens to catalyze the biogenic production of methane as the final step of organic matter decomposition in natural and anthropogenic anaerobic environments with limited availability of external electron acceptors, such as nitrate, sulfate, manganese, and iron [2]. Thus, methanogens are found in various anaerobic environments, including peatlands, rice paddy fields, hydrothermal vents, and the digestive tracts of ruminants and termites [1]. This renders methanogens important for the global carbon cycle [2]. Genetic tools have been established for some methanogenic species, which was initially driven by an interest in fundamental research to understand the unique physiology and biochemistry of the different methanogenic pathways [6].

However, as early as 1982, methanogens have been recognized as potential biocatalysts to replace fossil natural gas with biogenic methane in the search for renewable energy sources [7]. In this regard, methanogens are vital components in anaerobic digestion, which is widely applied for methane production from organic (waste) material [8]. More recently, the biological upgrading of carbon dioxide with pure cultures of hydrogenotrophic methanogens has attracted increasing attention [9, 10]. Species of the genus *Methanothermobacter* have been adopted for the power-to-gas technology with biomethanation due to their high biomass yields with high methane production rates [11, 12]. This technology platform converts hydrogen from water splitting with renewable electric power and carbon dioxide from industrial off-gases into methane [9, 10, 12].

In this chapter, we discuss novel developments that allow to consider methanogens for bioproduction beyond methane as the sole product. For the fundamental details of methanogenesis, we refer to previous reviews on methanogenic physiology and biochemistry [1, 2, 13]. We will (1) provide an overview of recent developments in genetic tools for methanogens; (2) discuss previous achievements on bioproduction with mesophilic methanogens, thermophilic non-methanogens, and a selection of putative target products beyond methane for thermophilic methanogens; and (3) give estimates of the potential for bioproduction with a thermophilic methanogen by employing a manually curated genome-scale metabolic model for *Methanothermobacter thermautotrophicus*. In the discussion of the genetic tools, we focus on the accomplishments after the review by Enzmann et al. [6], which gives an overview until 2018. We emphasize the potential of thermophilic methanogens in bioproduction and discuss the connected limitations because thermophilic methanogens have been adopted for large-scale bioproduction already [14]. Finally, we summarize and give an outlook on future prospects.

2 Genetic Tools for Methanogens

Methanogens are not a monophyletic group but form at least two groups with eight distinct orders in the kingdom of Euryarchaeota [15, 16]. Despite this diversity of methanogens, until recently, genetic tools were only available for mesophilic members of the two orders Methanococcales (*Methanococcus* spp.) and Methanosarcinales (*Methanosarcina* spp.), which were already made genetically tractable decades ago [17, 18]. Thus, genetic work in methanogens focused on these two genera with many genetic tools established over the years [6]. The limitation to a few genetically tractable model microbes restricted genetic investigations of the physiology and metabolism within the diverse group of methanogens. In the last 4 years, researchers around the world changed this by adding genetic tools for thermophilic species from three orders of methanogens: Methanococcales (*Methanocaldococcus* spp.), Methanomicrobiales (*Methanoculleus thermophilus*), and Methanobacteriales (*Methanothermobacter thermautotrophicus*) [19–22]. This now enables genetic studies in thermophilic methanogens with distinct physiological traits. Furthermore, as thermophilic non-methanogenic microbes have already been adopted for biotechnological purposes, genetic tools will allow considering metabolic engineering strategies for bioproduction with thermophilic methanogens.

2.1 Novel Tools for Existing Genetically Tractable Model Systems: Mesophilic Methanogens

In recent years, state-of-the-art genetic tools were developed for species of both the genera *Methanosarcina* and *Methanococcus*, transitioning these genetically tractable model systems to the modern genome-engineering era. In 2017, the CRISPR/Cas9 system from *Streptococcus pyogenes* was adapted for genome editing in *Methanosarcina acetivorans* [23]. The Cas9 enzyme is guided to a specific genomic location by an engineered single-guide RNA and introduces a double-strand break [24]. For the correct recognition of the target sequence, a protospacer adjacent motif on the genome is required that is specific to the Cas enzyme. For *M. acetivorans*, the repair of the resulting double-strand breaks was mediated either by a homology-directed repair with homologous repair templates or by a non-homologous end joining machinery from a related *Methanocella* strain [23]. In addition, a CRISPR interference (CRISPRi) system, which was based on a deactivated Cas9 (dCas9), was adopted for *M. acetivorans* [25]. The system was used to repress the expression of different genes in the nitrogenase gene cluster ($nifHI_1I_2DKEN$). The resulting strains showed impaired or no growth with molecular nitrogen as the sole nitrogen source. Furthermore, the replicative plasmid vector pWM321, which is routinely applied in *Methanosarcina* genetics, was reduced in size, still maintaining the functionality for *Methanosarcina mazei* [26].

Several new genetic tools were developed for *Methanococcus maripaludis*. In 2020, the same CRISPR/Cas9 system as for *M. acetivorans* was implemented for genome editing in *M. maripaludis* [27]. However, another CRISPR system, which was based on Cas12a from *Lachnospiraceae bacterium* ND2006, yielded higher efficiencies and provided a protospacer adjacent motif sequence better suited for the AT-rich *M. maripaludis* genome [28]. In addition, an improved random-mutagenesis protocol, based on the Himar1 transposase, was established for *M. maripaludis* [29]. Certain strains of *M. maripaludis* had been described as naturally competent for DNA uptake from the environment [30]. With the help of the random-mutagenesis protocol, natural competence-deficient mutants were screened. This allowed the identification of several genes required for the natural competence of *M. maripaludis*, thus demonstrating the method to be suitable for functional screening of mutants and identification of previously unknown gene functions [29].

Finally, the fluorescence-activating and absorption-shifting tag (FAST) system was adopted for *M. maripaludis* and *M. acetivorans* [29, 31]. The FAST system is based on a small protein tag that is non-fluorescent by itself. In combination with fluorogenic chemicals that specifically bind the small protein tag, light of a specific wavelength is emitted upon excitation [32]. Notably, the FAST system is active under anaerobic conditions, enabling in vivo experiments [33]. Previously, only in vitro experiments were possible because of the oxygen requirement of fluorescent markers, such as green fluorescent protein, its derivatives, and other alternatives [34, 35]. Furthermore, the FAST system overcomes limitations by the strong autofluorescence of all methanogens caused by the cofactor F_{420} [36]. In *M. maripaludis*, the FAST system was suitable for live-cell fluorescent imaging, studying protein interactions in vivo, and investigating the intracellular location of proteins [31]. For *M. acetivorans*, a FAST-labeled strain could be identified in a mixed microbial population [37]. Recently, the function of the FAST system was demonstrated also for a thermophilic non-methanogenic microbe [38].

2.2 New Genetically Tractable Model Systems: Thermophilic Methanogens

The previously described genetic tools were only applied in mesophilic methanogens. This only recently changed as we discuss in this section. Fundamental limitations had to be overcome to make thermophilic methanogens genetically tractable. The most prominent issue with genetically manipulating thermophilic methanogens was the limited options for the choice of selective markers. Usually, antibiotic resistances are employed as selective markers, but all methanogens (including mesophiles) are resistant to many antibiotics that are typically used in bacteria, such as ampicillin, kanamycin, and tetracycline [39]. Chloramphenicol inhibits methanogens efficiently, but the product of the resistance-conferring enzyme, chloramphenicol acetyltransferase (i.e., acetyl-chloramphenicol), is still toxic to methanogens [40]. Antibiotics that inhibit methanogens include puromycin and neomycin [6]. For thermophiles, in addition, the antibiotic needs to be thermostable. This is the case for both puromycin and neomycin. However, the antibiotic resistance must also be thermostable, which further limits the options. Puromycin is the first choice for mesophilic methanogens, and the puromycin resistance cassette is routinely used in both *Methanococcus* and *Methanosarcina* [6]. However, no reports are available adopting the puromycin resistance for any thermophilic microbe. For three groups of thermophilic methanogens, the limitations of available selective markers could be overcome, which led to the genetic tractability of those microbes (Table 1).

Table 1 Overview of novel genetic systems for thermophilic methanogens

Genus	DNA transfer method	Selective marker	Plasmid vectors	Tools for genome editing	Other tools	References
Methanocaldococcus	Heat shock	Simvastatin/ Mevinolin	Non-replicating plasmid vectors	Gene deletion via homology-directed recombination	Different promoters, purification of tagged proteins	[19]
Methanocaldococcus	PEG-mediated	Neomycin, simvastatin	Non-replicating plasmid vectors	Traceless mutagenesis with *hpt* counterselection	Different promoters	[22]
Methanoculleus	Natural competence	Neomycin	Non-replicating plasmid vectors	Traceless mutagenesis with *upt* counterselection	Purification of tagged proteins	[20, 47]
Methanothermobacter	Conjugation	Neomycin	Shuttle vectors and non-replicating plasmid vectors	Gene deletion via homology-directed recombination	Different promoters, quantified by thermostable ß-galactosidase	[21, 56]

2.2.1 Methanocaldococcus

The first example of how the described limitations for thermophiles were overcome was the presentation of the genetic tools for *Methanocaldococcus jannaschii* in 2019 [19]. *M. jannaschii* is phylogenetically closely related to mesophilic *Methanococcus* spp. but has an optimal growth temperature of 85°C [41]. The limitation of the selective marker was circumvented by applying thermostable mevinolin or simvastatin as selective agents. Those substances are statins and inhibitors of the 3-hydroxymethylglutaryl (HMG)-CoA reductase, a key enzyme of cell-membrane biosynthesis in all archaea, including methanogens [42]. Resistance is not achieved via a classical resistance gene as with antibiotics but by overexpression of the native HMG-CoA reductase-encoding gene (*hmgA*). This selective marker strategy was previously employed in several thermophilic non-methanogenic archaea [43].

DNA transfer into *M. jannaschii* was facilitated by heat-shock (85°C) treatment of previously cold-adapted (65°C) cells. The transferred DNA was a non-replicative and linearized plasmid vector, replacing the F_{420}-dependent sulfite reductase (Fsr)-encoding gene with a mevinolin resistance cassette by homology-directed recombination to yield an allelic exchange mutant [19]. Fsr reduces sulfite, which is toxic to methanogens, to sulfide, and the resulting *M. jannaschii* Δ*fsr* strain was shown to be highly sensitive to sulfite. Thus, it was demonstrated that with heat shock as the transfer method and HMG-CoA reductase-overproduction as a selective marker, DNA could be transferred into *M. jannaschii* to generate allelic exchange mutants. In addition, the authors also demonstrated that the system could be used to tag and overproduce proteins for easier purification: the 5′ region of the gene encoding the cofactor F_{420}-hydrogen oxidase (*fprA*) was replaced by a strong constitutive promoter and a sequence encoding a translational fusion of a 3xFLAGtwin Strep-tag along with the mevinolin resistance cassette. The resulting strain was used to purify Strep-tagged Fpr, and the Fpr activity was confirmed in enzyme assays [19].

Independently from this work, an alternative strategy was employed to genetically manipulate the strain *Methanocaldococcus* sp. FS406-22 [22]. In this case, the authors adopted the existing genetic tools for *M. maripaludis*. In *M. maripaludis*, traceless gene deletion is achieved by a "pop-in/pop-out" strategy, which requires a selective marker and a counterselective marker, the latter conferring sensitivity to a counterselective agent [44]. In *M. maripaludis*, the hypoxanthine phosphoribosyl transferase (*hpt*)-encoding gene is employed as the counterselective marker. Cells that carry an *hpt* gene are sensitive, and cells that do not carry an *hpt* gene are resistant to toxic base analogs, such as 8-aza-hypoxanthine. Both *M. maripaludis* and *Methanocaldococcus* sp. FS406-22 possess an *hpt* gene and are sensitive to 8-azahypoxanthine. Thus, generating an *hpt*-deletion (Δ*hpt*) strain is a prerequisite for using the "pop-in/pop-out" strategy. In *Methanocaldococcus* sp. FS406-22, this was achieved by introducing a non-replicative plasmid vector for *hpt* gene deletion by homology-directed recombination. The DNA transfer was mediated by a polyethylene glycol (PEG)-transformation protocol, which is routinely used in *M. maripaludis* [45]. Selecting successful deletion mutants was accomplished by screening for resistance to 8-aza-hypoxanthine [22]. The resulting *Methanocaldococcus* sp. FS406-22 Δ*hpt* strain was used to establish a "pop-in/pop-out" strategy akin to *M. maripaludis*, with a thermostable neomycin resistance as a selective marker. The functionality of the thermostable neomycin resistance was demonstrated for thermophilic microbes before [46]. With the help of this technique, knock-out mutants were generated for the *nifD* gene, which encodes for a nitrogenase subunit, as well as several additional nitrogenase-associated genes. This allowed investigating which of those genes are required for diazotrophic growth. It was also demonstrated that simvastatin and *hmgA* expression could be used as a selective marker, similar to *M. jannaschii* [22].

2.2.2 Methanoculleus

In 2020, genetic tools for *Methanoculleus thermophilus* from the order Methanomicrobiales were presented [20]. *M. thermophilus* was found to be naturally competent for DNA uptake of non-replicative plasmid vectors that could be integrated into the genome by homology-directed recombination with the same thermostable neomycin resistance as a selective marker that had been applied in *Methanocaldococcus* sp. FS406-22 [20]. Furthermore, the authors adapted the "pop-in/pop-out" strategy from *M. maripaludis*. Instead of the *hpt* gene, the uracil phosphoribosyl transferase-encoding gene (*upt*) combined with the neomycin resistance marker was applied to generate traceless mutations. Uracil phosphoribosyl transferase confers sensitivity to uracil base analogs, such as 6-aza-uracil, and can be used as a counterselective marker in the same manner as the previously described *hpt* gene [20].

Interestingly, *M. thermophilus* was naturally resistant to 6-aza-uracil, circumventing the requirement of a prior deletion of the *upt* gene. With the thermostable neomycin resistance marker and the *upt* gene from *M. maripaludis* as the counterselective marker, a traceless mutant of *M. thermophilus* was generated, which was lacking the operon that encodes for a putative type IV pili system. The resulting strain was not naturally competent anymore, revealing an involvement of the type IV pili system in the uptake of foreign DNA. Furthermore, the authors demonstrated that by expressing the type IV pili system of naturally competent *M. maripaludis* strain JJ in the non-competent strain *M. maripaludis* S2, the latter strain acquired natural competence [20].

In a follow-up publication, the genetic system for *M. thermophilus* was applied to investigate formate metabolism in *M. thermophilus* by attempting to delete *fdhA* isogenes that encode homologs of the α-subunit of the formate dehydrogenase [47]. While the mutants lacking either *fdhA1*, *fdhA3*, or both genes did not have an observable phenotype, the deletion of *fdhA2* was unsuccessful, suggesting that *fdhA2* was essential even when *M. thermophilus* was grown on hydrogen and carbon dioxide. The reason for this was further investigated by translationally fusing a polyhistidine-tag to the heterodisulfide reductase and subsequently purifying the recombinant enzyme. Several other enzymes were copurified in this process, including Fdh2, but no hydrogenase. Activity assays of the enzyme preparation showed that the heterodisulfide reductase is active with formate but not hydrogen as an electron donor. This suggested that in *M. thermophilus*, formate instead of hydrogen is the electron donor for the heterodisulfide reductase [47].

2.2.3 Methanothermobacter

Species of the genus *Methanothermobacter* in the order Methanobacteriales, especially *M. marburgensis* and *M. thermautotrophicus*, have been studied as model microbes for methanogenesis and the involved enzymes [13]. For this reason, genetic tools for those microbes were long sought after [48–50]. Attempts to establish genetic tools for *Methanothermobacter* spp. were made as early as 1985 [51]. While there were reports about DNA transfer into *M. marburgensis* by either natural competence [52] or phage transduction [53], DNA was only transferred between different mutants of *M. marburgensis*. No uptake of heterologous DNA was ever demonstrated, and there were no follow-up publications on either method for the next 30 years.

Only in 2021, a genetic system for *M. thermautotrophicus* ΔH was published [21]. The genetic systems for *Methanocaldococcus* spp. and *M. thermophilus* rely on non-replicating plasmid vectors. In contrast, the first genetic tools for *M. thermautotrophicus* ΔH were established by employing a shuttle vector that can replicate in both *Escherichia coli* and *M. thermautotrophicus* ΔH, similar to the shuttle vectors used in *Methanococcus* spp. and *Methanosarcina* spp. [6]. In the case of *M. thermautotrophicus* ΔH, the origin of replication was the complete cryptic plasmid pME2001 from *M. marburgensis* [49]. DNA transfer was realized by interdomain conjugation from *E. coli* to *M. thermautotrophicus* ΔH, which had been reported for *M. maripaludis* before [54]. Exconjugants were selected employing the same thermostable neomycin resistance that was utilized in *Methanocaldococcus* sp. FS406-22 and *M. thermophilus*. Heterologous gene expression was verified by producing a thermostable β-galactosidase, which was then applied to test the strength of different constitutive promoters. In addition, the potential for metabolic engineering was demonstrated by expressing the formate dehydrogenase (*fdh*) operon from the closely related strain *M. thermautotrophicus* Z-245. *M. thermautotrophicus* Z-245 can grow with formate as an alternative electron donor instead of hydrogen [55]. It was confirmed that a strain of *M. thermautotrophicus* ΔH carrying the *fdh* operon from *M. thermautotrophicus* Z-245 can also grow with formate as the sole electron donor [21].

A follow-up publication demonstrated that the genetic tools could be applied to delete genes in *M. thermautotrophicus* ΔH by an allelic exchange strategy [56]. For this, non-replicating plasmid vectors were designed. These plasmids did not carry the origin of replication for *M. thermautotrophicus* ΔH but instead encoded the neomycin resistance marker flanked by 1-kb homologous regions up- and downstream of the target site. Such a plasmid was used to delete the operon that encodes for so-called Mth60 fimbriae, which are cell appendages influencing interactions of *M. thermautotrophicus* ΔH with abiotic surfaces and other cells [57, 58]. The resulting deletion strain showed a significant decrease in intercellular interactions, further substantiating the importance of the Mth60 fimbriae in forming cell-cell connections [56].

3 Putative Products with Methanogens

Biogenic methane has been recognized as a valuable methanogenic product [9, 10]. Recently, secreted amino acids from wild-type strains have been identified as potential methanogenic products [59]. The previously discussed genetic tools for various methanogens opened the door to consider the recombinant production of value-added chemicals beyond methane in methanogen-based bioproduction [14, 60]. In this section, we will (1) discuss the achievements with mesophilic methanogens, (2) provide an overview of a selection of products with thermophilic non-methanogens, and (3) discuss putative products with the thermophilic methanogen *M. thermautotrophicus*.

3.1 Previous Achievements with Methanogens

As reviewed above, genetic tools for alternative methanogens only became available recently. Thus, it comes with no surprise that efforts to generate methanogenic strains with the capacity to produce additional chemicals besides methane recombinantly are, thus far, only available for the orders Methanococcales and Methanosarcinales (Table 2).

The first success in recombinant chemical production with a methanogen was in 2016 when *M. maripaludis* was engineered to produce the monoterpenoid geraniol

Table 2 Overview of previous achievements in recombinant production of chemicals with methanogens

Organism	Product	Substance class	Titer [g/L]	References
Methanosarcina acetivorans	Isoprene	Hemiterpene	0.06–0.07	[61, 62]
Methanosarcina barkeri	Isoprene	Hemiterpene	0.002–0.003	[61]
Methanococcus maripaludis	Geraniol	Monoterpenoid	0.0005	[63]
Methanosarcina acetivorans	Acetate	Carboxylic acid	0.61	[64]
Methanosarcina acetivorans	Acetate	Carboxylic acid	0.05–0.06	[5]
Methanosarcina acetivorans	Lactate	Hydroxy acid	0.10	[65]
Methanococcus maripaludis	3-HB	Hydroxy acid	0.15	[66]
Methanococcus maripaludis	PHB	Polyhydroxyalkanoate	0.17	[66]

3-HB 3-hydroxybutyrate, *PHB* poly(3-hydroxybutyrate)

(Table 2) [63]. Methanogens, like all archaea, possess cell membranes that are composed of terpenoid-based lipids. These terpenoids are biosynthesized via the mevalonate pathway with isopentenyl pyrophosphate and dimethylallyl pyrophosphate as essential intermediates [67]. Thus, the capability for terpenoid production is natively present in methanogens. Because of growth requirements, the metabolic flux through the mevalonate pathway is naturally high [68]. Thus, during growth, the intermediates isopentenyl pyrophosphate and dimethylallyl pyrophosphate and their derivatives geranyl pyrophosphate, farnesyl pyrophosphate, and geranylgeranyl pyrophosphate should be abundant in archaea. However, the pathway's genetic and enzymatic feedback control orchestrates the lipid metabolism and may limit precursor availability for bioproduction [67]. Dimethylallyl pyrophosphate is the precursor for the simplest terpenoid, the hemiterpene isoprene, while geraniol can be derived from geranyl pyrophosphate [68]. This makes archaea, and thus methanogens, ideal chassis strains for terpenoid production because only a few heterologous genes are required for terpenoid production. Indeed, geraniol production with *M. maripaludis* was achieved by expressing the geraniol synthase gene from *Ocimum basilicum* [63]. Following this first demonstration of terpenoid production, *M. acetivorans* and *Methanosarcina barkeri* were engineered to produce isoprene by expressing the isoprene synthase gene from *Populus alba* [61, 62].

Recently, *M. acetivorans* was engineered to produce acetate as the main metabolic product, with only small amounts of methane and formate remaining as side products with carbon monoxide as the sole substrate [5]. *M. acetivorans* can utilize acetate as a growth substrate in aceticlastic methanogenesis. Thus, the genetic equipment for acetate production by reversal of the pathway is already present. However, the predominant production of acetate with a methanogen was a milestone for potential bioproduction and provides further experimental insight into the evolutionary origin of methanogens [5].

Two more studies reported acetate and lactate production with *M. acetivorans*, respectively [64, 65]. Both studies reported that this was realized by hijacking methanogenesis and reverting the carbon flux from methane toward acids in reverse methanogenesis in a strain that expresses the methyl-coenzyme M reductase from anaerobic methane-oxidizing archaea. For lactate production, the recombinant expression of a 3-hydroxybutyryl-CoA dehydrogenase (Hbd)-encoding gene was necessary in addition [65]. In both cases, the conversion of a fraction of the provided methane to acetate or lactate in batch cultivations was observed. However, the inoculation with high cell density cultures that were pre-grown with methanol was required, the conversion of methane to acids was limited, and substantiated growth was not achieved [69].

Finally, a recent publication demonstrated the recombinant production of 3-hydroxybutyrate and the derived polymer, poly(3-hydroxybutyrate), with *M. maripaludis* [66]. This milestone can be considered the first production of the bioplastic poly(3-hydroxybutyrate) with a methanogen. After achieving 3-hydroxybutyrate production, the authors modified the established synthetic pathway to achieve the conversion of acetyl-CoA to poly(3-hydroxybutyrate) with the enzymes acetoacetyl-CoA thiolase (PhaA, from *Clostridium acetobutylicum*), *3R*-

hydroxybutyryl-CoA dehydrogenase (PhaB, from *Halomonas bluephagenesis*), and poly(3-hydroxybutyrate) polymerase (PhaC, from *Cupriavidus necator*). In addition, the authors implemented a pathway optimizing the regeneration of the NADH pool, which provided the reducing equivalents for the 3-hydroxybutyrate and poly (3-hydroxybutyrate) production pathways. This was required because methanogens evolved to utilize ferredoxin and F_{420} as the major redox cofactors. Increasing NADH availability yielded up to 0.171 ± 0.004 g/L poly(3-hydroxybutyrate) from hydrogen and carbon dioxide [66].

The list of achievements in metabolic engineering of methanogenic archaea is still relatively short, and the titers of the produced compounds are low. However, they highlight the potential of methanogens as alternative production hosts, with a large area of possible applications remaining unexplored. This includes adopting thermophilic methanogens as production hosts with the novel genetic tools (Sect. 2). In the following section, we discuss selected recombinant products in non-methanogenic thermophiles, which provides a foundation for the discussion of potential products with thermophilic methanogens.

3.2 Previous Achievements with Thermophilic Non-methanogenic Microbes

For biotechnological purposes, enzymes from thermophilic microbes are widely applied in different fields, such as molecular biology (e.g., polymerase chain reaction, PCR), food industry (e.g., baking), and laundry (e.g., lipases and proteases) [70]. However, thermophilic microbes also often have higher growth and metabolic rates than mesophilic microbes, making them promising biotechnological production hosts [71]. In addition, thermophilic processes are less susceptible to contamination [72]. Metabolic engineering of thermophilic microbes led to promising results regarding recombinant chemical production (Table 3). Various products were targeted, including (as a non-exclusive list): (1) the ketones/alcohols acetone, ethanol, butanol, isobutanol, 2,3-butanediol, and acetoin; (2) the terpenoid phytoene; and (3) the hydroxy acids lactate and 3-hydroxypropionate (Table 3). These compounds are relevant platform chemicals as they can be used as solvents, food and fuel additives, fragrances, flavor enhancers, and precursors or building blocks for other compounds, including polymers that have potential applications as bioplastics [73–75]. The recombinant production of these compounds with thermophiles indicates the availability of thermostable production routes with respective thermostable enzymes for heterologous production. Most of these recombinant production routes were based on heterotrophic microbes, mainly with sugars as substrates. One exception is the production of acetone with *Moorella thermoacetica* from carbon dioxide and hydrogen, carbon monoxide, and syngas (Table 3).

Most of the displayed results were achieved with bacteria, with the archaea *Thermococcus kodakarensis* and *Pyrococcus furiosus* as exemptions. In addition,

Table 3 Non-exclusive selection of previous achievements in metabolic engineering of thermophilic microbes. The given titers are the reported maxima presented in the respective reference

Organism	Domain	Substrate(s)	Product(s)	Titer [g/L]	References
Pyrococcus furiosus	Archaea	Maltose	Lactate	0.27	[76]
Pyrococcus furiosus	Archaea	Maltose + H_2 + HCO_3^-	3-HP	0.05	[77]
Pyrococcus furiosus	Archaea	Maltose	Butanol	0.06	[77]
Thermococcus kodakarensis	Archaea	Pyruvate	Phytoene	0.003	[74]
Geobacillus thermoglucosidasius	Bacteria	Cellobiose	Ethanol	14.83	[78]
Geobacillus thermoglucosidasius	Bacteria	Glucose	Isobutanol	3.30	[79]
Parageobacillus thermoglucosidasius	Bacteria	Glucose	2,3-BD	7.20	[80]
Parageobacillus thermoglucosidasius	Bacteria	Glucose	Acetoin 2,3-BD	7.60 15.60	[75]
Thermoanaerobacterium saccharolyticum	Bacteria	Glucose + xylose + mannose + galactose	Ethanol	37.00	[81]
Thermoanaerobacterium saccharolyticum	Bacteria	Xylose	Butanol	1.05	[82]
Thermoanaerobacterium thermosaccharolyticum	Bacteria	Cellobiose	Butanol	0.38	[82]
Caldicellulosiruptor bescii	Bacteria	Cellobiose	Ethanol	0.68	[83]
Moorella thermoacetica	Bacteria	Glucose	Ethanol	9.21	[84]
Moorella thermoacetica	Bacteria	Fructose	Lactate	3.60	[72]
Moorella thermoacetica	Bacteria	Fructose H_2 + CO_2 CO Syngas	Acetone	0.58 0.10 0.06 0.19	[85]

3-HP 3-hydroxypropionate, *2,3-BD* 2,3-butanediol

genetic tools exist for archaea of the genera *Sulfolobus* and *Saccharolobus*, although those have not yet been applied for bioproduction [86]. With the recent achievements in genetic tools for thermophilic methanogens, as discussed above, the adoption of thermophilic archaea for bioproduction can be expected to continue.

3.3 *Putative Products with* M. thermautotrophicus

From the discussed thermophilic methanogens, only species of the genus *Methanothermobacter* are already applied on a large scale in power-to-gas technology with biomethanation [21, 86–88]. The genetic tools that are now available make *Methanothermobacter* spp. attractive hosts for producing value-added chemicals.

M. thermautotrophicus is a hydrogenotrophic methanogen and a true autotroph [48]. Those microbes require only mineral salt medium, carbon dioxide, and hydrogen for growth. Thus, bioproduction with *M. thermautotrophicus* has minimal requirements. Based on the available literature for mesophilic methanogens and bioproduction in thermophilic non-methanogens, in this section, we discuss putative bioproduction routes that branch off from the central metabolic intermediates acetyl-CoA and pyruvate in *M. thermautotrophicus* (Fig. 1). The discussion will primarily focus on selected products previously achieved with thermophilic microbes (Sect. 3.2). We further restrict the discussion to pathways for which no more than three heterologous genes must be introduced. This should keep the metabolic burden as low as possible. Furthermore, we will focus on products that could potentially compete with bacterial production strains when thermophilic methanogens are employed for production. Thus, not all products from Sect. 3.2 (Table 3) are discussed.

3.3.1 Isoprene and Terpenoids

Isoprene is an important platform chemical as a precursor for producing rubber, pesticides, biofuels, and pharmaceuticals [89]. As a methanogenic archaeon, *M. thermautotrophicus* natively produces isoprene precursors (and precursors for more complex terpenoids) in the mevalonate pathway (Sect. 3.1). Since recombinant isoprene and geraniol production were demonstrated in mesophilic methanogens, isoprene and terpenoids are promising candidates for recombinant chemical production with *M. thermautotrophicus*.

Bacteria do not synthesize terpenoid-based lipids. However, some bacteria contain the deoxyxylulose 5-phosphate pathway as an alternative pathway for the biosynthesis of isopentenyl pyrophosphate/dimethylallyl pyrophosphate [90]. Several species have been reported to natively produce isoprene, including *Bacillus subtilis*, *Micrococcus luteus*, *Bacillus amyloliquefaciens*, and *Acinetobacter calcoaceticus* [91]). Bacteria have been engineered to express parts of the mevalonate pathway in addition to isoprene synthases for recombinant isoprene production. This way, isoprene production was accomplished heterotrophically with bacteria, such as *E. coli* (25–60 g/L from glucose; [90]), *Clostridium cellulolyticum* (0.41×10^{-4} g/L from cellulose [92]), and *C. necator* (1.5×10^{-6}–3.8×10^{-6} g/L from fructose [93]). Autotrophic isoprene production was achieved with bacteria, such as the acetogen *Clostridium ljungdahlii* (1.5×10^{-6} g/L from syngas; 1.4×10^{-4} g/L with external mevalonate addition [94]), and the cyanobacterium *Synechococcus elongatus* (1.26 g/L from carbon dioxide [95]).

Thus, the highest isoprene production was reported with a heterotrophic bacterium from sugars, followed by the photoautotrophic cyanobacterial production from carbon dioxide. Using sugars is not ideal, as this poses questions on land use for food vs chemical production [96]. For the phototrophic production with cyanobacteria, the limitation of light distribution in large-scale bioreactors presents a technical issue [97]. Instead, isoprene production with hydrogenotrophic methanogens, such as

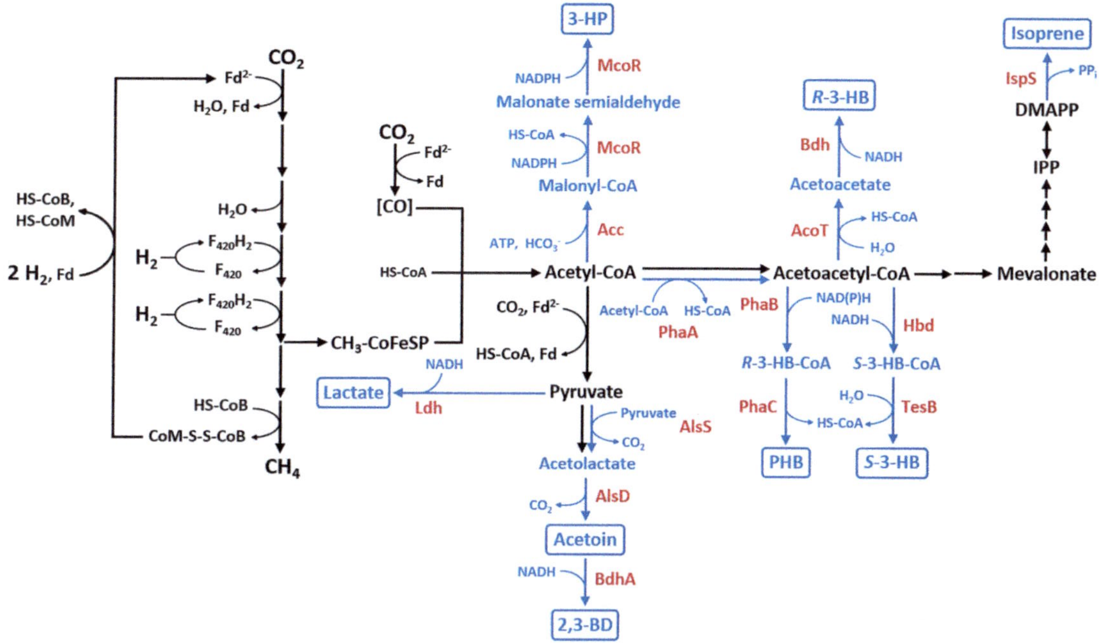

Fig. 1 Schematic overview of metabolic pathways of *M. thermautotrophicus* leading to the natural product methane and several potential recombinant products (framed). Native pathways are shown in black, and synthetic pathways are displayed in blue. *3-HP* 3-hydroxypropionate, *R-/S-3-HB R-/S*-3-hydroxybutyrate, *IPP* isopentenyl pyrophosphate, *DMAPP* dimethylallyl pyrophosphate, *PHB* poly(3-hydroxybutyrate), *2,3-BD* 2,3-butanediol, *Acc* acetyl-CoA carboxylase, *McoR* malonyl-CoA reductase (also has a 3-hydroxypropionate dehydrogenase activity leading to 3-HP), *AcoT* acetoacetyl-CoA hydrolase (acyl-CoA thioesterase), *Bdh* 3-hydroxybutyrate dehydrogenase, *IspS* isoprene synthase, *Ldh* lactate dehydrogenase, *AlsS* acetolactate synthase, *AlsD* acetolactate decarboxylase, *BdhA* acetoin reductase (2,3-BD dehydrogenase), *PhaA* β-ketothiolase, *PhaB* acetoacetyl-CoA reductase, *PhaC* PHB synthase, *Hbd* 3-hydroxybutyryl-CoA dehydrogenase, *TesB* acyl-CoA thioesterase

M. thermautotrophicus, can be achieved with hydrogen and carbon dioxide as the substrates. Furthermore, adding an isoprene synthase in methanogens suffices without the need to express additional parts of the mevalonate pathway, as is the case for most bacterial hosts (Fig. 1). As opposed to mesophilic methanogens, in *M. thermautotrophicus* isoprene would readily be separated from the culture broth because it would be gaseous at a cultivation temperature of approximately 65°C as it has a boiling point of 34°C.

However, for the thermophilic production of isoprene with *M. thermautotrophicus*, a thermostable isoprene synthase would be necessary. Isoprene synthases predominantly occur in chloroplasts of green plants [98]. Although most isoprene synthases have reported in vitro temperature optima around 40–50°C [99], thermo-adaptation might be required for isoprene (or terpenoid) production in *M. thermautotrophicus*. Furthermore, most isoprene synthases have low activity as the K_m for the substrate dimethylallyl pyrophosphate is relatively high (in the millimolar range), resulting in low catalytic activity (k_{cat}) values [99].

Lastly, introducing an isoprene synthase or terpenoid biosynthesis pathway would likely stress the cells, as carbon is diverted from cell-membrane metabolism to recombinant bioproduction. At the same time, regulatory mechanisms may keep precursors and, eventually, isoprene titers low. Thus, further metabolic engineering would likely be necessary to optimize isoprene production. Once established and optimized, the isoprene production from carbon dioxide could be a valuable and more sustainable alternative to current isoprene production, which relies on fossil sources [90, 94].

3.3.2 Lactate

Lactate is utilized in the pharmaceutical, food, and cosmetics industries and is the precursor for polylactic acid, which is considered an important bioplastic [96]. For microbial lactate production, lactic acid bacteria are hard to beat with maximum theoretical yields as high as 2 mol of lactate per mole of glucose [100]. However, these heterotrophic bacteria need glucose or other sugars as substrates. Only a few reports on autotrophic lactate production are available, with acetogens from hydrogen and carbon dioxide, carbon monoxide, or syngas. Some acetogens naturally produce lactate at low levels, for example, *Clostridium autoethanogenum* (0.02 g/L), *Butyribacterium methylotrophicum* (0.41 g/L), and *Clostridium ragsdalei* (9×10^{-3} g/L) [96, 101–103]. In those microbes, lactate reutilization so far prevented higher lactate yields. *Acetobacterium woodii* was engineered to prevent lactate reutilization, which resulted in increased lactate production (up to 0.9 g/L) [104, 105].

Implementing a recombinant lactate production pathway in *M. thermautotrophicus* could be an alternative for lactate production from carbon dioxide, as *M. thermautotrophicus* does not utilize lactate. Thus, reutilization should not be an issue. Metabolic engineering would include the introduction of a thermostable lactate dehydrogenase that can reduce pyruvate to lactate (Fig. 1). With

NADH as the electron donor, the reaction is exergonic ($\Delta G^{\circ\prime} = -25$ kJ/mol lactate) and could serve as an electron sink [96]. However, similar to NADH-dependent bioproduction in *M. maripaludis* (Sect. 3.1), implementing NADH-regenerating systems might be necessary for *M. thermautotrophicus* [66]. If the lactate dehydrogenase were electron-bifurcating, such as the one from *Acetobacterium woodii* [106], additional reduced ferredoxin might be produced [96]. This, in turn, might reduce the metabolic cost of lactate bioproduction, potentially increasing the yield. In terms of enzyme selection, several options from thermophilic microbes with reported lactate dehydrogenases can be considered, exemplified by *Thermoanaerobacter ethanolicus* [107], *Thermotoga maritima* [108], *Caldicellulosiruptor bescii* [76], and *Streptococcus thermophilus* [109].

3.3.3 Acetoin and 2,3-Butanediol

Acetoin is used as a fragrance and flavor enhancer [110], while 2,3-butanediol has applications as a solvent and fuel additive [96]. Many bacteria have been reported that naturally produce acetoin and 2,3-butanediol with varying yields, such as *Klebsiella* spp., *Bacillus* spp., *Paenibacillus* spp., *Serratia* spp., and *Clostridium* spp. [110–112]. Most studies focused on the production of 2,3-butanediol as titers of the natural producers are already high. However, additional native products, such as acetate, ethanol, and lactate, reduced the specificity (i.e., 2,3-butanediol production over all other products) that was achieved [112]. Thermophilic production of 2,3-butanediol has been realized recombinantly with *Parageobacillus thermoglucosidasius* and glucose as a substrate [75, 80]. Acetoin is the precursor of 2,3-butanediol and, thus far, has not received much attention in bioproduction attempts [110].

One advantage of methanogenic production of these compounds would be the facilitated separation from other fermentation products, which in the case of *M. thermautotrophicus* (and other methanogens) would only be the gaseous side-product methane. For acetoin production, which starts from pyruvate, two enzymes are required: acetolactate synthase and acetolactate decarboxylase (Fig. 1). Both enzymes produce carbon dioxide, which could be refixated by the microbes given sufficient hydrogen supply, providing additional reducing equivalents. 2,3-Butanediol production requires a third enzyme, which reduces acetoin to the alcohol (i.e., acetoin reductase/2,3-butanediol dehydrogenase). Many thermophilic sources are available for these enzymes, for example, *Bacillus* spp. and *Streptococcus thermophilus* [75, 109].

3.3.4 3-Hydroxypropionate

3-Hydroxypropionate is a precursor for chemicals, such as acrylates, 1,3-propanediol, malonic acid, and the polymer poly(3-hydroxypropionic acid) [111, 113]. 3-Hydroxypropionate is natively produced by different microbes, including *Metallosphaera sedula*, *Sulfolobus* spp., and *Chloroflexus aurantiacus* [111]. Although high titers of 3-hydroxypropionate can already be produced with different production strains, such as *Klebsiella pneumoniae*, *Lactobacillus reuteri*, and *E. coli* [114], all rely on heterotrophic routes utilizing sugars or glycerol as substrates. Only a few studies have explored the possibility of producing 3-hydroxypropionate autotrophically, including with *C. necator* (0.05 g/L [115]), *Synechocystis* sp. PCC 6803 (0.85 g/L [116]), and *A. woodii* (0.08 g/L [117]). For the implementation of 3-hydroxypropionate production in *M. thermautotrophicus*, the introduction of three enzymes would be required, namely, acetyl-CoA carboxylase, malonyl-CoA reductase, and malonate semialdehyde reductase (i.e., 3-hydroxypropionate dehydrogenase) [111]. In some microbes, such as *C. aurantiacus*, the malonyl-CoA reductase and 3-hydroxypropionate dehydrogenase reactions are catalyzed by the same enzyme (McoR), which could limit the introduction of recombinant enzymes to two (Fig. 1). Because the conversion of acetyl-CoA to malonyl-CoA includes the consumption of one ATP, the introduction of the 3-hydroxypropionate pathway would pose an extra energetic burden to *M. thermautotrophicus*. The malonyl-CoA reductase and 3-hydroxypropionate dehydrogenase reactions are NADPH-dependent. According to literature, $NADP^+$ can be reduced directly with reduced cofactor F_{420} in *M. thermautotrophicus* [118]. Thus, the introduction of a synthetic pathway to 3-hydroxypropionate might not require further cofactor engineering. However, it cannot be ruled out that a cofactor regeneration system might be required to optimize the reducing equivalent pools, similar to what has been demonstrated for NADH-dependent poly (3-hydroxybutyrate) production in *M. maripaludis* [66]. Thermophilic 3-hydroxypropionate production was successfully established in *P. furiosus*, utilizing parts of the 3-hydroxypropionate/4-hydroxybutyrate pathway, which generates 3-hydroxypropionate from acetyl-CoA [77], providing a source for thermostable enzymes.

3.3.5 3-Hydroxybutyrate and Poly(3-Hydroxybutyrate)

3-Hydroxybutyrate is a building block for cosmetics, pharmaceuticals, and the bioplastic poly(3-hydroxybutyrate) [119]. Many native 3-hydroxybutyrate/poly (3-hydroxybutyrate) producers, such as *Burkholderia* spp., *Clostridium* spp., *Chromobacterium* spp., and *C. necator*, have been described [119–121]. *C. necator* can accumulate polyhydroxyalkanoates, including poly (3-hydroxybutyrate), with up to 96% of its cell dry weight. It is further natively able to produce poly(3-hydroxybutyrate) autotrophically, whereas the other microbes produce poly(3-hydroxybutyrate) from carbohydrates or organic acids [119–121]. For these reasons, *C. necator* has been most extensively studied concerning poly(3-hydroxybutyrate) production [121].

Furthermore, recombinant production with cyanobacteria or autotrophic clostridia, such as *Clostridium coskatii*, *C. ljungdahlii*, and *A. woodii*, was made, however, with limited success [119, 121–123]. One remaining challenge for the polymer poly(3-hydroxybutyrate) is the need for extraction from the cells because this polymer is stored within the cells. Therefore, proper extraction methods are required. Recent protocols realize the extraction with lyophilization followed by lysis of the cells [119, 123]. The efficiency of such extraction methods would need to be evaluated for methanogens.

The implementation of 3-hydroxybutyrate production by metabolic engineering of *M. thermautotrophicus* would require the addition of β-ketothiolase (PhaA), 3-hydroxybutyryl-CoA dehydrogenase (Hbd), and acyl-CoA thioesterase (TesB), yielding *R*-3-hydroxybutyrate as it was performed for *M. maripaludis* (Fig. 1) [66]. Alternatively, the pathway could proceed from acetoacetyl-CoA via acetoacetate by the action of acetoacetyl-CoA hydrolase (AcoT) and 3-hydroxybutyrate dehydrogenase (Bdh), yielding *S*-3-hydroxybutyrate [119, 122]. The bioproduction of poly(3-hydroxybutyrate) would include PhaA, acetoacetyl-CoA reductase (PhaB), and poly(3-hydroxybutyrate) synthase (PhaC) [121].

To implement 3-hydroxybutyrate or poly(3-hydroxybutyrate) production in *M. thermautotrophicus,* suitable thermostable enzymes with high catalytic efficiency must be identified. For 3-hydroxybutyrate, to our knowledge, no thermophilic producer is reported, but respective pathways could be pieced together using enzymes from different thermophiles in combination with thermo-adapted homologs from mesophilic microbes. Potential candidates for poly(3-hydroxybutyrate) are the *pha* genes from thermophilic polyhydroxyalkanoate producers such as *Thermus thermophilus*, *Bacillus licheniformis*, and *Caldimonas taiwanensis* [124].

4 Theoretical Yields for Products with *M. thermautotrophicus*

We discussed several putative production routes with *M. thermautotrophicus* in the previous section. While those hypothetic production routes await experimental implementation in *M. thermautotrophicus* or alternative thermophilic methanogens, the theoretical potential can be further assessed. For this purpose, bioinformatic tools, such as genome-scale metabolic models, can provide insight. Of course, those modeling results need to be validated experimentally. Genome-scale metabolic models use the reconstruction of an organism's metabolism derived from the genomically encoded metabolic potential to simulate the metabolic fluxes through the metabolism. Genes encoding metabolic enzymes are assigned metabolic reactions, which then reconstitute the metabolic network of an organism [125]. The reconstruction can be automated with existing databases, but for high-quality models that accurately represent the metabolic network of an organism, manual curation of the assigned reactions, integration of experimentally derived data [126], and thermodynamic calculations to determine the reversibility of each reaction [127, 128] are required. The resulting genome-scale metabolic model is the basis for methods such as flux balance analysis, which simulates steady-state metabolic fluxes within the cell based on defined constraints [125].

Such a manually curated genome-scale metabolic model was recently published for *M. thermautotrophicus* ΔH and the related strains *M. thermautotrophicus* Z-245 and *M. marburgensis* [128]. Based on the genome-scale metabolic model for *M. thermautotrophicus* ΔH, we provide estimations of the bioproduction potential for the products that we discussed before (Table 4). These estimations must be taken cautiously: the genome-scale metabolic model was based on curated genomic information and includes thermodynamic considerations. Further, it was constrained with transcriptomic, proteomic, and steady-state fermentation data (specifically, growth rates and gas consumption/production rates) [128]. However, it does not consider any kinetic limitations for the involved enzymes but reflects steady-state metabolic fluxes. Furthermore, while the model was constrained with proteome and transcriptome data, it cannot reflect regulation on the enzyme level and regulatory changes (e.g., caused by the presence of a recombinant metabolic pathway, which is not present in the wild-type strain). Therefore, the presented values should not be seen as predicted production values but as theoretical estimates of the metabolic potential. The calculations are further based on the constraints from the original publication [128]. Thus, variations in the modeling parameters with experimental constraints from bioreactors that are operated in different conditions will likely change the outcome of the modeling. However, the flux balance analyses with maximization of biomass and products, respectively, give a range of the ability to convert carbon from carbon dioxide into the various products (Table 4).

Table 4 Theoretical yields calculated with the genome-scale metabolic model for *M. thermautotrophicus* ΔH, given as percent values of total carbon consumed (in the form of carbon dioxide). We constrained the model based on the experimental data ± error rates from Casini et al. [128]. Non-growth-associated ATP maintenance (NGAM, reaction "ATPM" in the model) of *M. thermautotrophicus* ΔH was calculated by maximizing the ATPM reaction as the objective function and performing a flux balance analysis. The resulting value was then applied as a constraint for all further calculations. For each product, a new instance of the genome-scale metabolic model was created for which the specific reactions for the respective product were added. Eventually, two independent flux balance analyses were performed for each model instance, one with maximizing biomass and one with maximizing the product as the objective function; the range gives the estimation with maximization of biomass to the estimation with maximization of the respective product

Product	Biomass	Methane	Product yield
Isoprene	4.9–3.1%	95.1–84.1%	0–12.1%
Lactate	4.4–2.7%	95.4–80.4%	0.24–16.9%
Acetoin	4.3–3.8%	93.5–82.7%	2.17–14.6%
2,3-BD	4.4–2.7%	93.9–82.7%	1.75–14.6%
3-HP	4.9–2.7%	95.1–83.6%	0–13.6%
3-HB	4.4–2.7%	95.2–80.1%	0.40–17.1%
PHB	4.4–2.7%	95.2–79.0%	0.40–18.1%

2,3-BD 2,3-butanediol, *3-HP* 3-hydroxypropionate, *3-HB* 3-hydroxybutyrate, *PHB* poly (3-hydroxybutyrate)

Adding the reactions for the bioproduction of the chemicals to the model provided valuable insight. First, the condensation of two molecules of pyruvate to acetolactate, which is the first reaction in the pathway to acetoin and 2,3-butanediol, is already present in *M. thermautotrophicus* ΔH because this is part of valine biosynthesis. Second, the first reaction toward 3-hydroxypropionate production is missing. This is surprising at the first glance because the carboxylation of acetyl-CoA to malonyl-CoA is the initial reaction of fatty acid biosynthesis. However, the cell membrane of *M. thermautotrophicus* ΔH, as discussed before, is composed of terpenoid-based lipids instead of fatty acid-based lipids [129]. A closer look into the model and the genome of *M. thermautotrophicus* ΔH revealed the absence of any genes required for fatty acid biosynthesis.

In our modeling, with the maximization of biomass as the objective function, no production of isoprene and 3-hydroxybutyrate was predicted, and the production for other products was considerably lower than with the maximization of the product as the objective function (Table 4). Under these constraints, the model predicts metabolic fluxes for the highest possible flux toward biomass production. However, none of the recombinant products would contribute to biomass, and thus, the predicted fluxes are low. With the maximization of the respective products as the objective function, flux toward all products was predicted (Table 4). More complex and energy-intensive molecules, such as isoprene, resulted in lower theoretical yields, which might have been expected intuitively. Instead, lactate, 3-hydroxybutyrate, and poly(3-hydroxybutyrate) resulted in higher theoretical yields than the other chemicals. Considering the required precursors may explain this observation. The hydroxybutyrates are synthesized from acetoacetyl-CoA, which itself is synthesized from acetyl-CoA. Acetyl-CoA is derived from the reduction of carbon dioxide in the Wood-Ljungdahl pathway and requires 2 mol of reduced ferredoxin. The third mole of reduced ferredoxin is required when acetyl-CoA is carboxylated to pyruvate (Fig. 1) [130]. The oxidized ferredoxins are reduced with hydrogen by the energy-converting hydrogenases (Eha or Ehb) [48]. This, however, requires the translocation of two sodium ions across the membrane, and the sodium-ion gradient also powers the A_1/A_0-ATPase (four sodium ions per ATP), which is the only means of ATP production in *M. thermautotrophicus* ΔH [130]. Thereby, the consumption of reduced ferredoxin indirectly drains ATP from the cell. $NADP^+$, on the other hand, can be directly reduced with reduced cofactor F_{420} [118] and, thus, does not cause additional ATP costs. Notably, the estimations from the modeling are given in percent of the carbon that was consumed as carbon dioxide. Therefore, the energy demands of the products have to be calculated considering this. 3-Hydroxybutyrate and poly(3-hydroxybutyrate) are synthesized from 2 mol of acetyl-CoA. This conversion demands 1 mol of reduced ferredoxin per mole of carbon. The same is true for lactate, which is synthesized from pyruvate, requiring 3 mol of reduced ferredoxin. Thus, the energy cost is 1 mol of reduced ferredoxin per mole of carbon for lactate production. Furthermore, in *M. thermautotrophicus*, no system for direct NAD^+ reduction with either the oxidation of reduced ferredoxin or cofactor F_{420} is known. This suggests that recombinantly producing NADH-dependent enzymes might cause imbalances in the redox metabolism of *M. thermautotrophicus*, and rebalancing mechanisms might be required as discussed in Sects. 3.1 and 3.3.2.

But why is the theoretical poly(3-hydroxybutyrate) yield higher than for lactate and 3-hydroxybutyrate? Another factor that must be considered is that several of the produced compounds are acids. Their production would drop the intracellular pH, which *M. thermautotrophicus* ΔH has to balance again. The genome-scale metabolic model does that with a sodium ion-proton antiporter, which contributes to draining the sodium-ion gradient and, thereby, the ATP pool [48]. Therefore, the number of protons that the products release must also be considered. Lactate production generates 1 mol of protons per three carbon atoms, while 3-hydroxybutyrate releases 1 mol of protons per four carbon atoms. However, poly(3-hydroxybutyrate) is an ester, and no protons are released. Thus, the yield is the highest for poly (3-hydroxybutyrate). It should be noted, however, that poly(3-hydroxybutyrate) is an insoluble polymer, and the model cannot account for the effects of poly (3-hydroxybutyrate) accumulation within the cells.

The lower yields for the other products can be explained similarly. Acetoin and 2,3-butanediol are synthesized from 2 mol of pyruvate (3 mol of reduced ferredoxin each), thus requiring 1.5 mol of reduced ferredoxin per carbon, which results in lower yields. 3-Hydroxypropionate and isoprene directly require ATP in their respective biosynthetic pathways, which results in comparatively lower theoretical yields [67, 111, 114]. This is in line with the experimental results from other methanogens, where isoprene and geraniol production yielded lower titers than lactate, 3-hydroxybutyrate, and poly(3-hydroxybutyrate) (Table 2).

5 Summary and Outlook

In recent years, a considerable progress has been made regarding genetic tools for methanogens, including the development of CRISPR/Cas-based methods for mesophilic species and several new sets of genetic tools for thermophilic species. However, especially for the thermophilic methanogens, many challenges remain. No replicative vectors have been described for the genetic systems of *Methanocaldococcus* spp. and *M. thermophilus*. Non-replicative plasmid vectors that rely on homology-based recombination always bear the risk of unwanted polar effects when used for heterologous expression of genes from the genome. In contrast, a replicative vector is available for *M. thermautotrophicus* ΔH. However, no system for traceless deletion of genes, such as the "pop-in/pop-out" strategy or CRISPR/Cas-based methods, has been reported. The allelic exchange system for *M. thermautotrophicus* ΔH has the considerable drawback that the resistance marker remains in the genome, which can lead to unwanted polar effects and restricts the reuse of the selective marker for additional deletions.

With the novel genetic tools, the range of genetically tractable methanogens now includes members of the orders (1) Methanococcales (*Methanococcus* spp. and *Methanocaldococcus* spp.), (2) Methanosarcinales (*Methanosarcina* spp.),

(3) Methanomicrobiales (*M. thermophilus*), and (4) Methanobacteriales (*M. thermautotrophicus*). Genetic tools are unavailable for the described orders Methanopyrales, Methanocellales, Methanomassiliicoccales, and Methanonatronarchaeia, which might be due to their growth requirements and recent time of discovery. The order Methanopyrales contains only the extreme hyperthermophilic species *Methanopyrus kandleri*, which grows at temperatures up to 110°C [131]. This makes handling in the lab and genetic work particularly demanding. The orders Methanocellales, Methanomassiliicoccales, and Methanonatronarchaeia were only described relatively recently, with only a few isolated representative species [16, 132, 133], which might explain that genetic tools have not yet been established. It should be noted that recent discoveries suggest that methanogenesis, or at least the genes that encode involved enzymes, is by no means limited to these eight orders but can be found all over the archaeal domain in metagenome-assembled genomes from environmental samples [4]. This does not immediately imply functional relevance, but it would not be surprising that many methanogens out there await to be investigated genetically.

For bioproduction, the question can be posed of why to use *M. thermautotrophicus* or other (thermophilic) methanogens for archaea-based production, as opposed to mesophilic or thermophilic autotrophic bacteria, such as *C. autoethanogenum*, *C. ljungdahlii*, *M. thermoacetica*, and *Thermoanaerobacter kivui*. Those microbes grow with gaseous substrates, including carbon dioxide, hydrogen, and carbon monoxide, and more sophisticated genetic tools already exist for several species [134]. In addition, products, including acetate, butanol, ethanol, acetone, and isopropanol, can be produced at high titers and, thus, might not be outcompeted by recombinant production in methanogens. Acetate, ethanol, acetone, and isopropanol are already produced at an industrial scale by LanzaTech using proprietary *C. autoethanogenum* strains [135]. Despite these facts, many chemicals have the potential to be produced with methanogens because of metabolic properties, such as the native presence of the mevalonate pathway. Those production pathways could potentially outcompete respective bacterial producers. Furthermore, unlike with acetogens, the only metabolic byproduct of a recombinant methanogenic producer strain would be methane, which is easier to separate from the fermentation broth than, for example, acetate and is itself of high value. However, our calculations show that the main metabolic product of hydrogenotrophic methanogenesis will always be methane. Thus, all processes that produce additional recombinant products should be considered an upgrade to the existing power-to-gas process to increase the economic value of the process (power-to-x). Furthermore, it is still largely unknown how the methanogenic host strains react to genetic or metabolic changes and which possible obstacles would have to be overcome (e.g., genetic instability, regulatory mechanisms, the accumulation of insoluble polymers, and the extraction/purification of potential products).

The genetic tools for thermophilic methanogens allow considering metabolic engineering with those microbes. While species of the genus *Methanothermobacter* were found to have beneficial properties for power-to-gas technology with biomethanation [11, 12], this does not rule out that other methanogens may be adopted for biotechnological purposes in the years to come [10]. In any case, metabolic engineering strategies for bioproduction will benefit from the implementation of modern genetic engineering technologies in combination with mathematical modeling, such as with genome-scale metabolic models. In conclusion, bioprocessing with thermophilic methanogens for bioproduction of value-added products is expected to increase in importance.

Acknowledgments The authors would like to thank Dr. Isabella Casini for her assistance and input in setting up and constraining the genome-scale metabolic model. This work was supported by the German Federal Ministry of Education and Research (MethanoPEP, 031B0851C). Additional funding was from the Deutsche Forschungsgemeinschaft (DFG, German Research Foundation) under Germany's Excellence Strategy-EXC 2124-390838134.

References

1. Lyu Z, Shao N, Akinyemi T, Whitman WB (2018) Methanogenesis. Curr Biol 28(13):R727–RR32. https://doi.org/10.1016/j.cub.2018.05.021
2. Thauer RK, Kaster A-K, Seedorf H, Buckel W, Hedderich R (2008) Methanogenic archaea: ecologically relevant differences in energy conservation. Nat Rev Microbiol 6(8):579–591. https://doi.org/10.1038/nrmicro1931
3. Kurth JM, Op den Camp HJM, Welte CU (2020) Several ways one goal – methanogenesis from unconventional substrates. Appl Microbiol Biotechnol 104(16):6839–6854. https://doi.org/10.1007/s00253-020-10724-7
4. Garcia PS, Gribaldo S, Borrel G (2022) Diversity and evolution of methane-related pathways in archaea. Annu Rev Microbiol 76(1):727–755. https://doi.org/10.1146/annurev-micro-041020-024935
5. Schöne C, Poehlein A, Jehmlich N, Adlung N, Daniel R, von Bergen M et al (2022) Deconstructing *Methanosarcina acetivorans* into an acetogenic archaeon. Proc Natl Acad Sci U S A 119(2):e2113853119. https://doi.org/10.1073/pnas.2113853119
6. Enzmann F, Mayer F, Rother M, Holtmann D (2018) Methanogens: biochemical background and biotechnological applications. AMB Express 8(1):1. https://doi.org/10.1186/s13568-017-0531-x
7. Reeve JN, Trun NJ, Hamilton PT (1982) Beginning genetics with methanogens. Basic Life Sci 19:233–244. https://doi.org/10.1007/978-1-4684-4142-0_19
8. Angenent LT, Usack JG, Sun T, Fink C, Molitor B, Labatut R et al (2022) Upgrading anaerobic digestion within the energy economy – the methane platform. In: Pikaar I, Guest J, Ganigué R, Jensen P, Rabaey K, Seviour T et al (eds) Resource recovery from water: principles and application. IWA Publishing
9. Thema M, Weidlich T, Hörl M, Bellack A, Mörs F, Hackl F et al (2019) Biological CO_2-methanation: an approach to standardization. Energies 12(9):1670
10. Pfeifer K, Ergal İ, Koller M, Basen M, Schuster B, Rittmann SKR (2021) Archaea biotechnology. Biotechnol Adv 47:107668. https://doi.org/10.1016/j.biotechadv.2020.107668
11. Martin MR, Fornero JJ, Stark R, Mets L, Angenent LT (2013) A single-culture bioprocess of *Methanothermobacter thermautotrophicus* to upgrade digester biogas by CO_2-to-CH_4 conversion with H_2. Archaea 2013:157529. https://doi.org/10.1155/2013/157529
12. Thema M, Weidlich T, Kaul A, Böllmann A, Huber H, Bellack A et al (2021) Optimized biological CO_2-methanation with a pure culture of thermophilic methanogenic archaea in a trickle-bed reactor. Bioresour Technol 333:125135. https://doi.org/10.1016/j.biortech.2021.125135

13. Costa KC, Whitman WB (2023) Model organisms to study methanogenesis, a uniquely archaeal metabolism. J Bacteriol 205(8):e00115–e00123. https://doi.org/10.1128/jb.00115-23
14. Contreras G, Thomsen J, Pfitzer M, Hafenbradl D, Kostner D, Holtmann D et al (2022) New perspectives for biotechnological applications of methanogens. Curr Res Biotechnol 4:468–474. https://doi.org/10.1016/j.crbiot.2022.10.001
15. Adam PS, Borrel G, Brochier-Armanet C, Gribaldo S (2017) The growing tree of archaea: new perspectives on their diversity, evolution and ecology. ISME J 11(11):2407–2425. https://doi.org/10.1038/ismej.2017.122
16. Sorokin DY, Merkel AY, Abbas B (2022) Ecology of *Methanonatronarchaeia*. Environ Microbiol 24(11):5217–5229. https://doi.org/10.1111/1462-2920.16108
17. Bertani G, Baresi L (1987) Genetic transformation in the methanogen *Methanococcus voltae* PS. J Bacteriol 169(6):2730–2738. https://doi.org/10.1128/jb.169.6.2730-2738.1987
18. Metcalf WW, Zhang JK, Apolinario E, Sowers KR, Wolfe RS (1997) A genetic system for archaea of the genus *Methanosarcina*: liposome-mediated transformation and construction of shuttle vectors. Proc Natl Acad Sci U S A 94(6):2626–2631. https://doi.org/10.1073/pnas.94.6.2626
19. Susanti D, Frazier MC, Mukhopadhyay B (2019) A genetic system for *Methanocaldococcus jannaschii*: an evolutionary deeply rooted hyperthermophilic methanarchaeon. Front Microbiol 10(1256). https://doi.org/10.3389/fmicb.2019.01256
20. Fonseca DR, Halim MFA, Holten MP, Costa KC (2020) Type IV-like pili facilitate transformation in naturally competent archaea. J Bacteriol 202(21):e00355–e00320. https://doi.org/10.1128/jb.00355-20
21. Fink C, Beblawy S, Enkerlin AM, Mühling L, Angenent LT, Molitor B (2021) A shuttle-vector system allows heterologous gene expression in the thermophilic methanogen *Methanothermobacter thermautotrophicus* ΔH. MBio 12(6):e02766–e02721. https://doi.org/10.1128/mBio.02766-21
22. Lie TJ, Kuo YP, Leite M, Costa KC, Harwood CS, Leigh JA (2022) A genetic study of *nif*-associated genes in a hyperthermophilic methanogen. Microbiol Spectr 10(1):e02093–e02021. https://doi.org/10.1128/spectrum.02093-21
23. Nayak DD, Metcalf WW (2017) Cas9-mediated genome editing in the methanogenic archaeon *Methanosarcina acetivorans*. Proc Natl Acad Sci U S A 114(11):2976–2981. https://doi.org/10.1073/pnas.1618596114
24. Jinek M, Chylinski K, Fonfara I, Hauer M, Doudna JA, Charpentier E (2012) A programmable dual-RNA–guided DNA endonuclease in adaptive bacterial immunity. Science 337(6096):816–821. https://doi.org/10.1126/science.1225829
25. Dhamad AE, Lessner DJ (2020) A CRISPRi-dCas9 system for archaea and its use to examine gene function during nitrogen fixation by *Methanosarcina acetivorans*. Appl Environ Microbiol 86(21):e01402–e01420. https://doi.org/10.1128/AEM.01402-20
26. Thomsen J, Schmitz RA (2022) Generating a small shuttle vector for effective genetic engineering of *Methanosarcina mazei* allowed first insights in plasmid replication mechanism in the methanoarchaeon. Int J Mol Sci 23(19). https://doi.org/10.3390/ijms231911910
27. Li J, Zhang L, Xu Q, Zhang W, Li Z, Chen L et al (2022) CRISPR-Cas9 toolkit for genome editing in an autotrophic CO_2-fixing methanogenic archaeon. Microbiol Spectr 10(4):e01165–e01122. https://doi.org/10.1128/spectrum.01165-22
28. Bao J, de Dios ME, Scheller S (2022) Efficient CRISPR/Cas12a-based genome-editing toolbox for metabolic engineering in *Methanococcus maripaludis*. ACS Synth Biol 11(7):2496–2503. https://doi.org/10.1021/acssynbio.2c00137
29. Fonseca DR, Loppnow MB, Day LA, Kelsey EL, Abdul Halim MF, Costa KC (2023) Random transposon mutagenesis identifies genes essential for transformation in *Methanococcus maripaludis*. Mol Genet Genomics 298(3):537–548. https://doi.org/10.1007/s00438-023-01994-7

30. Tumbula DL, Makula RA, Whitman WB (1994) Transformation of *Methanococcus maripaludis* and identification of a Pst I-like restriction system. FEMS Microbiol Lett 121(3):309–314. https://doi.org/10.1111/j.1574-6968.1994.tb07118.x
31. Hernandez E, Costa KC (2022) The fluorescence-activating and absorption-shifting tag (FAST) enables live-cell fluorescence imaging of *Methanococcus maripaludis*. J Bacteriol 204(7):e00120–e00122. https://doi.org/10.1128/jb.00120-22
32. Plamont M-A, Billon-Denis E, Maurin S, Gauron C, Pimenta FM, Specht CG et al (2016) Small fluorescence-activating and absorption-shifting tag for tunable protein imaging *in vivo*. Proc Natl Acad Sci U S A 113(3):497–502. https://doi.org/10.1073/pnas.1513094113
33. Streett HE, Kalis KM, Papoutsakis ET (2019) A strongly fluorescing anaerobic reporter and protein-tagging system for *Clostridium* organisms based on the fluorescence-activating and absorption-shifting tag protein (FAST). Appl Environ Microbiol 85(14):e00622–e00619. https://doi.org/10.1128/AEM.00622-19
34. Heim R, Prasher DC, Tsien RY (1994) Wavelength mutations and posttranslational autoxidation of green fluorescent protein. Proc Natl Acad Sci U S A 91(26):12501–12504. https://doi.org/10.1073/pnas.91.26.12501
35. Chia HE, Marsh ENG, Biteen JS (2019) Extending fluorescence microscopy into anaerobic environments. Curr Opin Chem Biol 51:98–104. https://doi.org/10.1016/j.cbpa.2019.05.008
36. Lambrecht J, Cichocki N, Hübschmann T, Koch C, Harms H, Müller S (2017) Flow cytometric quantification, sorting and sequencing of methanogenic archaea based on F_{420} autofluorescence. Microb Cell Fact 16(1):180. https://doi.org/10.1186/s12934-017-0793-7
37. Adlung N, Scheller S (2023) Application of the fluorescence-activating and absorption-shifting tag (FAST) for flow cytometry in methanogenic archaea. Appl Environ Microbiol 89(4):e01786–e01722. https://doi.org/10.1128/aem.01786-22
38. Hocq R, Bottone S, Gautier A, Pflügl S (2023) A fluorescent reporter system for anaerobic thermophiles. Front Bioeng Biotechnol:11. https://doi.org/10.3389/fbioe.2023.1226889
39. Hilpert R, Winter J, Hammes W, Kandler O (1981) The sensitivity of archaebacteria to antibiotics. Zentralblatt für Bakteriologie Mikrobiologie und Hygiene: I Abt Originale C: Allgemeine, angewandte und ökologische Mikrobiologie 2(1):11–20. https://doi.org/10.1016/S0721-9571(81)80014-2
40. Beckler GS, Hook LA, Reeve JN (1984) Chloramphenicol acetyltransferase should not provide methanogens with resistance to chloramphenicol. Appl Environ Microbiol 47(4): 868–869
41. Jones WJ, Leigh JA, Mayer F, Woese CR, Wolfe RS (1983) *Methanococcus jannaschii* sp. nov., an extremely thermophilic methanogen from a submarine hydrothermal vent. Arch Microbiol 136(4):254–261. https://doi.org/10.1007/BF00425213
42. Jain S, Caforio A, Driessen AJM (2014) Biosynthesis of archaeal membrane ether lipids. Front Microbiol:5. https://doi.org/10.3389/fmicb.2014.00641
43. Farkas JA, Picking JW, Santangelo TJ (2013) Genetic techniques for the archaea. Annu Rev Genet 47(1):539–561. https://doi.org/10.1146/annurev-genet-111212-133225
44. Moore BC, Leigh JA (2005) Markerless mutagenesis in *Methanococcus maripaludis* demonstrates roles for alanine dehydrogenase, alanine racemase, and alanine permease. J Bacteriol 187(3):972–979. https://doi.org/10.1128/jb.187.3.972-979.2005
45. Li J, Akinyemi TS, Shao N, Chen C, Dong X, Liu Y et al (2023) Genetic and metabolic engineering of *Methanococcus* spp. Curr Res Biotechnol 5:100115. https://doi.org/10.1016/j.crbiot.2022.11.002
46. Hoseki J, Yano T, Koyama Y, Kuramitsu S, Kagamiyama H (1999) Directed evolution of thermostable kanamycin-resistance gene: a convenient selection marker for *Thermus thermophilus*. J Biochem 126(5):951–956. https://doi.org/10.1093/oxfordjournals.jbchem.a022539
47. Abdul Halim MF, Day LA, Costa KC (2021) Formate-dependent heterodisulfide reduction in a *Methanomicrobiales* archaeon. Appl Environ Microbiol 87(6). https://doi.org/10.1128/aem.02698-20

48. Kaster A-K, Goenrich M, Seedorf H, Liesegang H, Wollherr A, Gottschalk G et al (2011) More than 200 genes required for methane formation from H_2 and CO_2 and energy conservation are present in *Methanothermobacter marburgensis* and *Methanothermobacter thermautotrophicus*. Archaea 2011:973848. https://doi.org/10.1155/2011/973848
49. Luo Y, Leisinger T, Wasserfallen A (2001) Comparative sequence analysis of plasmids pME2001 and pME2200 of *Methanothermobacter marburgensis* strains Marburg and ZH3. Plasmid 45(1):18–30. https://doi.org/10.1006/plas.2000.1493
50. Leisinger T, Meile L (1993) Plasmids, phages, and gene transfer in methanogenic bacteria. In: Sebald M (ed) Genetics and molecular biology of anaerobic bacteria. Springer, New York, pp 1–12
51. Meile L, Reeve JN (1985) Potential shuttle vectors based on the methanogen plasmid pME2001. Nat Biotechnology 3(1):69–72. https://doi.org/10.1038/nbt0185-69
52. Worrell VE, Nagle Jr DP, McCarthy D, Eisenbraun A (1988) Genetic transformation system in the archaebacterium *Methanobacterium thermoautotrophicum* Marburg. J Bacteriol 170(2): 653–656. https://doi.org/10.1128/jb.170.2.653-656.1988
53. Meile L, Abendschein P, Leisinger T (1990) Transduction in the archaebacterium *Methanobacterium thermoautotrophicum* Marburg. J Bacteriol 172(6):3507–3508. https://doi.org/10.1128/jb.172.6.3507-3508.1990
54. Dodsworth JA, Li L, Wei S, Hedlund BP, Leigh JA, Figueiredo P (2010) Interdomain conjugal transfer of DNA from bacteria to archaea. Appl Environ Microbiol 76(16):5644–5647. https://doi.org/10.1128/AEM.00967-10
55. Nölling J, Reeve JN (1997) Growth- and substrate-dependent transcription of the formate dehydrogenase (*fdhCAB*) operon in *Methanobacterium thermoformicicum* Z-245. J Bacteriol 179(3):899–908. https://doi.org/10.1128/jb.179.3.899-908.1997
56. Fink C, Martinez-Cano G, Shuster J, Panzera A, Rennhack KE, Rohbohm N et al (2023) The targeted deletion of genes responsible for expression of the Mth60 fimbriae leads to loss of cell-cell connections in *Methanothermobacter thermautotrophicus* ΔH. Appl Environ Microbiol 89(7):e00575–e00523. https://doi.org/10.1128/aem.00575-23
57. Doddema HJ, Derksen JWM, Vogels GD (1979) Fimbriae and flagella of methanogenic bacteria. FEMS Microbiol Lett 5(3):135–138. https://doi.org/10.1111/j.1574-6968.1979.tb03264.x
58. Thoma C, Frank M, Rachel R, Schmid S, Näther D, Wanner G et al (2008) The Mth60 fimbriae of *Methanothermobacter thermoautotrophicus* are functional adhesins. Environ Microbiol 10(10):2785–2795. https://doi.org/10.1111/j.1462-2920.2008.01698.x
59. Taubner R-S, Baumann LMF, Steiner M, Pfeifer K, Reischl B, Korynt K et al (2023) Lipidomics and comparative metabolite excretion analysis of methanogenic archaea reveal organism-specific adaptations to varying temperatures and substrate concentrations. mSystems 8(2):e01159–e01122. https://doi.org/10.1128/msystems.01159-22
60. Lyu Z, Whitman WB (2019) Transplanting the pathway engineering toolbox to methanogens. Curr Opin Biotechnol 59:46–54. https://doi.org/10.1016/j.copbio.2019.02.009
61. Aldridge J, Carr S, Weber KA, Buan NR (2021.:AEM.02417-20) Anaerobic production of isoprene by engineered *Methanosarcina* spp. archaea. Appl Environ Microbiol. https://doi.org/10.1128/aem.02417-20
62. Carr S, Aldridge J, Buan NR (2021) Isoprene production from municipal wastewater biosolids by engineered archaeon. Appl Sci 11(8):3342
63. Lyu Z, Jain R, Smith P, Fetchko T, Yan Y, Whitman WB (2016) Engineering the autotroph *Methanococcus maripaludis* for geraniol production. ACS Synth Biol 5(7):577–581. https://doi.org/10.1021/acssynbio.5b00267
64. Soo VWC, McAnulty MJ, Tripathi A, Zhu F, Zhang L, Hatzakis E et al (2016) Reversing methanogenesis to capture methane for liquid biofuel precursors. Microb Cell Fact 15(1):11. https://doi.org/10.1186/s12934-015-0397-z

65. McAnulty MJ, Poosarla VG, Li J, Soo VW, Zhu F, Wood TK (2017) Metabolic engineering of *Methanosarcina acetivorans* for lactate production from methane. Biotechnol Bioeng 114(4): 852–861. https://doi.org/10.1002/bit.26208
66. Thevasundaram K, Gallagher JJ, Cherng F, Chang MCY (2022) Engineering nonphotosynthetic carbon fixation for production of bioplastics by methanogenic archaea. Proc Natl Acad Sci U S A 119(23):e2118638119. https://doi.org/10.1073/pnas.2118638119
67. Koga Y, Morii H (2007) Biosynthesis of ether-type polar lipids in archaea and evolutionary considerations. Microbiol Mol Biol Rev 71(1):97–120. https://doi.org/10.1128/mmbr.00033-06
68. Carr S, Buan NR (2022) Insights into the biotechnology potential of *Methanosarcina*. Front Microbiol:13. https://doi.org/10.3389/fmicb.2022.1034674
69. Chadwick GL, Nayak DD, Rother M, Metcalf WW (2024) No evidence for methanotrophic growth of diverse marine methanogens. Proc Natl Acad Sci U S A 121(20):e2404143121. https://doi.org/10.1073/pnas.2404143121
70. Rigoldi F, Donini S, Redaelli A, Parisini E, Gautieri A (2018) Review: engineering of thermostable enzymes for industrial applications. APL Bioeng 2(1):011501. https://doi.org/10.1063/1.4997367
71. Cowan DA, Albers SV, Antranikian G, Atomi H, Averhoff B, Basen M et al (2024) Extremophiles in a changing world. Extremophiles 28(2):26. https://doi.org/10.1007/s00792-024-01341-7
72. Iwasaki Y, Kita A, Yoshida K, Tajima T, Yano S, Shou T et al (2017) Homolactic acid fermentation by the genetically engineered thermophilic homoacetogen *Moorella thermoacetica* ATCC 39073. Appl Environ Microbiol 83(8). https://doi.org/10.1128/aem.00247-17
73. Jang YS, Kim B, Shin JH, Choi YJ, Choi S, Song CW et al (2012) Bio-based production of C2-C6 platform chemicals. Biotechnol Bioeng 109(10):2437–2459. https://doi.org/10.1002/bit.24599
74. Fuke T, Sato T, Jha S, Tansengco ML, Atomi H (2018) Phytoene production utilizing the isoprenoid biosynthesis capacity of *Thermococcus kodakarensis*. Extremophiles 22(2): 301–313. https://doi.org/10.1007/s00792-018-0998-7
75. Sheng L, Madika A, Lau MSH, Zhang Y, Minton NP (2023) Metabolic engineering for the production of acetoin and 2,3-butanediol at elevated temperature in *Parageobacillus thermoglucosidasius* NCIMB 11955. Front Bioeng Biotechnol 11:1191079. https://doi.org/10.3389/fbioe.2023.1191079
76. Basen M, Sun J, Adams MWW (2012) Engineering a hyperthermophilic archaeon for temperature-dependent product formation. MBio 3(2):e00053–e00e12. https://doi.org/10.1128/mBio.00053-12
77. Keller MW, Schut GJ, Lipscomb GL, Menon AL, Iwuchukwu IJ, Leuko TT et al (2013) Exploiting microbial hyperthermophilicity to produce an industrial chemical, using hydrogen and carbon dioxide. Proc Natl Acad Sci U S A 110(15):5840–5845. https://doi.org/10.1073/pnas.1222607110
78. Cripps RE, Eley K, Leak DJ, Rudd B, Taylor M, Todd M et al (2009) Metabolic engineering of *Geobacillus thermoglucosidasius* for high yield ethanol production. Metab Eng 11(6): 398–408. https://doi.org/10.1016/j.ymben.2009.08.005
79. Lin PP, Rabe KS, Takasumi JL, Kadisch M, Arnold FH, Liao JC (2014) Isobutanol production at elevated temperatures in thermophilic *Geobacillus thermoglucosidasius*. Metab Eng 24:1–8. https://doi.org/10.1016/j.ymben.2014.03.006
80. Zhou J, Lian J, Rao CV (2020) Metabolic engineering of *Parageobacillus thermoglucosidasius* for the efficient production of (2R, 3R)-butanediol. Appl Microbiol Biotechnol 104(10):4303–4311. https://doi.org/10.1007/s00253-020-10553-8
81. Shaw AJ, Podkaminer KK, Desai SG, Bardsley JS, Rogers SR, Thorne PG et al (2008) Metabolic engineering of a thermophilic bacterium to produce ethanol at high yield. Proc Natl Acad Sci U S A 105(37):13769–13774. https://doi.org/10.1073/pnas.0801266105

82. Bhandiwad A, Shaw AJ, Guss A, Guseva A, Bahl H, Lynd LR (2014) Metabolic engineering of *Thermoanaerobacterium saccharolyticum* for n-butanol production. Metab Eng 21:17–25. https://doi.org/10.1016/j.ymben.2013.10.012
83. Chung D, Cha M, Guss AM, Westpheling J (2014) Direct conversion of plant biomass to ethanol by engineered *Caldicellulosiruptor bescii*. Proc Natl Acad Sci U S A 111(24): 8931–8936. https://doi.org/10.1073/pnas.1402210111
84. Rahayu F, Kawai Y, Iwasaki Y, Yoshida K, Kita A, Tajima T et al (2017) Thermophilic ethanol fermentation from lignocellulose hydrolysate by genetically engineered *Moorella thermoacetica*. Bioresour Technol 245:1393–1399. https://doi.org/10.1016/j.biortech.2017.05.146
85. Kato J, Takemura K, Kato S, Fujii T, Wada K, Iwasaki Y et al (2021) Metabolic engineering of *Moorella thermoacetica* for thermophilic bioconversion of gaseous substrates to a volatile chemical. AMB Express 11(1):59. https://doi.org/10.1186/s13568-021-01220-w
86. Wagner M, Berkner S, Ajon M, Driessen Arnold JM, Lipps G, Albers S-V (2009) Expanding and understanding the genetic toolbox of the hyperthermophilic genus *Sulfolobus*. Biochem Soc Trans 37(1):97–101. https://doi.org/10.1042/bst0370097
87. Sato T, Fukui T, Atomi H, Imanaka T (2003) Targeted gene disruption by homologous recombination in the hyperthermophilic archaeon *Thermococcus kodakaraensis* KOD1. J Bacteriol 185(1):210–220. https://doi.org/10.1128/jb.185.1.210-220.2003
88. Lipscomb GL, Stirrett K, Schut GJ, Yang F, Jenney Jr FE, Scott RA et al (2011) Natural competence in the hyperthermophilic archaeon *Pyrococcus furiosus* facilitates genetic manipulation: construction of markerless deletions of genes encoding the two cytoplasmic hydrogenases. Appl Environ Microbiol 77(7):2232–2238. https://doi.org/10.1128/AEM.02624-10
89. Li M, Liu C, Chen H, Deng L, Zhang H, Nian R et al (2019) Biochemical characterization of isoprene synthase from *Ipomoea batatas*. J Biosci Bioeng 127(2):138–144. https://doi.org/10.1016/j.jbiosc.2018.07.022
90. Isar J, Jain D, Joshi H, Dhoot S, Rangaswamy V (2022) MICROBIAL isoprene production: an overview. World J Microbiol Biotechnol 38(7). https://doi.org/10.1007/s11274-022-03306-4
91. Kuzma J, Nemecek-Marshall M, Pollock WH, Fall R (1995) Bacteria produce the volatile hydrocarbon isoprene. Curr Microbiol 30(2):97–103. https://doi.org/10.1007/bf00294190
92. Janke C, Gaida S, Jennewein S (2020) The production of isoprene from cellulose using recombinant *Clostridium cellulolyticum* strains expressing isoprene synthase. Microbiology 9(4):e1008-e. https://doi.org/10.1002/mbo3.1008
93. Lee H-W, Park J-H, Lee H-S, Choi W, Seo S-H, Anggraini ID et al (2019) Production of bio-based isoprene by the mevalonate pathway cassette in *Ralstonia eutropha*. J Microbiol Biotechnol 29(10):1656–1664. https://doi.org/10.4014/jmb.1909.09002
94. Diner BA, Fan J, Scotcher MC, Wells DH, Whited GM (2017) Synthesis of heterologous mevalonic acid pathway enzymes in *Clostridium ljungdahlii* for the conversion of fructose and of syngas to mevalonate and isoprene. Appl Environ Microbiol 84(1):e01723–e01717. https://doi.org/10.1128/AEM.01723-17
95. Gao X, Gao F, Liu D, Zhang H, Nie X, Yang C (2016) Engineering the methylerythritol phosphate pathway in cyanobacteria for photosynthetic isoprene production from CO_2. Energ Environ Sci 9(4):1400–1411. https://doi.org/10.1039/c5ee03102h
96. Bertsch J, Müller V (2015) Bioenergetic constraints for conversion of syngas to biofuels in acetogenic bacteria. Biotechnol Biofuels 8:210. https://doi.org/10.1186/s13068-015-0393-x
97. Hobisch M, Spasic J, Malihan-Yap L, Barone GD, Castiglione K, Tamagnini P et al (2021) Internal illumination to overcome the cell density limitation in the scale-up of whole-cell photobiocatalysis. ChemSusChem 14(15):3219–3225. https://doi.org/10.1002/cssc.202100832
98. Sharkey TD, Yeh S, Wiberley AE, Falbel TG, Gong D, Fernandez DE (2005) Evolution of the isoprene biosynthetic pathway in kudzu. Plant Physiol 137(2):700–712. https://doi.org/10.1104/pp.104.054445

99. Li M, Nian R, Xian M, Zhang H (2018) Metabolic engineering for the production of isoprene and isopentenol by *Escherichia coli*. Appl Microbiol Biotechnol 102(18):7725–7738. https://doi.org/10.1007/s00253-018-9200-5
100. Eiteman MA, Ramalingam S (2015) Microbial production of lactic acid. Biotechnol Lett 37(5):955–972. https://doi.org/10.1007/s10529-015-1769-5
101. Liew F, Henstra AM, Köpke M, Winzer K, Simpson SD, Minton NP (2017) Metabolic engineering of *Clostridium autoethanogenum* for selective alcohol production. Metab Eng 40:104–114. https://doi.org/10.1016/j.ymben.2017.01.007
102. Humphreys JR, Hebdon SD, Rohrer H, Magnusson L, Urban C, Chen Y-P et al (2022) Establishing *Butyribacterium methylotrophicum* as a platform organism for the production of biocommodities from liquid C(1) metabolites. Appl Environ Microbiol 88(6):e0239321-e. https://doi.org/10.1128/aem.02393-21
103. Drake HL, Daniel SL (2004) Physiology of the thermophilic acetogen *Moorella thermoacetica*. Res Microbiol 155(10):869–883. https://doi.org/10.1016/j.resmic.2004.10.002
104. Mook A, Beck MH, Baker JP, Minton NP, Dürre P, Bengelsdorf FR (2022) Autotrophic lactate production from $H_2 + CO_2$ using recombinant and fluorescent FAST-tagged *Acetobacterium woodii* strains. Appl Microbiol Biotechnol 106(4):1447–1458. https://doi.org/10.1007/s00253-022-11770-z
105. Moon J, Waschinger LM, Müller V (2023) Lactate formation from fructose or C1 compounds in the acetogen *Acetobacterium woodii* by metabolic engineering. Appl Microbiol Biotechnol 107(17):5491–5502. https://doi.org/10.1007/s00253-023-12637-7
106. Weghoff MC, Bertsch J, Müller V (2015) A novel mode of lactate metabolism in strictly anaerobic bacteria. Environ Microbiol 17(3):670–677. https://doi.org/10.1111/1462-2920.12493
107. Zhou Q, Shao WL (2010) Molecular genetic characterization of the thermostable L-lactate dehydrogenase gene (ldhL) of *Thermoanaerobacter ethanolicus* JW200 and biochemical characterization of the enzyme. Biochemistry 75(4):526–530. https://doi.org/10.1134/s0006297910040188
108. Ostendorp R, Auerbach G, Jaenicke R (1996) Extremely thermostable L-lactate dehydrogenase from *Thermotoga maritima*: cloning, characterization, and crystallization of the recombinant enzyme in its tetrameric and octameric state. Protein Sci 5(5):862–873. https://doi.org/10.1002/pro.5560050508
109. Akyol I, Ozcelik FG, Karakas-Sen A, Ozkose E, Gezginc Y, Ekinci MS (2015) Cloning and overexpression of the *als*, *pflA*, and *adhB* genes in *Streptococcus thermophilus* and their effects on metabolite formation. Mol Biotechnol 57(10):923–930. https://doi.org/10.1007/s12033-015-9882-1
110. Xiao Z, Lu JR (2014) Strategies for enhancing fermentative production of acetoin: a review. Biotechnol Adv 32(2):492–503. https://doi.org/10.1016/j.biotechadv.2014.01.002
111. Kumar V, Ashok S, Park S (2013) Recent advances in biological production of 3-hydroxypropionic acid. Biotechnol Adv 31(6):945–961. https://doi.org/10.1016/j.biotechadv.2013.02.008
112. Köpke M, Mihalcea C, Liew F, Tizard JH, Ali MS, Conolly JJ et al (2011) 2,3-butanediol production by acetogenic bacteria, an alternative route to chemical synthesis, using industrial waste gas. Appl Environ Microbiol 77(15):5467–5475. https://doi.org/10.1128/AEM.00355-11
113. Werpy T, Petersen G (2004) Top value added chemicals from biomass: volume I – results of screening for potential candidates from sugars and synthesis gas. Office of Scientific and Technical Information (OSTI)
114. Wang X, Cui Z, Sun X, Wang Z, Chen T (2023) Production of 3-hydroxypropionic acid from renewable substrates by metabolically engineered microorganisms: a review. Molecules 28(4):1888. https://doi.org/10.3390/molecules28041888
115. Salinas A, McGregor C, Irorere V, Arenas-López C, Bommareddy RR, Winzer K et al (2022) Metabolic engineering of *Cupriavidus necator* H16 for heterotrophic and autotrophic

production of 3-hydroxypropionic acid. Metab Eng 74:178–190. https://doi.org/10.1016/j.ymben.2022.10.014
116. Wang Y, Sun T, Gao X, Shi M, Wu L, Chen L et al (2016) Biosynthesis of platform chemical 3-hydroxypropionic acid (3-HP) directly from CO_2 in cyanobacterium *Synechocystis sp.* PCC 6803. Metab Eng 34:60–70. https://doi.org/10.1016/j.ymben.2015.10.008
117. Beck MH (2020) Biological conversion of CO_2 into the platform chemicals lactate and 3-hydroxypropionate using recombinant strains of *Acetobacterium woodii*. Dissertation
118. Dudley Eirich L, Dugger RS (1984) Purification and properties of an F420-dependent $NADP^+$ reductase from *Methanobacterium thermoautotrophicum*. Biochim Biophys Acta Gen Subj 802(3):454–458. https://doi.org/10.1016/0304-4165(84)90364-7
119. Flüchter S, Follonier S, Schiel-Bengelsdorf B, Bengelsdorf FR, Zinn M, Dürre P (2019) Anaerobic production of poly(3-hydroxybutyrate) and its precursor 3-hydroxybutyrate from synthesis gas by autotrophic *Clostridia*. Biomacromolecules 20(9):3271–3282. https://doi.org/10.1021/acs.biomac.9b00342
120. Rodríguez-Contreras A, Koller M, Miranda-de Sousa Dias M, Calafell-Monfort M, Braunegg G, Marqués-Calvo MS (2015) Influence of glycerol on poly(3-hydroxybutyrate) production by *Cupriavidus necator* and *Burkholderia sacchari*. Biochem Eng J 94:50–57. https://doi.org/10.1016/j.bej.2014.11.007
121. Lee J, Park HJ, Moon M, Lee J-S, Min K (2021) Recent progress and challenges in microbial polyhydroxybutyrate (PHB) production from CO_2 as a sustainable feedstock: a state-of-the-art review. Bioresour Technol 339:125616. https://doi.org/10.1016/j.biortech.2021.125616
122. Jia D, He M, Tian Y, Shen S, Zhu X, Wang Y et al (2021) Metabolic engineering of gas-fermenting *Clostridium ljungdahlii* for efficient co-production of isopropanol, 3-hydroxybutyrate, and ethanol. ACS Synth Biol 10(10):2628–2638. https://doi.org/10.1021/acssynbio.1c00235
123. Höfele F, Dürre P (2023) Production of potential substitutes for conventional plastics using metabolically engineered *Acetobacterium woodii*. Fermentation 9(9):799. https://doi.org/10.3390/fermentation9090799
124. Chavan S, Yadav B, Tyagi RD, Drogui P (2021) A review on production of polyhydroxyalkanoate (PHA) biopolyesters by thermophilic microbes using waste feedstocks. Bioresour Technol 341:125900. https://doi.org/10.1016/j.biortech.2021.125900
125. Gu C, Kim GB, Kim WJ, Kim HU, Lee SY (2019) Current status and applications of genome-scale metabolic models. Genome Biol 20(1):121. https://doi.org/10.1186/s13059-019-1730-3
126. Simeonidis E, Price ND (2015) Genome-scale modeling for metabolic engineering. J Ind Microbiol Biotechnol 42(3):327–338. https://doi.org/10.1007/s10295-014-1576-3
127. Mahamkali V, McCubbin T, Beber ME, Noor E, Marcellin E, Nielsen LK (2021) multiTFA: a Python package for multi-variate thermodynamics-based flux analysis. Bioinformatics 37(18):3064–3066. https://doi.org/10.1093/bioinformatics/btab151
128. Casini I, McCubbin T, Esquivel-Elizondo S, Luque GG, Evseeva D, Fink C et al (2023) An integrated systems biology approach reveals differences in formate metabolism in the genus *Methanothermobacter*. iScience 26(10):108016. https://doi.org/10.1016/j.isci.2023.108016
129. Nováková Z, Blaško J, Hapala I, Šmigáň P (2010) Effects of 3-hydroxy-3-methylglutaryl-coenzyme a reductase inhibitor pravastatin on membrane lipids and membrane associated functions of *Methanothermobacter thermautotrophicus*. Folia Microbiol 55(4):359–362. https://doi.org/10.1007/s12223-010-0058-x
130. Fuchs G (2011) Alternative pathways of carbon dioxide fixation: insights into the early evolution of life? Annu Rev Microbiol 65(1):631–658. https://doi.org/10.1146/annurev-micro-090110-102801
131. Oren A (2014) The family *Methanopyraceae*. In: Rosenberg E, DeLong EF, Lory S, Stackebrandt E, Thompson F (eds) The prokaryotes: other major lineages of bacteria and the archaea. Springer, Berlin, pp 247–252
132. Sakai S, Imachi H, Hanada S, Ohashi A, Harada H, Kamagata Y (2008) *Methanocella paludicola* gen. nov., sp. nov., a methane-producing archaeon, the first isolate of the lineage 'Rice Cluster I', and proposal of the new archaeal order *Methanocellales* ord. nov. Int J Syst Evol Microbiol 58(Pt 4):929–936. https://doi.org/10.1099/ijs.0.65571-0

133. Dridi B, Fardeau ML, Ollivier B, Raoult D, Drancourt M (2012) *Methanomassiliicoccus luminyensis* gen. nov., sp. nov., a methanogenic archaeon isolated from human faeces. Int J Syst Evol Microbiol 62(Pt 8):1902–1907. https://doi.org/10.1099/ijs.0.033712-0
134. Fackler N, Heijstra BD, Rasor BJ, Brown H, Martin J, Ni Z et al (2021) Stepping on the gas to a circular economy: accelerating development of carbon-negative chemical production from gas fermentation. Annu Rev Chem Biomol Eng 12:439–470. https://doi.org/10.1146/annurev-chembioeng-120120-021122
135. Liew FE, Nogle R, Abdalla T, Rasor BJ, Canter C, Jensen RO et al (2022) Carbon-negative production of acetone and isopropanol by gas fermentation at industrial pilot scale. Nat Biotechnol 40(3):335–344. https://doi.org/10.1038/s41587-021-01195-w

Adv Biochem Eng Biotechnol (2025) 192: 59–84
https://doi.org/10.1007/10_2024_269

Published online: 4 October 2024

From Knallgas Bacterium to Promising Biomanufacturing Host: The Evolution of *Cupriavidus necator*

Daniel Casey, Laura Diaz-Garcia, Mincen Yu, Kang Lan Tee, and Tuck Seng Wong

Contents

D. Casey, L. Diaz-Garcia, and M. Yu
School of Chemical, Materials and Biological Engineering, The University of Sheffield, Sheffield, UK

K. L. Tee
School of Chemical, Materials and Biological Engineering, The University of Sheffield, Sheffield, UK

Evolutor Ltd, The Innovation Centre, Sheffield, UK

T. S. Wong (✉)
School of Chemical, Materials and Biological Engineering, The University of Sheffield, Sheffield, UK

Evolutor Ltd, The Innovation Centre, Sheffield, UK

National Center for Genetic Engineering and Biotechnology (BIOTEC), National Science & Technology Development Agency (NSTDA), Khlong Nueng, Khlong Luang, Pathum Thani, Thailand

School of Pharmacy, Bandung Institute of Technology, Bandung, West Java, Indonesia
e-mail: t.wong@sheffield.ac.uk

Abstract The expanding field of synthetic biology requires diversification of microbial chassis to expedite the transition from a fossil fuel-dependent economy to a sustainable bioeconomy. Relying exclusively on established model organisms such as *Escherichia coli* and *Saccharomyces cerevisiae* may not suffice to drive the profound advancements needed in biotechnology. In this context, *Cupriavidus necator*, an extraordinarily versatile microorganism, has emerged as a potential catalyst for transformative breakthroughs in industrial biomanufacturing. This comprehensive book chapter offers an in-depth review of the remarkable technological progress achieved by *C. necator* in the past decade, with a specific focus on the fields of molecular biology tools, metabolic engineering, and innovative fermentation strategies. Through this exploration, we aim to shed light on the pivotal role of *C. necator* in shaping the future of sustainable bioprocessing and bioproduct development.

Keywords Biological CO_2 capture and utilisation (bioCCU), Chemolithotrophic bacterium, CO_2 bioconversion, *Cupriavidus necator*, *Ralstonia eutropha*

1 Introduction

Cupriavidus necator, a Gram-negative bacterium classified within the β-subclass of *Proteobacteria*, is a ubiquitous inhabitant of soil and freshwater ecosystems. Initially isolated in 1961 [1], this microorganism is renowned for its innate capacity to accumulate poly(3-hydroxybutyrate) (PHB), constituting as much as 90% of its cell dry weight (CDW). Over time, *C. necator* has undergone several nomenclatural revisions. In earlier scientific literature and publications, it was referred to as *Hydrogenomonas eutrophus*, *Alcaligenes eutrophus*, *Ralstonia eutropha*, and *Wautersia eutropha*.

C. necator serves as a paradigmatic example of an aerobic lithoautotroph and formatotroph. Its proficiency in utilising H_2 and CO_2 has earned it the moniker 'Knallgas bacterium'. In recent times, it has emerged as a compelling platform for metabolic engineering aimed at the bioconversion of CO_2, without being constrained by the availability of light. Moreover, its exceptional adaptability to dynamic environmental shifts and its capacity to seamlessly transition between various metabolic strategies (including heterotrophic, autotrophic, and mixotrophic) position this non-pathogenic microorganism as a prospective catalyst for revolutionary advancements in the realm of industrial biomanufacturing.

In this book chapter, our aim is to provide an extensive overview of the advancements in *C. necator* within the past 10 years. Our focus encompasses three key areas: molecular biology tools, metabolic engineering, and fermentation strategies.

2 Molecular Biology Tools for *Cupriavidus necator*

Over the past decade, a notable proliferation of specialised genetic tools has emerged, purposefully tailored to meet the needs of the *C. necator* research community. This surge in resources stands as a resounding testament to the microbe's growing potential in industrial biotechnology. In this section, we offer an extensive exploration of genetic toolkits encompassing five critical domains: transformation methodologies, plasmid systems, promoter elements, genome editing techniques, and random mutagenesis approaches.

2.1 *Transformation Techniques*

Bacterial conjugation has stood as the primary method for introducing plasmids into *C. necator*. Despite its time-intensive nature, it remains a favoured approach, especially when handling sizable or potentially toxic plasmids [2]. In 2017, Tee et al. pioneered a highly efficient electroporation-based transformation technique for *C. necator* H16 [3]. This method yielded a remarkable transformation efficiency of 3.86×10^5 cfu/μg of plasmid pBHR1, representing a 10^3-fold enhancement compared to a previously reported value for the identical plasmid [4]. Other research groups also sought to optimise the electroporation protocol by systematically comparing different growth media, the cell growth phase at the time of harvest, and the composition of the electroporation buffer employed [5]. The electroporation efficiency of *C. necator* was constrained by its restriction-modification (RM) systems. Notably, the deletion of H16_A0006, encoding a type I restriction enzyme R subunit, resulted in a striking 1,658-fold increase in electroporation efficiency. Similarly, the removal of H16_A0008-9, which encodes 5-methylcytosine-specific restriction enzymes B and C, led to a 4-fold improvement in electroporation efficiency [6]. Plasmids can be further improved for transformation into *C. necator* by propagating and purifying them from an *E. coli* strain that is deficient in both Dam and Dcm DNA methylation, such as C2925 from New England Biolabs [7]. In addition to electroporation, an alternative method involving heat shock transformation of *C. necator* was explored. Nevertheless, it's worth noting that the transformation efficiency achieved through this approach was approximately 40-fold lower when compared to the results obtained through electroporation [8]. Beyond the influence of the transformation protocol and RM systems, the transformation efficiency of *C. necator* is intricately linked to the design and sequence of the plasmid, a topic we delve into more detail below. As an example, the elimination of non-essential DNA sequences, redundant for plasmid replication, resulted in a fourfold enhancement in electroporation efficiency [9].

2.2 *Plasmid Systems*

When crafting plasmid systems tailored for *C. necator*, it is essential to account for three pivotal factors: the choice of antibiotic selection markers, plasmid stability, and transformation efficiency. Notably, *C. necator* H16 exhibits inherent resistance to gentamicin, spectinomycin, and kanamycin at low concentration [8, 10].

A diverse array of extensively characterised broad-host-range plasmids, originally isolated from Gram-negative bacteria, have demonstrated compatibility with *C. necator* (Table 1). These plasmids encompass various replication elements, including RP4 (belonging to the IncP incompatibility group), pSa (IncW), RSF1010 (IncQ), pBBR1 (incompatibility undefined), and pMOL28 (incompatibility undefined). However, even when these plasmids were successfully introduced into *C. necator* H16, a notable plasmid loss was observed, notwithstanding the application of antibiotic selection [11]. In addressing the plasmid loss challenge, a proactive approach was taken by fortifying the plasmid design through the integration of the RP4 partition sequence. This pivotal enhancement significantly augmented plasmid stability, resulting in fermentation processes characterised by negligible plasmid loss in recombinant *C. necator* H16, maintaining stability for a minimum of 96 h [12]. The RP4 partitioning system not only encodes a post-segregational killing system (*parDE*) but also features a site-specific recombination system (*parCBA*), which facilitates the resolution of plasmid multimers [12]. The same strategy was applied in the development of an inducible plasmid system for *C. necator*, yielding promising results. Notably, no plasmid loss was observed over the course of at least four consecutive cultivations, underscoring the system's robust stability [13]. The choice of plasmid replicon was observed to exert a significant influence on plasmid stability [5, 14]. Notably, plasmids relying on the *Pseudomonas* pVS1 replicon exhibited the highest stability, with over 97% of cells retaining the plasmids throughout a 9-day period [5]. In contrast, plasmids harbouring the

Table 1 Plasmids developed for *C. necator*

Plasmid	Genotype/relevant characteristics	References
pKRSF1010	Km^r, *par*, RSF1010 *mob*, and origin of replication	[12]
pKRTc	Tc^r, *par*, RSF1010 *mob*, and origin of replication	[13]
pE-BR20	Km^r, *mob*, f1 ori, *RP4*, pBBR1 origin of replication	[15]
pMPJAS03	Cm^r, Km^r, pBBR1 origin of replication	[16]
pCAT201	Km^r, pBBR1 OriV-Rep	[8]
pMTL71101	Cm^r, pBBR1 origin of replication	[5]
pMTL75111	Cm^r, pVS1 origin of replication	[5]

Cm^r, chloramphenicol resistance; Km^r, kanamycin resistance, Tc^r, tetracycline resistance

pMOL28 and pBBR1 replicons displayed considerably lower stability, with only 9% and 74% of cells, respectively, retaining the plasmids over the same duration.

The transformation efficiency of *C. necator* has been shown to be influenced by both the kanamycin resistance (KanR) cassette and the replication origin sequence within a plasmid [8]. Specifically, a KanR gene encoding an aminoglycoside *O*-phosphotransferase class I, subtype 'a', demonstrated a twofold increase in transformation efficiency compared to KanR gene of the class II, subtype 'a'. Furthermore, plasmids harbouring the complete OriV (origin of vegetative replication) and Rep sequences exhibited transformation efficiencies ranging from 3,000 to approximately 5,000 times higher than those carrying Rep sequences paired with truncated OriV sequences.

2.3 Promoter Elements

Several renowned and well-characterised promoters, including P_{lac}, P_{lacUV5}, P_{tac}, P_{trp}, P_{BAD}, and P_{T7}, were shown to be active in *C. necator* H16 [12, 17, 18]. Over the last decade, a diverse array of constitutive and inducible promoters, each varying in strength, has been developed, thus enriching the continually expanding genetic toolkit available for research and engineering purposes.

Numerous native promoters sourced from *C. necator* H16 have found utility in facilitating constitutive gene expression. These promoters are harnessed from operons involved in metabolic pathways such as PHB biosynthesis (P_{phaC}), acetoin metabolism (P_{acoD}, P_{acoX}, P_{acoE}), or pyruvate metabolism (P_{pdhE}) [19]. In an endeavour to expand the promoter repertoire, Gruber et al. tested and characterised a suite of constitutive promoters (P_{de33}, P_{f30}, P_{g25}, P_{h207}, P_{h22b}, P_{j5}, P_{k28a}, P_{k28b}, P_{n25}, P_{n26}, and P_{T5}) sourced from the bacteriophage T5 [12]. Among these 11 promoters, P_{j5} was identified as the strongest in driving gene expression. In a comprehensive study conducted by Johnson et al., the architecture and genetic components of the four most prevalent constitutive promoters in *C. necator* H16 (viz., P_{phaC1}, P_{rrsC}, P_{j5}, and P_{g25}) were meticulously examined. Leveraging these four promoters as foundational templates and employing a spectrum of genetic modifications, including point mutations, adjustments in length, integration of regulatory genetic elements, promoter hybridisation, and configuration alterations, they constructed a diverse library of 42 functional constitutive promoters tailored for use in *C. necator* H16 [20]. The promoter engineering approach put forth by Johnson et al. was applied to augment the expression of *Vitreoscilla* hemoglobin in *C. necator* H16, employing the engineered $P_{phaC1\text{-}j5}$ promoter [21]. Further, the engineered promoter ($P_{j5[C2]}$) was used to construct the pCAT vector series tailored for *C. necator* H16 [8].

Similar to constitutive promoters, a diverse array of inducible promoters has emerged in the past decade, affording precise control over the timing and magnitude of gene expression. Promoter P_{BAD} is perhaps the most frequently used for inducible gene expression in *C. necator* [6, 16, 18, 20]. A frequently used promoter design

Fig. 1 Inducible promoter design that combined constitutive promoters and regulatory elements. The example illustrated here used the CymR repressor [17] to regulate GFP expression driven by the constitutive P_{T5} promoter. GFP expression was induced by addition of *p*-cumate

approach involves the fusion of a constitutive promoter (e.g. P_{rrsC}, P_{T5}, P_{j5}) with regulatory elements (e.g. operator sites), as illustrated in Fig. 1. In simpler terms, this strategy employs a repressor protein that responds to an inducer to enable controlled and inducible gene expression. These inducible systems encompass a variety of repressor-inducer pairs. Examples include the TetR repressor, which responds to the inducer anhydrotetracycline (aTc) [22]; the AraC repressor and L-arabinose [23]; the LacI repressor, which relies on IPTG induction with transport facilitated by the integrated lactose permease (LacY) in *C. necator* H16 [17]; the XylS repressor responding to *m*-toluic acid [23]; the CymR repressor and *p*-cumate [17]; and the AcuR repressor and acrylate [18]. Furthermore, hybrid promoters have been developed, incorporating the native *tolC* promoter of *C. necator* in conjunction with a synthetic tetO-operator. These promoters are contingent upon the introduction of the exogenous inducer, doxycycline (dc), for gene expression [15]. Additional systems worth noting encompass the L-rhamnose inducible system [13, 18] and the T7 expression system [9]. The latter was established through the integration of a T7 RNA polymerase gene under the control of the P_{BAD} promoter into the *C. necator* genome, along with the introduction of a P_{T7} promoter upstream of the target gene.

Exploration was also carried out on the promoter (P_{SH}) associated with the soluble hydrogenase (SH) operon within *C. necator* [24]. Intriguingly, this promoter remained dormant during hydrogenase-repressing conditions, such as growth on fructose, while becoming active under hydrogenase-derepressing conditions, as seen during growth on glycerol. This promoter therefore holds a significant potential for facilitating growth-related optimisations in *C. necator*.

2.4 Genome Editing Methods

In the past, genome editing in *C. necator* relied on a suicide vector incorporating the *sacB* gene from *Bacillus subtilis*, which becomes activated in the presence of sucrose and proves lethal when expressed in Gram-negative bacteria [25]. In a nutshell, this process involved transferring the suicide plasmid to *C. necator* through conjugation from *E. coli* S17-1. Transconjugants carrying the integrated plasmid were selectively isolated using appropriate antibiotics, and strains that had lost the integration vector via a second single crossover were identified in a nutrient-rich medium containing sucrose, utilising *sacB* as the negative selection marker. However, this method, which typically took an average of 2–3 weeks to delete a single gene, was not only time-consuming but also exhibited relatively low efficiency.

In 2018, a CRISPR-Cas9 genome editing method was introduced for *C. necator*. This technique involved driving the expression of Cas9 from the editing plasmid using the P_{BAD} promoter, while the corresponding sgRNA was transcribed from a constitutive promoter. To mitigate unintended expression from the arabinose-inducible P_{BAD} promoter, which was used to optimise Cas9 expression, fructose was employed to minimise leaky expression. This setup facilitated genome editing through homologous recombination, resulting in the successful editing of a total of five genes, with editing efficiencies ranging from 78.3% to 100% [6]. The P_{BAD} promoter is known for its inherent leakiness. Studies have shown that the presence of fructose [6] or glucose [26] effectively suppress its leaky expression.

2.5 *Random Mutagenesis*

Transposon-directed insertion-site sequencing (TraDIS), as developed by Langridge et al. [27], has proven effective in generating transposon mutants for *C. necator* H16, facilitating the assessment of gene essentiality [7]. Transposon mutagenesis stands out as a potent and practical approach for creating single-gene disruption mutants in *C. necator* [28].

Adaptive laboratory evolution (ALE) emerges as a powerful and robust technique for creating *C. necator* variants tailored to specific cultivation conditions or the utilisation of distinct carbon sources, all without prior knowledge of its metabolic network. This methodology has proven successful in adapting *C. necator* to efficiently harness both laboratory-grade glycerol and crude glycerol derived from the fat-splitting process [29].

The genome sequence of *C. necator* H16 was initially published in 2006 [30], followed by a comprehensive re-sequencing effort in 2019 employing a combination of PacBio and Illumina sequencing technologies [31]. Additionally, there are several genome-scale models available for this organism, namely, RehMBEL1391 [32] and *i*CN1361 [7]. The convergence of these invaluable resources, coupled with the genetic tools discussed herein, undeniably represents a major leap forward in the realm of *C. necator* engineering. This progress is evident from the extensive array of bioproducts achievable through the manipulation of *C. necator*, as elaborated upon in the subsequent section.

3 Products Manufactured by Engineered *C. necator*

The current market volume for bio-based chemical production is estimated at US$ 80 billion, projected to grow annually by approximately 10% [33]. With its remarkable metabolic versatility, *C. necator* stands poised to significantly bolster this growth. This bacterium has a diverse range of metabolic pathways, including carbon fixation via the Calvin-Benson-Bassham (CBB) cycle, hydrogen and formate oxidation, utilisation of nitrate/nitrite as alternative electron acceptors, and xenobiotic degradation [34]. While *C. necator* is renowned for its polyhydroxyalkanoate (PHA) production, the prevailing metabolic engineering strategy for generating value-added products involves disrupting this pathway through the deletion of the *phaCAB* operon, concurrently boosting the activity of heterologous pathways to redirect carbon flux [35]. This section delves into the robust and versatile capabilities of *C. necator* in utilising various carbon sources and metabolic pathways to synthesise a diverse array of value-added chemicals, as succinctly depicted in Fig. 2. Rather than enumerating products achievable through engineered *C. necator*, this chapter adopts a distinctive approach by spotlighting the frequently targeted pathways for product synthesis.

3.1 Carbohydrate Metabolism

Carbohydrate metabolism not only serves as the primary energy source but also provides essential intermediates for PHB synthesis within *C. necator*. Carbohydrates undergo metabolic processing through various pathways, including, but not limited to, the glycolytic pathway, tricarboxylic acid (TCA) cycle, the Entner-Doudoroff (ED) pathway, the pentose phosphate (PP) pathway, and gluconeogenesis. Intermediates generated within these pathways, such as glucose-6-phosphate and fructose-6-phosphate, can be strategically redirected to facilitate the production of value-added products. For instance, the production of sugar from CO_2 has emerged as a compelling strategy to transition from a fossil fuel-dependent economy to a sustainable one [36]. By predominantly harnessing gluconeogenesis while obstructing the glucose catabolic, ED, and PHB biosynthesis pathways, a yield of 253.3 mg/L glucose was synthesised from CO_2 [37]. A comparable approach was employed in the quest for generating monosaccharide derivatives of glucose, such as *N*-acetylglucosamine (GlcNAc) and *myo*-inositol. Through the strategic disruption of GlcNAc import and its intracellular metabolic pathways, blocking the ED and PHB biosynthesis pathways, and expressing the GlcNAc-6-phosphate *N*-acetyltransferase gene (*gna1*), a GlcNAc titre of 75.3 mg/L was attained in autotrophic fermentation [38]. Through the deliberate overexpression of *myo*-inositol-3-phosphate synthase (IPS) from *S. cerevisiae* and inositol monophosphatase (IMP) from *E. coli*, in conjunction with the blocking of the ED and PHB biosynthesis pathways, an impressive yield of 1054.8 mg/L of *myo*-inositol was attained from CO_2. Similarly, the expression of

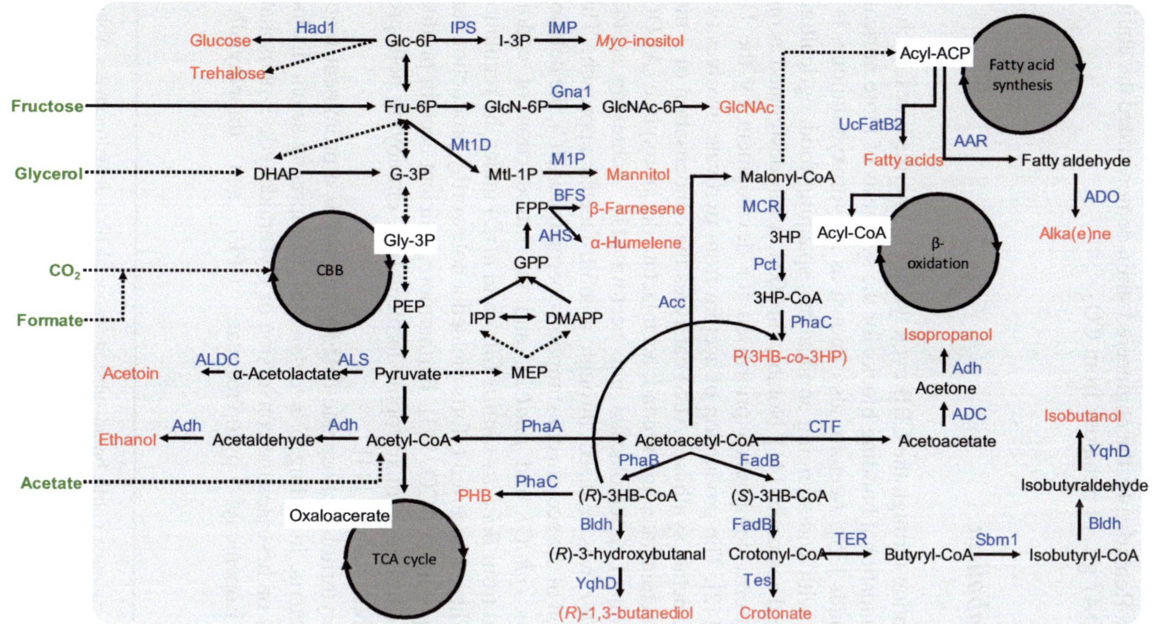

Fig. 2 The native and engineered metabolic pathways in *C. necator* for the manufacture of diverse products. The carbon feedstocks are in green, products are in red, and enzymes catalysing the reactions are in blue. Dotted lines represent multiple reactions. Abbreviations of enzymes: AAR, acyl-ACP reductase; Acc, acetyl-CoA carboxylase; ADC, acetoacetate decarboxylase; Adh, alcohol dehydrogenase; ADO, aldehyde decarbonylase oxygenase; AHS, α-humulene synthase; ALDC, acetolactate decarboxylase; ALS, acetolactate synthase; BFS, β-farnesene synthase; Bldh, CoA-acylating aldehyde dehydrogenase; CTF, acetoacetyl-CoA reductase; FadB, enoyl-CoA hydratase/3-hydroxyacyl-CoA dehydrogenase; Gna1, *N*-acetylglucosamine-6-phosphate *N*-acetyltransferase; Had1, haloacid dehalogenase-like phosphatase; IMP, inositol monophosphatase; IPS, inositol-3-phosphate synthase; M1P, mannitol-1-phosphate phosphatase; MCR, malonyl-CoA reductase; Mt1D, mannitol-1-phosphate dehydrogenase; Pct, propionyl-CoA transferase; PhaA, β-ketothiolase; PhaB, acetoacetyl-CoA reductase; PhaC, PHA synthase; Sbm1, acyl-CoA mutase; TER, trans-2-enoyl-CoA reductase; Tes, thioesterase; UcFatB2, medium-chain-length specific acyl-

mannitol-1-phosphate dehydrogenase (MtlD) and mannitol-1-phosphate phosphatase (M1P) sourced from the brown alga *Ectocarpus siliculosus* enabled the production of 3.9 g/L of mannitol from CO_2 [39]. Under stress conditions induced by NaCl, *C. necator* naturally produces trehalose, as elucidated by Lowe et al. [40]. Introducing the sugar efflux transporter A (*setA*) from *E. coli* resulted in a trehalose-leaky phenotype in *C. necator*. Remarkably, the engineered strain demonstrated the ability to produce as much as 0.47 g/L of trehalose from CO_2.

3.2 Pyruvate Metabolism

Glycolysis and CO_2 fixation through the CBB cycle yield C3 intermediates, such as pyruvate, serving as foundational building blocks for the synthesis of more intricate and diversified compounds [35]. Acetoin, also known as 3-hydroxybutanone, is a valuable C4 compound with wide-ranging applications spanning food, agriculture, the chemical industry, and cosmetics [41]. Notably, in 2004, it was designated as one of the top 30 key platform compounds prioritised for development by the US Department of Energy [42]. The production of acetoin from pyruvate involves two crucial steps: first, acetolactate synthase (ALS) catalyses the condensation of 2 mol of pyruvate to yield α-acetolactate, and second, acetolactate decarboxylase (ALDC) facilitates the conversion of α-acetolactate into acetoin [41]. Achieving acetoin production in *C. necator* was realised through a multi-faceted approach. This encompassed the deletion of the acetoin consumption operon (*acoABC*), the deletion of competing pathways (*phaC1* and *phaC2*), and the introduction of codon-optimised genes sourced from *Bacillus subtilis* (*alsS* and *alsD*) under the regulation of the PHB promoter. These concerted efforts yielded a noteworthy production rate of 0.32 mol of acetoin per mol of CO_2 [43]. Furthermore, efficient acetoin production can be accomplished through mixotrophic fermentation, employing both propionate and CO_2 as carbon sources [44].

The 2-C-methyl-D-erythritol-4-phosphate (MEP) pathway is a prominent metabolic route found in bacteria, including *C. necator* [45]. Comprising seven distinct steps, the MEP pathway orchestrates the conversion of glyceraldehyde-3-phosphate and pyruvate into isopentenyl-5-pyrophosphate (IPP) and dimethylallyl-

Fig. 2 (continued) ACP thioesterase; YqhD, broad-substrate-range alcohol dehydrogenase. Abbreviations of metabolites: 3HP, 3-hydroxypropionate; 3HP-CoA, 3-hydroxypropionyl-CoA; DHAP, dihydroxyacetone phosphate; DMAPP, dimethylallyl pyrophosphate; FPP, farnesyl pyrophosphate; Fru-6P, fructose-6-phosphate; G-3P, glyceraldehyde-3-phosphate; Glc-6P, glucose-6-phosphate; GlcN-6P, glucosamine-6-phosphate; GlcNAc, *N*-acetylglucosamine; GlcNAc-6P, *N*-acetylglucosamine-6-phosphate; GPP, geranyl pyrophosphate; Gly-3P, glycerate-3-phosphate; I-3P, myo-inositol-3-phosphate; IPP, isopentenyl pyrophosphate; MEP, 2-C-methyl-D-erythritol-4-phosphate; Mtl-1P, mannitol-1-phosphate; PEP, phosphoenolpyruvate; (*R*)-3HB-CoA, (*R*)-3-hydroxybutyryl-CoA; (*S*)-3HB-CoA, (*S*)-3-hydroxybutyryl-CoA

pyrophosphate (DMAPP). These compounds serve as essential building blocks for the synthesis of longer-chain precursors required for terpenoid production. Terpenoids hold a significant industrial value, finding applications in perfumes, pharmaceuticals, as eco-friendly alternatives to chlorinated solvents, and as potential substitutes for petroleum-derived fuels [46]. Among terpenoids, sesquiterpenes (C15), including β-farnesene and α-humulene, constitute the largest subgroup [47], each offering distinct industrial applications. β-Farnesene serves as a precursor for fuel additives [48], while α-humulene holds a promise as a potential anticancer drug [49]. The introduction of β-farnesene synthase from *Artemisia annua* into *C. necator* H16 PHB^{-}4 resulted in a β-farnesene product yield of 26.3 ± 1.3 μM through a fed-batch process [50]. Furthermore, the augmentation of the native MEP pathway in *C. necator* with a heterologous mevalonate (MVA) pathway, as demonstrated by Krieg et al. [51], led to the production of 17 mg of α-humulene per gram of cell dry weight (CDW) under autotrophic conditions. This same engineered strain achieved a substantial production rate of 2 g/L of α-humulene during fed-batch fermentation, utilising fructose as the carbon source in conjunction with *n*-dodecane as an extracting organic phase for in situ product removal [47].

3.3 Acetyl-CoA Metabolism

Much like pyruvate mentioned earlier, acetyl-CoA serves as a pivotal hub in metabolism, owing to its central role in interfacing with numerous metabolic pathways and transformations. Acetyl-CoA also functions as a foundational precursor for the synthesis of fatty acids, PHB, and a plethora of other essential molecules. Consequently, it is a logical focal point for endeavours aimed at generating value-added compounds, exemplified by the production of crotonate and similar molecules. Crotonate, a short-chain unsaturated carboxylic acid, holds a significant industrial promise as a fundamental building block for the synthesis of copolymers and organic chemical intermediates [52]. Collas et al. [33] devised a heterologous crotonate pathway geared towards the conversion of acetyl-CoA into crotonate, employing intermediates such as acetoacetyl-CoA and crotonyl-CoA. This approach led to the successful production of 148 mg/L of crotonate from formate.

Countries such as the United States of America, Brazil, and various European nations have embraced ethanol as a prominent fuel source [53]. Introduction of an alcohol dehydrogenase (Adh) from *E. coli* into *C. necator* has enabled the conversion of acetyl-CoA into ethanol, passing through the intermediary acetaldehyde. This led to the synthesis of 0.35 g/L of ethanol from acetate [54]. It is worth noting that the established fuel distribution infrastructure in many countries readily accommodates C4 alcohols, allowing for their direct utilisation as vehicle fuels without the necessity for blending with petrol [45]. While prior research endeavours have primarily focused on engineering the Ehrlich pathway for isobutanol production, an intriguing alternative has emerged through the development of a CoA-dependent pathway. This novel approach entails elongating acetyl-CoA into butyryl-CoA, followed by a rearrangement from butyryl-CoA to isobutyryl-CoA facilitated by acyl-CoA mutase (Sbm1) and the conversion of isobutyryl-CoA into isobutanol. The engineered *C. necator* strain was capable of producing 32 mg/L of isobutanol from fructose [55].

Leveraging acetyl-CoA or its precursors, notably pyruvate, offers a promising avenue for the synthesis of (*R*)-1,3-butanediol (1,3-BDO) and isopropanol. 1,3-BDO features diverse applications across food, chemical, and cosmetic sectors [56]. As a naturally occurring C4 diol found in bell peppers, it imparts a delicate fruity aroma. It serves as a pivotal intermediate or solvent in the production of paints, coatings, inks, plasticisers, 1,3-butadiene, and azetidinone derivatives. In personal care, 1,3-BDO acts as an effective humectant and emollient. In the quest for (*R*)-1,3-BDO production, two distinct pathways were engineered [57]. The first pathway facilitates the conversion of acetyl-CoA into (*R*)-1,3-BDO via intermediates like acetoacetyl-CoA, (*R*)-3-hydroxybutyryl-CoA, and (*R*)-3-hydroxybutanal, while the second pathway employs pyruvate as a precursor to generate (*R*)-1,3-BDO through the intermediate (*R*)-3-hydroxybutanal. By amalgamating these pathways, coupled with the strategic elimination of PHB biosynthesis and reducing flux through the TCA cycle, the engineered strain produced 2.97 g/L of (*R*)-1,3-BDO from CO_2. Isopropanol is used as a fuel additive, primarily attributed to its high-octane rating [53]. In the pursuit of isopropanol production, a similar strategy was employed, wherein acetyl-CoA was transformed into isopropanol, traversing through intermediary compounds such as acetoacetyl-CoA, acetoacetate, and acetone [58]. Overexpression of native *groEL* and *groES* genes was found to enhance *C. necator*'s tolerance to exogenous isopropanol [59]. The tolerance enhancement facilitated the attainment of an impressive final isopropanol concentration of 9.8 g/L in a fed-batch culture, employing fructose as the exclusive carbon source.

3.4 Malonyl-CoA Pathway

3-Hydroxypropionate (3HP) is a compelling platform chemical that holds an immense promise as a precursor for a diverse range of essential compounds, including acrylate, acrylamide, and even as a monomer for biodegradable plastics. The synthesis of 3HP can be accomplished through various intermediates, such as glycerol, malonyl-CoA, and β-alanine. Among these bioengineering pathways, the malonyl-CoA route presents distinctive advantages, including its versatility in accommodating a wide array of feedstocks, favourable thermodynamics, and inherent redox neutrality [60]. An enticing approach for achieving autotrophic 3-HP production involves the implementation of the 3-HP bicycle pathway, recognised for its remarkable efficiency in aerobic CO_2 fixation. This pathway leverages the malonyl-CoA reductase (MCR) as one of its key enzymatic components [61]. Through the strategic expression of an MCR responsible for the conversion of malonyl-CoA into 3HP and a propionyl-CoA transferase tasked with converting

3HP into 3HP-CoA, engineered *C. necator* could produce 3.11 g/L of [P(3HB-*co*-3HP)] with a remarkable 32.25% molar fraction of 3-HP [62].

3.5 Fatty Acid Biosynthesis and β-Oxidation Pathways

Fatty acids, a diverse and vital class of molecules, have garnered substantial attention in the realm of metabolic engineering due to their potential as precursors for biofuels [45]. In the biosynthesis of fatty acids, acetyl-CoA is iteratively condensed on an acyl carrier protein (ACP). Fatty acids are also subject to consumption through the competitive β-oxidation pathway [63]. The introduction of UcFatB2, a thioesterase specific to medium-chain-length acyl-ACP, led to the generation of medium-chain-length fatty acids within *C. necator*. This achievement was further enhanced by strategically eliminating PHB synthesis and selectively deleting acyl-CoA ligases, which serve as entry points for fatty acids into the β-oxidation pathway. These combined efforts yielded a production of up to 62 mg/L of free fatty acids [64]. Likewise, in a study conducted by Li et al. [65], a notable synthesis of fatty acids was achieved, with yields of 128.48 mg/g CDW using fructose as the carbon source and 60.64 mg/g CDW when CO_2 was employed. This accomplishment was realised through the strategic overexpression of endogenous acetyl-CoA carboxylase and cytoplasmic thioesterase enzymes, complemented by the introduction of exogenous type I polyketide synthase and holo-ACP synthase.

Alkanes with defined carbon chain lengths offer promising alternatives to petroleum-based fuels. To realise alkane synthesis, Crepin et al. [66] introduced a pathway from *Synechococcus elongatus* consisting of two genes encoding an acyl-ACP reductase (*aar*) and an aldehyde-deformylating oxygenase (*ado*) into a *C. necator* mutant strain deficient in the PHB synthesis pathway. Under heterotrophic conditions with fructose as the carbon source and nitrogen limitation, in the presence of an organic phase (decane), the strain was capable of producing 670 mg/L of total hydrocarbons, consisting of 286 mg/L of pentadecane, 131 mg/L of heptadecene, 18 mg/L of heptadecane, and 236 mg/L of hexadecanal.

Metabolic engineering has empowered *C. necator* to produce a wide array of valuable products. Nonetheless, effective strain engineering should seamlessly integrate with advanced fermentation strategies to not only create economically viable bioprocesses but also enhance product yields.

4 Fermentation Methods

C. necator exhibits remarkable versatility in harnessing a wide array of carbon feedstocks, such as carbohydrates, glycerol, organic acids, and C1 compounds [67]. Its genetic adaptability has significantly broadened its capacity to utilise diverse

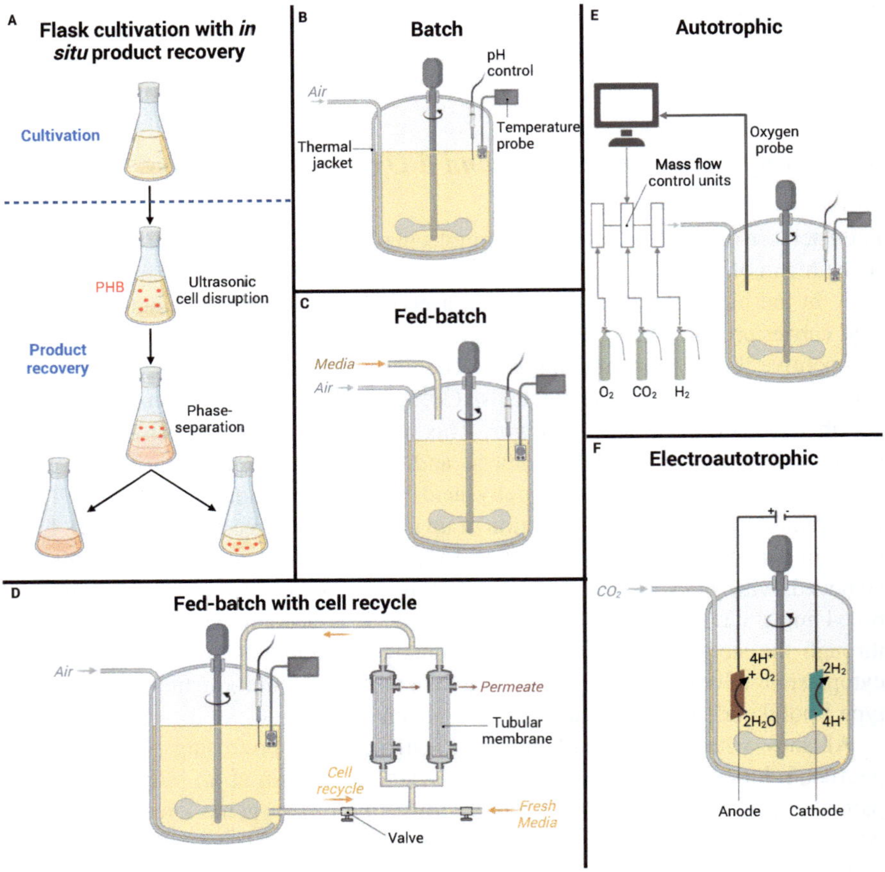

Fig. 3 Various fermentation strategies using *Cupriavidus necator* as a biomanufacturing host. (**a**) Flask cultivation with or without in situ product recovery [68], (**b**) batch fermentation, (**c**) fed-batch fermentation, (**d**) fed-batch fermentation with cell recycle [69], (**e**) autotrophic fermentation [70], and (**f**) electroautotrophic fermentation [71]. Created in BioRender. Casey, D. (2023) BioRender.com/s12q850

carbon sources [29], rendering it an exceedingly appealing candidate as a biomanufacturing host for the burgeoning biotech industry. Numerous fermentation strategies have been innovatively devised to unlock its full biomanufacturing potential (Fig. 3), with a growing focus on sustainable feedstock utilisation (Table 2). While polyhydroxybutyrate (PHB) remains the predominant product of *C. necator*, we are witnessing a noteworthy diversification in its product portfolio (Table 2), facilitated by the organism's amenability to genetic manipulation and the continuous expansion of its genetic toolkit.

Table 2 Biomanufacturing with *C. necator* using different fermentation strategies, along with the products created and the feedstocks used. Yield is reported in terms of gram product per gram substrate consumed

Fermentation strategies	Product	Feedstock	Titre (g/L)	Productivity (g/L/h)	Yield (g/g)	References
Flask culture	PHB	Fructose	4.6	0.06	0.31	[72]
Flask culture	PHB	Glucose	11.83	0.25[a]	0.91[a]	[73]
Flask culture	PHB	Glucose	2.155	–	–	[74]
Flask culture	P(3HB-*co*-3HV)	Volatile fatty acids from broken rice	0.95	0.01[a]	–	[75]
Flask culture with in situ product recovery	PHB	Glycerol	0.44[a]	–	–	[68]
Co-culture	P(3HB-*co*-3HV)	Sucrose	2.30	–	0.08	[76]
Batch	PHB	Milled corn hydrolysate and crude glycerol	14.17	0.197	0.43	[77]
Batch	PHB	Molasses	1.31	0.012[a]	–	[78]
Batch	P(3HB-*co*-3HV)	Synthetic volatile fatty acids	1.60	0.044[a]	0.25	[79]
Fed-batch	PHB	Acetic acid	4.95	0.17[a]	0.33	[80]
Fed-batch	PHB	Acetic acid	58.5	0.93	–	[81]
Fed-batch	PHB	Waste makgeolli lees hydrolysate and glucose	24.1	0.33[a]	–	[82]
Fed-batch	PHB	Corn stover alkaline pretreatment liquor	4.47	0.080	–	[83]
Fed-batch	PHB	Fructose	25.7	0.43	–	[84]
Fed-batch	PHB	Crude glycerol and rapeseed meal hydrolysate	24.75	0.21	0.32	[85]
Fed-batch	PHB	Crude glycerol and sunflower meal hydrolysate	27	0.28	0.32	[86]
Fed-batch	PHB	Waste glycerol	65.6	1.36	–	[87]
Fed-batch	PHB	Glycerol	85.8	1.83	–	[88]
Fed-batch	PHB	Banana pulp juice and fructose	2.816	–	–	[89]

(continued)

Table 2 (continued)

Fermentation strategies	Product	Feedstock	Titre (g/L)	Productivity (g/L/h)	Yield (g/g)	References
Fed-batch	PHB	Used cooking oil	19.8	0.53	0.52	[90]
Fed-batch	P(3HB-*co*-3HV)	Volatile fatty acids from cheese whey and glucose	10.6[a]	–	0.60	[91]
Fed-batch	P(3HB-*co*-3HV)	Waste vegetable oil and propionic acid	121.7	2.03	0.67[b]	[92]
Fed-batch	P(3HB-*co*-3HHx)	Fructose and rapeseed oil	–	1.45	–	[93]
Fed-batch	P(3HB-*co*-3HD)	Beeswax hydrolysate and glucose	8.91	–	–	[94]
Fed-batch with cell recycle	PHB	Glucose	113	3.10	0.33[b]	[95]
Fed-batch with cell recycle	PHB	Grass silage	9.4	0.31	0.15	[96]
Fed-batch with cell recycle	P(3HB-*co*-3HV)	Synthetic volatile fatty acids	2.8	–	–	[97]
Fed-batch with cell recycle	P(3HB-*co*-3HV)	Glucose and propionic acid	64.6	1.24	–	[69]
Autotrophic	PHB	CO_2	10.2	–	–	[70]
Autotrophic	PHB	CO_2	11.6[a]	–	–	[98]
Autotrophic	Isopropanol	CO_2	3.5	–	–	[99]
Heterotrophic-autotrophic	PHB	Waste glycerol and CO_2	28	0.168	–	[100]
Heterotrophic-autotrophic	Isopropanol	Fructose and CO_2	0.25	0.021[a]	0.093	[59]
Electroautotrophic	Isopropanol	CO_2	0.216	–	–	[71]
Electroautotrophic	α-Humulene	CO_2	0.0108	0.00008	–	[51]
Mixotrophic	P(3HB-*co*-3HV)	Synthetic volatile fatty acids and CO_2	3.1[a]	0.026[a]	–	[101]

PHB, poly(3-hydroxybutyrate); P(3HB-*co*-3HV), poly(3-hydroxybutyrate-*co*-3-hydroxyvalerate); P(3HB-*co*-3HHx), poly(3-hydroxybutyrate-*co*-3-hydroxyhexanoate); P(3HB-*co*-3HD), poly(3-hydroxybutyrate-*co*-3-hydroxydecanoate)

[a] Value was not directly given in paper, but calculated from other data within the study

[b] Yield is reported in terms of gram product per gram substrate fed

4.1 Heterotrophic Cultivation

PHB production in *C. necator* is typically achieved in media rich in carbon, often with either nitrogen or phosphorus limitation [102]. The simplest mode of cultivating *C. necator* is a shake flask culture. This approach is often used to assess the impact of different culture conditions on the bacteria. For example, response surface methodology (RSM) was applied to optimise culture media for PHB production by *C. necator* ATCC 17697 [72]. With a carbon concentration of 20 g/L, nitrogen limited to 1.5 g/L, phosphate levels at 8.75 g/L, and pH 7.5, a maximum PHB production of 4.6 g/L was achieved [72]. Another study, using the engineered strain *C. necator* NSDG-GG, showed that optimal conditions differed whether the focus was on cell growth or PHB production [73]. This suggests a two-stage approach would be advantageous, where biomass is initially maximised before PHB production prioritised.

Flask cultures can also be employed to establish innovative fermentation strategies, including integrated biorefinery concepts [75]. The use of ultrasound during the cultivation of *C. necator* NRRL B14690 was found to increase PHB yield by a factor of almost 2 compared to culture without ultrasonication [74]. Extractive bioconversion that integrates fermentation and downstream PHB purification has also been investigated in flask cultures [68]. An aqueous two-phase system was applied, which utilises heat and a thermo-separating polymer to split the culture into a water-rich top phase containing PHB and a polymer-rich bottom phase containing cells that can be recycled (Fig. 3a). Co-culture of *C. necator* with another microorganism has been demonstrated at an even smaller scale. In 25 mL tubes containing 10 mL of media, 2.30 g/L poly(3-hydroxybutyrate-*co*-3-hydroxyvalerate) (P(3HB-*co*-3HV)) was produced from sucrose by a culture of *C. necator* 5119 and *Bacillus subtilis* [76]. *B. subtilis* hydrolysed sucrose, which cannot otherwise be utilised as a carbon source by *C. necator*.

At an increased scale, batch fermentations are often used to assess the viability of new carbon sources (Fig. 3b). P(3HB-*co*-3HV) production from volatile fatty acids (VFA) at different concentrations and with varying compositions has been investigated, with a highest titre of 1.60 g/L achieved despite the focus being on optimising growth [79]. de Mello et al. [77] studied PHB accumulation using reducing sugars from milled corn hydrolysate and crude glycerol, with the aim of coupling bioplastic and bioethanol production in an integrated biorefinery. In an 8 L stirred tank reactor, 14.17 g/L PHB was achieved with 0.197 g/L/h productivity [77]. Another innovation trialled in batch cultivation is the reuse of filtrate from cell recovery as substrate [78]. Following initial cultivation of *C. necator* TISTR 1095 on molasses, cells were recovered and PHB extracted. The filtrate from this process contained sugar and organic acid; hence it could be recycled as the carbon source in a second reactor. While the production was lower in the second reactor (1.13 g/L vs 1.31 g/L) [78], this may be a viable strategy to reduce waste.

The most common strategy for polyhydroxyalkanoate (PHA) production from *C. necator* is in a fed-batch bioreactor [69, 80–97], which allows the substrate concentration to be controlled throughout fermentation (Fig. 3c). Kedia et al. [80] continually fed VFA as both carbon source and pH control, meaning NaOH was not required to adjust pH. Sodium accumulation has been shown to negatively affect *C. necator* [103]; hence the change in feeding regime increased PHA accumulation almost twofold [80]. Another benefit of fed-batch fermentation is the ability to control the carbon-to-nitrogen (C:N) ratio. Makgeolli lees enzymatic hydrolysate was used as feedstock for PHB production by *C. necator* H16 [82]. During fed-batch cultivation with the C:N ratio controlled to 20:1 by the periodic addition of glucose, 30 times more PHB was synthesised compared with batch fermentation [82]. This was a significant increase in product yield, despite only approximately doubling the carbon source supplied and increasing cell growth by 1.5-fold [82]. However, fed-batch fermentation is not always more efficient than batch. Li and Wilkins [83] compared two fed-batch approaches to batch fermentation for PHB production by *C. necator* DSM 545 using lignin from corn stover alkaline pretreatment liquor. A higher concentration of PHB was reached during fed-batch fermentation but over a longer period of time, with productivity being highest in batch cultivation [83].

High cell density fermentation has been used with fed-batch feeding to increase PHA production further. Under oxygen-limited conditions, 121.7 g/L P(3HB-*co*-3HV) was produced from waste frying oil and propionic acid by *C. necator* H16 [92]. In an attempt to overcome substrate inhibition, an immersed membrane was used to maintain high cell density in fed-batch fermentation of *C. necator* DSM 545 on VFA [97]. Following 128 h of cultivation, a maximum PHB titre of 2.8 g/L was reached, a 1.75-fold increase on the concentration achieved by the same authors with a conventional batch method [79]. The use of an external membrane to recycle cells (Fig. 3d) has also been employed to increase PHA accumulation. Haas et al. [95] produced 113 g/L PHB from glucose with a productivity of 3.10 g/L/h, while Schmidt et al. [69] added propionic acid and achieved 64.6 g/L P(3HB-*co*-3HV) at 1.24 g/L/h.

Given the difference in optimal conditions for growth and product formation by *C. necator*, fermentation is often split into biomass and product accumulation phases [88–91, 96]. For example, to increase P(3HB-*co*-3HV) accumulation from VFA to 10.6 g/L, glucose was used as the carbon source for the cell growth phase [91]. Cruz et al. [90] used batch fermentation for biomass accumulation and assessed different feeding strategies for the polymer-producing fed-batch stage. Used cooking oil was supplied either through exponential feeding or DO-stat mode. The DO-stat mode, where nutrient feed is controlled based on dissolved oxygen concentration, gave the highest PHB productivity (0.53 g/L/h) [90]. A two-stage fed-batch strategy has recently been assessed at a 150 L pilot scale, with purified glycerol used to produce 85.8 g/L PHB at 1.83 g/L/h [88]. The *C. necator* B-10646 cells were initially grown under nitrogen deficiency, before PHB synthesis was induced by the use of a nitrogen-free medium [88]. The two-stage approach can also facilitate more complete use of feedstock. Based on an integrated biorefinery concept, different products from processing of grass silage were employed for PHB production by *C. necator* DSM 531 [96]. Greater control has been applied to fermentation in various 'three-stage' approaches, where biomass accumulation is split into two phases with distinct feeding strategies [84, 87, 94]. For growth of *C. necator* DSM 545 on waste glycerol, exponential feeding was initially employed before switching to feeding based on alkali addition monitoring [87]. Following this, PHB synthesis took place under constant substrate feeding with 65.6 g/L produced at 1.36 g/L/h [87].

Separating cell growth and PHA synthesis into two bioreactors can improve productivity further. For production of poly(3-hydroxybutyrate-*co*-3-hydroxyhexanoate) from fructose and rapeseed oil, Santolin et al. [93] recycled 10% of the culture broth in the first reactor and transferred the remainder to a second for PHA accumulation. This repeated fed-batch procedure meant the slow lag phase could be avoided, as well as downtime between cultures for cleaning, sterilisation, and filling of the reactor. This downtime can also be avoided in a one bioreactor set-up. When cultivating *C. necator* DSM 545 with acetic acid to produce PHB, Vlaeminck et al. [81] retained 0.5% v/v of the broth between fed-batch fermentations to act as inoculum for the next. A titre of 58.5 g/L PHB was reached, while productivity (0.93 g/L/h) doubled compared with a single fed-batch culture [81].

4.2 *Autotrophic Cultivation*

When using carbon dioxide as feedstock in autotrophic fermentation of *C. necator*, the explosivity of oxygen and hydrogen mixture is a significant hurdle. One way to avoid this is to keep the gas levels outside the explosion limits (Table 3); however this can limit the availability of oxygen to the bacteria. Lambauer and Kratzer [70] instead cultivated *C. necator* H16 in an explosion-proof bioreactor inside a grounded fume hood. They also assessed different O_2 supply strategies, finding a stepwise increase guided by dissolved oxygen levels gave the highest cell growth and a PHB titre of 10.2 g/L from CO_2 [70]. This was achieved with manual gas dosage, and the value was increased to 11.6 g/L using automated gas control [98]. Another way to improve autotrophic fermentation of *C. necator* is to increase gas transfer by

Table 3 Autotrophic fermentation of *C. necator*

C. necator strain	Initial H_2:CO_2:O_2 (vol%)	Control	Pressure	References
H16	85:10:2	Guided by dissolved O_2 concentration	Atmospheric	[70]
H16	40:5:2	Guided by dissolved O_2 and CO_2 concentration	Atmospheric	[98]
H16 PHB$^-$4	Electroautotrophic	–	Atmospheric	[51]
Re2133	87:3:2	Guided by dissolved O_2 concentration	4.2 bar	[99]
Re2133	60:10:2	Constant gas composition	Atmospheric	[59]
Re2133	Electroautotrophic	–	Atmospheric	[71]
DSM 545	84:13.2:2.8	Constant gas composition	Atmospheric	[100]

operating at high pressure. This approach was applied to cultivation of engineered *C. necator* Re2133 to produce 3.5 g/L isopropanol [99].

Another way to overcome the limitations in gas transfer, and therefore cell growth, at low oxygen levels is to complete the biomass accumulation phase under heterotrophic conditions first. Garcia-Gonzalez et al. [100] assessed this approach using either glucose or glycerol as carbon source before initiating autotrophic PHB production under nitrogen and oxygen limitation. They found that if biomass concentration was too high, PHB accumulation decreased significantly, likely due to poor O_2 transfer [100]. The highest PHB titre was 28 g/L, achieved when waste glycerol was used to accumulate 13 g/L cell mass before PHB production [100]. A similar approach was taken to produce isopropanol from engineered *C. necator* Re2133, with an additional stage to allow cells to adapt to gaseous substrates [59]. Only 0.25 g/L isopropanol was synthesised, likely due to production being conducted at a low cell density of around 1 g/L [59].

A more innovative method of autotrophic fermentation is the use of a bioelectrochemical cell to split water into hydrogen and oxygen (Fig. 3f). Water oxidation occurs at the anode, producing O_2 and four protons that are then reduced to $2H_2$ at the cathode. This system has been used to produce isopropanol and the terpene α-humulene from CO_2 with engineered strains of *C. necator* [51, 71]. While production of isopropanol was around 16 times lower than the titre reached in a high-pressure autotrophic culture [71, 99], the system has the benefit of not requiring H_2 and O_2 gas supplies. It can also be incorporated with photovoltaic cells to utilise solar energy [71].

4.3 Mixotrophic Cultivation

C. necator is able to utilise heterotrophic and autotrophic mechanisms simultaneously, known as mixotrophic cultivation. Mixotrophic growth and PHA production by *C. necator* H16 have been investigated using VFA and CO_2 [101]. Compared to VFA-only cultures, a higher optical density and PHA titre were reached in the presence of CO_2. To further improve mixotrophic production, H_2 was supplied as an additional energy source. Initially this had a negative impact, with uptake of both VFA and CO_2 being limited [101]. The process was then changed to have a lower starting VFA concentration before addition of more VFA at the 48-h mark. This system led to a P(3HB-*co*-3HV) accumulation of 3.1 g/L, higher than under heterotrophic conditions or mixotrophic conditions without H_2 [101].

In the past decade, there has been a significant innovation in the fermentation processes involving *C. necator*. Methods such as external cell recycling and reuse of culture broth have increased the productivity of fed-batch cultures significantly, while smaller-scale approaches have been used to optimise and assess new processes. For autotrophic fermentation, the use of a bioelectrochemical cell is a promising technology, although production is currently low. Techniques such as operating under pressure have also improved gas availability in traditional autotrophic cultivation. If the aim is to maximise production, separating fermentation into two stages is almost essential no matter the cultivation mode. This allows conditions to be optimised for biomass and product accumulation separately, as well as different feeding strategies to be applied. It can also enable a heterotrophic-autotrophic approach, which can combat some concerns around autotrophic production with *C. necator*.

5 Conclusion

The development of *C. necator* as a biomanufacturing host has made significant strides, as evident from the comprehensive review of advancements in this book chapter. However, several technological and knowledge gaps must be addressed before *C. necator* can reach its full potential as a biomanufacturing powerhouse.

Firstly, there is a pressing need to standardise and thoroughly characterise the biological parts designed for use in *C. necator*. This standardisation is essential to ensure reproducibility and wider adoption of these parts. Additionally, establishing an organised archive for these biological components would facilitate easy access and adoption, aligning with the principles of synthetic biology.

Secondly, while gas fermentation holds an immense promise for biological carbon dioxide capture and utilisation (bioCCU), it is crucial to address bioprocess safety concerns. Further technological developments are required, both in terms of hardware for gas fermentation and scaling up these processes, to make them more efficient and safer.

Thirdly, despite the significant progress in metabolic engineering and fermentation strategies, the choice of sustainable feedstock remains a top priority. *C. necator* should be comprehensively characterised for its compatibility with various potential feedstocks, especially after feedstock pretreatment, to ensure its versatility and adaptability.

With collaborative efforts from various scientific communities, including wet lab and dry lab scientists, *C. necator* has the potential to emerge as the next-generation biomanufacturing host, capable of meeting the diverse and evolving societal needs of a sustainable bioeconomy. The future holds a great promise for this versatile microorganism, and further research and innovation will undoubtedly propel it to the forefront of biotechnology.

Acknowledgement We express our sincere gratitude to the following organisations for their generous financial support: Department of Chemical and Biological Engineering at the University of Sheffield (studentships, to DC and LDG), the EPSRC New Investigator Award (EP/X025853/1, to KLT), the UKRI (Global Challenges Research Fellowship, to KLT), the RAEng|The Leverhulme Trust Senior Research Fellowship (LTSRF1819\15\21, to TSW), the National Research Council of Thailand (P2250317/3, to TSW), and BBSRC-IAA & C1Net (to TSW & KLT).

References

1. Schlegel HG, Kaltwasser H, Gottschalk G (1961) Arch Mikrobiol 38:209
2. Pavan M, Reinmets K, Garg S, Mueller AP, Marcellin E, Kopke M, Valgepea K (2022) Metab Eng 71:117. https://doi.org/10.1016/j.ymben.2022.01.015
3. Tee KL, Grinham J, Othusitse AM, Gonzalez-Villanueva M, Johnson AO, Wong TS (2017) Biotechnol J 12. https://doi.org/10.1002/biot.201700081
4. Solaiman DK, Swingle BM, Ashby RD (2010) J Microbiol Methods 82:120. https://doi.org/10.1016/j.mimet.2010.04.010
5. Ehsaan M, Baker J, Kovacs K, Malys N, Minton NP (2021) J Microbiol Methods 189:106323. https://doi.org/10.1016/j.mimet.2021.106323
6. Xiong B, Li Z, Liu L, Zhao D, Zhang X, Bi C (2018) Biotechnol Biofuels Bioprod 11:172. https://doi.org/10.1186/s13068-018-1170-4
7. Pearcy N, Garavaglia M, Millat T, Gilbert JP, Song Y, Hartman H, Woods C, Tomi-Andrino C, Reddy Bommareddy R, Cho BK, Fell DA, Poolman M, King JR, Winzer K, Twycross J, Minton NP (2022) PLoS Comput Biol 18:e1010106. https://doi.org/10.1371/journal.pcbi.1010106
8. Azubuike CC, Gatehouse AMR, Howard TP (2021) N Biotechnol 65:20. https://doi.org/10.1016/j.nbt.2021.07.003
9. Hu M, Xiong B, Li Z, Liu L, Li S, Zhang C, Zhang X, Bi C (2020) BMC Microbiol 20:121. https://doi.org/10.1186/s12866-020-01812-9
10. Park JM, Jang YS, Kim TY, Lee SY (2010) FEMS Microbiol Lett 309:193. https://doi.org/10.1111/j.1574-6968.2010.02041.x
11. Gruber S, Schwab H, Koefinger P (2015) N Biotechnol 32:552. https://doi.org/10.1016/j.nbt.2015.03.015
12. Gruber S, Hagen J, Schwab H, Koefinger P (2014) J Biotechnol 192(Pt B):410. https://doi.org/10.1016/j.jbiotec.2014.09.023
13. Sydow A, Pannek A, Krieg T, Huth I, Guillouet SE, Holtmann D (2017) J Biotechnol 263:1. https://doi.org/10.1016/j.jbiotec.2017.10.002
14. Sato S, Fujiki T, Matsumoto K (2013) J Biosci Bioeng 116:677. https://doi.org/10.1016/j.jbiosc.2013.05.026
15. Aboulnaga EA, Zou H, Selmer T, Xian M (2018) J Biotechnol 274:15. https://doi.org/10.1016/j.jbiotec.2018.03.007
16. Flores-Sanchez A, Rathinasabapathy A, Lopez-Cuellar MDR, Vergara-Porras B, Perez-Guevara F (2020) Int J Biol Macromol 164:1600. https://doi.org/10.1016/j.ijbiomac.2020.07.275
17. Gruber S, Schwendenwein D, Magomedova Z, Thaler E, Hagen J, Schwab H, Heidinger P (2016) J Biotechnol 235:92. https://doi.org/10.1016/j.jbiotec.2016.04.026
18. Alagesan S, Hanko EKR, Malys N, Ehsaan M, Winzer K, Minton NP (2018) Appl Environ Microbiol 84. https://doi.org/10.1128/AEM.00878-18
19. Delamarre SC, Batt CA (2006) Appl Microbiol Biotechnol 71:668. https://doi.org/10.1007/s00253-005-0217-1
20. Johnson AO, Gonzalez-Villanueva M, Tee KL, Wong TS (2018) ACS Synth Biol 7:1918. https://doi.org/10.1021/acssynbio.8b00136
21. Tang R, Weng C, Peng X, Han Y (2020) Metab Eng 61:11. https://doi.org/10.1016/j.ymben.2020.04.009
22. Li H, Liao JC (2015) ACS Synth Biol 4:101. https://doi.org/10.1021/sb4001189
23. Bi C, Su P, Muller J, Yeh YC, Chhabra SR, Beller HR, Singer SW, Hillson NJ (2013) Microb Cell Fact 12:107. https://doi.org/10.1186/1475-2859-12-107
24. Jugder BE, Welch J, Braidy N, Marquis CP (2016) PeerJ 4:e2269. https://doi.org/10.7717/peerj.2269
25. Quandt J, Hynes MF (1993) Gene 127:15. https://doi.org/10.1016/0378-1119(93)90611-6

26. Simcikova M, Prather KL, Prazeres DM, Monteiro GA (2014) Vaccine 32:2843. https://doi.org/10.1016/j.vaccine.2014.02.035
27. Langridge GC, Phan MD, Turner DJ, Perkins TT, Parts L, Haase J, Charles I, Maskell DJ, Peters SE, Dougan G, Wain J, Parkhill J, Turner AK (2009) Genome Res 19:2308. https://doi.org/10.1101/gr.097097.109
28. Raberg M, Heinrich D, Steinbuchel A (2015) McGenity T, Timmis K, Nogales B (eds) Hydrocarbon and lipid microbiology protocols. Springer, Berlin. https://doi.org/10.1007/8623_2015_110
29. Gonzalez-Villanueva M, Galaiya H, Staniland P, Staniland J, Savill I, Wong TS, Tee KL (2019) Int J Mol Sci 20. https://doi.org/10.3390/ijms20225737
30. Pohlmann A, Fricke WF, Reinecke F, Kusian B, Liesegang H, Cramm R, Eitinger T, Ewering C, Potter M, Schwartz E, Strittmatter A, Voss I, Gottschalk G, Steinbuchel A, Friedrich B, Bowien B (2006) Nat Biotechnol 24:1257. https://doi.org/10.1038/nbt1244
31. Little GT, Ehsaan M, Arenas-Lopez C, Jawed K, Winzer K, Kovacs K, Minton NP (2019) Microbiol Resour Announc 8. https://doi.org/10.1128/MRA.00814-19
32. Park JM, Kim TY, Lee SY (2011) BMC Syst Biol 5:101. https://doi.org/10.1186/1752-0509-5-101
33. Collas F, Dronsella BB, Kubis A, Schann K, Binder S, Arto N, Claassens NJ, Kensy F, Orsi E (2023) Metab Eng 79:49. https://doi.org/10.1016/j.ymben.2023.06.015
34. Jahn M, Crang N, Janasch M, Hober A, Forsstrom B, Kimler K, Mattausch A, Chen Q, Asplund-Samuelsson J, Hudson EP (2021) Elife:10. https://doi.org/10.7554/eLife.69019
35. Panich J, Fong B, Singer SW (2021) Trends Biotechnol 39:412. https://doi.org/10.1016/j.tibtech.2021.01.001
36. Nangle SN, Ziesack M, Buckley S, Trivedi D, Loh DM, Nocera DG, Silver PA (2020) Metab Eng 62:207. https://doi.org/10.1016/j.ymben.2020.09.002
37. Wang X, Luo H, Wang Y, Wang Y, Tu T, Qin X, Su X, Huang H, Bai Y, Yao B, Zhang J (2022) Bioresour Technol 362:127806. https://doi.org/10.1016/j.biortech.2022.127806
38. Wang X, Chang F, Wang T, Luo H, Su X, Tu T, Wang Y, Bai Y, Qin X, Zhang H, Wang Y, Yao B, Huang H, Zhang J (2023) Bioresour Technol 379:129024. https://doi.org/10.1016/j.biortech.2023.129024
39. Hanko EKR, Sherlock G, Minton NP, Malys N (2022) Metab Eng 72:24. https://doi.org/10.1016/j.ymben.2022.02.003
40. Lowe H, Beentjes M, Pfluger-Grau K, Kremling A (2021) Bioresour Technol:319. https://doi.org/10.1016/j.biortech.2020.124169
41. Cui ZZ, Wang ZW, Zheng MY, Chen T (2022) Crit Rev Biotechnol 42:1135. https://doi.org/10.1080/07388551.2021.1995319
42. Werpy T, Peterson G, Aden A, Holladay J, White J, Bozell J, Manheim A (2004) Top value added chemicals from biomass, vol 1. Pacific Northwest National Lab. https://doi.org/10.2172/926125
43. Windhorst C, Gescher J (2019) Biotechnol Biofuels 12:163. https://doi.org/10.1186/s13068-019-1512-x
44. Harrer D, Windhorst C, Bohner N, Ducassou JN, Coute Y, Gescher J (2021) Bioresour Technol 329. https://doi.org/10.1016/j.biortech.2021.124866
45. Pan HJ, Wang J, Wu HL, Li ZJ, Lian JZ (2021) Biotechnol Biofuels:14. https://doi.org/10.1186/s13068-021-02063-0
46. Peralta-Yahya PP, Ouellet M, Chan R, Mukhopadhyay A, Keasling JD, Lee TS (2011) Nat Commun 2. https://doi.org/10.1038/ncomms1494
47. Milker S, Sydow A, Torres-Monroy I, Jach G, Faust F, Kranz L, Tkatschuk L, Holtmann D (2021) Biotechnol Bioeng 118:2694. https://doi.org/10.1002/bit.27788
48. Rude MA, Schirmer A (2009) Curr Opin Microbiol 12:274. https://doi.org/10.1016/j.mib.2009.04.004
49. Wroblewska-Luczka P, Cabaj J, Bargiel J, Luszczki JJ (2023) Pharmacol Rep 75:1115. https://doi.org/10.1007/s43440-023-00512-1

50. Milker S, Holtmann D (2021) Microb Cell Fact:20. https://doi.org/10.1186/s12934-021-01562-x
51. Krieg T, Sydow A, Faust S, Huth I, Holtmann D (2018) Angew Chem Int Ed 57:1879. https://doi.org/10.1002/anie.201711302
52. Li Y, Yang SJ, Ma DL, Song W, Gao C, Liu LM, Chen XL (2021) Nat Prod Rep 38:1518. https://doi.org/10.1039/d0np00062k
53. Sohn YJ, Son J, Jo SY, Park SY, Yoo JI, Baritugo KA, Na JG, Choi JI, Kim HT, Joo JC, Park SJ (2021) Bioresour Technol 340:125693. https://doi.org/10.1016/j.biortech.2021.125693
54. Lee HM, Jeon BY, Oh MK (2016) Biotechnol Bioprocess Eng 21:402. https://doi.org/10.1007/s12257-016-0197-2
55. Black WB, Zhang L, Kamoku C, Liao JC, Li H (2018) ACS Synth Biol 7:794. https://doi.org/10.1021/acssynbio.7b00409
56. Islam T, Nguyen-Vo TP, Cho S, Lee J, Gaur VK, Park S (2023) Bioresour Technol 389:129814. https://doi.org/10.1016/j.biortech.2023.129814
57. Gascoyne JL, Bommareddy RR, Heeb S, Malys N (2021) Metab Eng 67:262. https://doi.org/10.1016/j.ymben.2021.06.010
58. Grousseau E, Lu JN, Gorret N, Guillouet SE, Sinskey AJ (2014) Appl Microbiol Biotechnol 98:4277. https://doi.org/10.1007/s00253-014-5591-0
59. Marc J, Grousseau E, Lombard E, Sinskey AJ, Gorret N, Guillouet SE (2017) Metab Eng 42:74. https://doi.org/10.1016/j.ymben.2017.05.007
60. Liu CS, Ding YM, Xian M, Liu M, Liu HZ, Ma QJ, Zhao G (2017) Crit Rev Biotechnol 37:933. https://doi.org/10.1080/07388551.2016.1272093
61. Liu Z, Wang K, Chen Y, Tan T, Nielsen J (2020) Nat Catal 3:274. https://doi.org/10.1038/s41929-019-0421-5
62. Li MD, Li W, Zhang TT, Guo KY, Feng DX, Liang FB, Xu C, Xian M, Zou HB (2023) Bioengineering 10. https://doi.org/10.3390/bioengineering10040446
63. Riedel SL, Lu J, Stahl U, Brigham CJ (2014) Appl Microbiol Biotechnol 98:1469. https://doi.org/10.1007/s00253-013-5430-8
64. Chen JS, Colon B, Dusel B, Ziesack M, Way JC, Torella JP (2015) PeerJ 3:e1468. https://doi.org/10.7717/peerj.1468
65. Li ZK, Xiong B, Liu L, Li SW, Xin XQ, Li Z, Zhang XL, Bi CH (2019) J Ind Microbiol Biotechnol 46:791. https://doi.org/10.1007/s10295-019-02164-8
66. Crepin L, Lombard E, Guillouet SE (2016) Metab Eng 37:92. https://doi.org/10.1016/j.ymben.2016.05.002
67. Volodina E, Raberg M, Steinbuchel A (2016) Crit Rev Biotechnol 36:978. https://doi.org/10.3109/07388551.2015.1079698
68. Leong YK, Show PL, Lan JCW, Krishnamoorthy R, Chu DT, Nagarajan D, Yen HW, Chang JS (2019) Bioresour Technol 287. https://doi.org/10.1016/j.biortech.2019.121474
69. Schmidt M, Ienczak JL, Quines LK, Zanfonato K, Schmidell W, de Aragao GMF (2016) Biochem Eng J 112:130. https://doi.org/10.1016/j.bej.2016.04.013
70. Lambauer V, Kratzer R (2022) Bioengineering:9. https://doi.org/10.3390/bioengineering9050204
71. Torella JP, Gagliardi CJ, Chen JS, Bediako DK, Colon B, Way JC, Silver PA, Nocera DG (2015) Proc Natl Acad Sci U S A 112:2337. https://doi.org/10.1073/pnas.1424872112
72. Nygaard D, Yashchuk O, Hermida EB (2019) Heliyon 5:e01374. https://doi.org/10.1016/j.heliyon.2019.e01374
73. Biglari N, Ganjali Dashti M, Abdeshahian P, Orita I, Fukui T, Sudesh K (2018) 3 Biotech 8:330. https://doi.org/10.1007/s13205-018-1351-7
74. Deshmukh AD, Pawar SV, Rathod VK (2020) Chem Eng Process:153. https://doi.org/10.1016/j.cep.2020.107923
75. Brojanigo S, Alvarado-Morales M, Basaglia M, Casella S, Favaro L, Angelidaki I (2022) Sci Total Environ 825:153931. https://doi.org/10.1016/j.scitotenv.2022.153931

76. Bhatia SK, Yoon JJ, Kim HJ, Hong JW, Gi Hong Y, Song HS, Moon YM, Jeon JM, Kim YG, Yang YH (2018) Bioresour Technol 257:92. https://doi.org/10.1016/j.biortech.2018.02.056
77. de Mello AFM, Vandenberghe LPD, Machado CMB, Valladares-Diestra KK, de Carvalho JC, Soccol CR (2023) Bioresour Technol:370. https://doi.org/10.1016/j.biortech.2022.128537
78. Keunun P, Rakkarn T, Yunu T, Paichid N, Prasertsan P, Sangkharak K (2018) J Polym Environ 26:2459. https://doi.org/10.1007/s10924-017-1140-0
79. Vu DH, Mahboubi A, Root A, Heinmaa I, Taherzadeh MJ, Akesson D (2022) Fermentation:8. https://doi.org/10.3390/fermentation8110605
80. Kedia G, Passanha P, Dinsdale RM, Guwy AJ, Esteves SR (2014) Biotechnol Bioprocess Eng 19:989. https://doi.org/10.1007/s12257-014-0144-z
81. Vlaeminck E, Quataert K, Uitterhaegen E, De Winter K, Soetaert WK (2022) J Biotechnol 343:102. https://doi.org/10.1016/j.jbiotec.2021.11.010
82. Gang S, Lee W, Kwon K, Kim T, Kim JS, Chung CW (2019) J Polym Environ 27:2182. https://doi.org/10.1007/s10924-019-01508-w
83. Li MX, Wilkins M (2020) Bioresour Technol:299. https://doi.org/10.1016/j.biortech.2019.122676
84. Nygaard D, Yashchuk O, Noseda DG, Araoz B, Hermida EB (2021) Heliyon:7. https://doi.org/10.1016/j.heliyon.2021.e05979
85. Salakkam A, Webb C (2018) Biochem Eng J 137:358. https://doi.org/10.1016/j.bej.2018.06.018
86. Kachrimanidou V, Kopsahelis N, Papanikolaou S, Kookos IK, De Bruyn M, Clark JH, Koutinas AA (2014) Bioresour Technol 172:121. https://doi.org/10.1016/j.biortech.2014.08.044
87. Mozumder MSI, De Wever H, Volcke EIP, Garcia-Gonzalez L (2014) Process Biochem 49:365. https://doi.org/10.1016/j.procbio.2013.12.004
88. Volova T, Demidenko A, Kiselev E, Baranovskiy S, Shishatskaya E, Zhila N (2019) Appl Microbiol Biotechnol 103:225. https://doi.org/10.1007/s00253-018-9460-0
89. Arias-Roblero M, Mora-Villalobos V, Velazquez-Carrillo C (2021) Front Sustain Food Syst:5. https://doi.org/10.3389/fsufs.2021.681596
90. Cruz MV, Gouveia AR, Dionisio M, Freitas F, Reis MAM (2019, 2019) Int J Polym Sci. https://doi.org/10.1155/2019/2191650
91. Domingos JMB, Puccio S, Martinez GA, Amaral N, Reis MAM, Bandini S, Fava F, Bertin L (2018) Chem Eng J 336:47. https://doi.org/10.1016/j.cej.2017.11.024
92. Kokpinar O, Altun M (2023) Prep Biochem Biotechnol 53:532. https://doi.org/10.1080/10826068.2022.2114009
93. Santolin L, Waldburger S, Neubauer P, Riedel SL (2021) Front Bioeng Biotechnol:9. https://doi.org/10.3389/fbioe.2021.623890
94. Quintanar-Gomez S, Abreu-Corona A, Zamudio-Perez E, Vargas-Hernandez G, Tellez-Jurado A, Gracida-Rodriguez J (2018) Rev Int De Contam Ambient 34:467. https://doi.org/10.20937/Rica.2018.34.03.09
95. Haas C, El-Najjar T, Virgolini N, Smerilli M, Neureiter M (2017) N Biotechnol 37:117. https://doi.org/10.1016/j.nbt.2016.06.1461
96. Schwarz D, Schoenenwald AKJ, Dorrstein J, Sterba J, Kahoun D, Fojtikova P, Vilimek J, Schieder D, Zollfrank C, Sieber V (2018) Bioresour Technol 269:237. https://doi.org/10.1016/j.biortech.2018.08.064
97. Vu DH, Mahboubi A, Root A, Heinmaa I, Taherzadeh MJ, Akesson D (2023) Membranes 13. https://doi.org/10.3390/membranes13060569
98. Lambauer V, Permann A, Petrasek Z, Subotic V, Hochenauer C, Kratzer R, Reichhartinger M (2023) Fermentation 9. https://doi.org/10.3390/fermentation9070619
99. Garrigues L, Maignien L, Lombard E, Singh J, Guillouet SE (2020) N Biotechnol 56:16. https://doi.org/10.1016/j.nbt.2019.11.005
100. Garcia-Gonzalez L, Mozumder MSI, Dubreuil M, Volcke EIP, De Wever H (2015) Catal Today 257:237. https://doi.org/10.1016/j.cattod.2014.05.025

101. Jawed K, Irorere VU, Bommareddy RR, Minton NP, Kovacs K (2022) Fermentation:8. https://doi.org/10.3390/fermentation8030125
102. Ryu HW, Hahn SK, Chang YK, Chang HN (1997) Biotechnol Bioeng 55:28. https://doi.org/10.1002/(SICI)1097-0290(19970705)55:1<28::AID-BIT4>3.0.CO;2-Z
103. Mozumder MS, Garcia-Gonzalez L, De Wever H, Volcke EI (2015) Bioresour Technol 191:213. https://doi.org/10.1016/j.biortech.2015.04.110

Adv Biochem Eng Biotechnol (2025) 192: 85–116
https://doi.org/10.1007/10_2024_271

Published online: 12 November 2024

Vibrio natriegens: Application of a Fast-Growing Halophilic Bacterium

Maurice Hädrich, Clarissa Schulze, Josef Hoff, and Bastian Blombach

Contents

Maurice Hädrich and Clarissa Schulze contributed equally to this work.

M. Hädrich
Microbial Biotechnology, Campus Straubing for Biotechnology and Sustainability, Technical University of Munich, Straubing, Germany

SynBiofoundry@TUM, Technical University of Munich, Straubing, Germany
e-mail: maurice.haedrich@tum.de

C. Schulze
Microbial Biotechnology, Campus Straubing for Biotechnology and Sustainability, Technical University of Munich, Straubing, Germany
e-mail: clarissa.schulze@tum.de

J. Hoff and B. Blombach (✉)
Microbial Biotechnology, Campus Straubing for Biotechnology and Sustainability, Technical University of Munich, Straubing, Germany

SynBiofoundry@TUM, Technical University of Munich, Straubing, Germany

Munich Institute of Integrated Materials, Energy and Process Engineering, Technical University of Munich, Garching, Germany
e-mail: josef.hoff@tum.de; bastian.blombach@tum.de

Abstract The fast growth accompanied with high substrate consumption rates and a versatile metabolism paved the way to exploit *Vibrio natriegens* as unconventional host for biotechnological applications. Meanwhile, a wealth of knowledge on the physiology, the metabolism, and the regulation in this halophilic marine bacterium has been gathered. Sophisticated genetic engineering tools and metabolic models are available and have been applied to engineer production strains and first chassis variants of *V. natriegens*. In this review, we update the current knowledge on the physiology and the progress in the development of synthetic biology tools and provide an overview of recent advances in metabolic engineering of this promising host. We further discuss future challenges to enhance the application range of *V. natriegens*.

Keywords Genetic engineering, Halophilic, Industrial biotechnology, Metabolic engineering, Synthetic biology, Systems biology, *Vibrio natriegens*

1 Introduction

Confronted with the ongoing challenges of climate change, pollution, and reliance on finite fossil resources, circular bioeconomy became a key research area to progress toward the United Nations' sustainable development goals. Microbial biotechnology plays a significant role within this field, contributing substantially to the production of chemicals and fuels from renewable resources. A critical hurdle, especially for bulk products, is the development of cost-competitive processes to established chemical ones. Therefore, key performance indicators (KPIs) such as titer, yield, and volumetric productivity of biotechnological processes have to be pushed to their limits. To tackle these challenges, the utilization of unconventional host strains with distinct superior intrinsic properties comes more and more into focus [1]. These considerations have led to the identification of *V. natriegens* as alternative host for biotechnological applications since this bacterium possesses extraordinary growth and substrate consumption rates. *V. natriegens* is the fastest-growing non-pathogenic bacterium known to date with a doubling time of less than 10 min in complex medium and 23 min in minimal medium with sucrose under aerobic conditions [2, 3]. This marine Gram-negative γ-proteobacterium is prototrophic and facultatively anaerobic and possesses a broad nutritional flexibility; however, it requires sufficient amounts of sodium ions for growth (halophilic) and shows a loss of viability upon long-term storage at 4°C and a distinct sensitivity toward oxidative stress [4–7]. Remarkably, the biomass-specific glucose uptake rates of 3.9 $g_{Glc}\ g_{CDW}\ h^{-1}$ and 7.8 $g_{Glc}\ g_{CDW}\ h^{-1}$ in minimal medium under aerobic and anaerobic conditions, respectively, exceed values of established organisms, such as *Escherichia coli*, *Corynebacterium glutamicum*, and *Bacillus subtilis*, at least by twofold and represent an excellent basis to develop production processes with improved productivity [3, 8–10].

Although *V. natriegens* was isolated in 1958 from salt marches in Georgia (USA), it was only in the 2010s that this bacterium gained significant attention for its beneficial characteristics [3–5]. These publications initiated further research activities with *V. natriegens* to rapidly expand the genetic engineering toolbox and the range of applications in biotechnology. Further significant research progress has been made since the last comprehensive reviews [11–14]. In this article, we provide an update on the physiology, tool development, and application of *V. natriegens* in biotechnology.

2 Systems-Level Analysis of *V. natriegens*' Metabolism

2.1 Genomic Background

The rapid advancement of molecular biology tools has facilitated precise metabolic engineering of *V. natriegens*. However, generation and application of production strains are currently limited by an incomplete comprehension of the system *V. natriegens* itself. As outlined below, first systems-level analyses by transcriptomics, proteomics, and metabolomics supported by metabolic models have been applied to gain more knowledge about the metabolism and its regulation. Fundamental information about the genetic repertoire is derived from several whole genome sequencing studies, reporting a total genome of 5.13 to 5.2 Mb [5, 15–17]. As a common trait of the genus *Vibrio*, the genome of *V. natriegens* consists of two chromosomes with a size of approximately 1.92 Mb and 3.24 Mb [15, 18] containing up to 4,788 annotated coding sequences [5, 15–17]. Multiple sequencing and resequencing results with individual gene annotations published in a condensed time period have given rise to a diversity of gene identifiers used in various publications. This emphasizes the critical significance of consistent and accurate labeling for precise gene identification. Notably, the most frequently referenced gene identifiers originate from the works of Lee et al. and Weinstock et al. [5, 15].

2.2 Oxidative Stress

Upon isolating *V. natriegens*, pronounced growth deficiencies at 4°C were observed [4]. This extends also to cultures grown on agar plates that did show a loss in viability after storage at 4°C for 3 weeks [5]. These viable but non-culturable (VBNC) states occur in a variety of *Vibrio* species and are described by cells losing the ability to grow on laboratory media despite being alive [19, 20]. The observed cold-induced loss of culturability (CILC) in *V. natriegens* is associated with oxidative stress, in particular by hydrogen peroxide [5, 21]. Weinstock et al. could reduce the CILC phenotype in *V. natriegens* by adding catalase to the medium or overexpression of the *E. coli* catalase gene *katG* or *E. coli* alkyl hydroperoxide reductase genes *ahpCF* [5]. Similar findings in *Vibrio vulnificus* show that OxyR-deficient strains have a pronounced VBNC phenotype [22]. The hydrogen peroxide-sensing transcriptional regulator OxyR induces expression of a number of genes in other bacteria, including the genes of *katG* and *ahpCF* [23–25]. Accordingly, OxyR-deficient *V. vulnificus* lacks catalase activity [22]. In *V. natriegens*, adaptive laboratory evolution (ALE) experiments in high iron environment resulted in constitutive *oxyR* expression mutants which showed higher hydrogen peroxide tolerance [26]. Deletion of *oxyR* in *V. natriegens* leads to a faster loss of viability of outgrown cells after exposure to hydrogen peroxide [27]. Transcriptomic analysis of cells stored at 4°C in liquid culture revealed an increased SOS response as a result of DNA damage compared to the control conditions at 25°C [21]. Cultivation in lower volumes led to accumulation of reactive oxygen species (ROS) and consequently accelerated CILC, likely due to increased oxygen availability. Moreover, inactivation of the Na^+-translocating NADH-ubiquinone oxidoreductase (Na^+-NQR) in *V. natriegens* also resulted in CILC [21]. The VBNC state of cells is still poorly understood, but ROS appear to be a major contributor to this dormant state in *V. natriegens*.

V. natriegens possesses the ability to naturally take up DNA from the environment which constitutes a central feature of several genetic engineering applications [28–30; see also section "Synthetic Biology Tools for *V. natriegens*"]. Besides the VBNC state, response to ROS was also found important for natural competence. Shin et al. started to unravel the regulatory relations behind the natural competency and its regulators TfoX and QstR [31]. By analyzing 104 RNA-seq samples from various conditions through machine learning, the authors identified 45 independently modulated gene groups (iModulons). Competency in *V. natriegens* is characterized by activation of the iModulons of TfoX and QstR and four additional iModulons of motility, arginine metabolism, ROS response, and one of unknown function, while the determined iModulons of iron metabolism and translation are repressed [31]. In the iron metabolism iModulon, genes encoding iron transporters are downregulated, and genes for iron storage proteins are upregulated, reducing oxidative stress by the Fenton reaction. Additionally, the activated arginine metabolism iModulon might indicate improved polyamine formation [31]. Polyamine deficiency leads in *E. coli* to reduced tolerance of oxidative stress which is connected to a lowered induction of *oxyR* and *ropS* expression [32, 33]. Similarly, *V. vulnificus* shows higher susceptibility to superoxide stress without the polyamine cadaverine [34]. The alkyl hydroperoxide reductase genes *ahpCF* are also upregulated by TfoX overexpression, further emphasizing the relevance of ROS mitigation during competency [31].

2.3 Hyperosmotic Stress

It is vital for most microorganisms to respond to changing environmental osmolarity by regulating their cytoplasmic osmotic potential. A common response to rising osmotic pressure is the production or import of compatible solutes, such as ectoine. Biosynthesis of ectoine from L-aspartate includes an aspartate kinase (Ask), aspartate-semialdehyde dehydrogenase as well as a L-2,4-diaminobutyrate aminotransferase (EctB), L-2,4-diaminobutyrate acyltransferase (EctA), and ectoine synthase (EctC, Fig. 1). These ectoine biosynthesis genes are found in a large number of bacteria across many phyla and habitats, including *Vibrionaceae* [35–41]. In *V. natriegens*, an operon containing *ectABC* and an *ask* gene is situated next to the operon *proWXV*, encoding the proline/glycine betaine ABC transporter ProU1 [38, 39]. The gene *ectD* coding for an ectoine dioxygenase producing 5-hydroxyectoine as another compatible solute is absent in *Vibrionaceae*

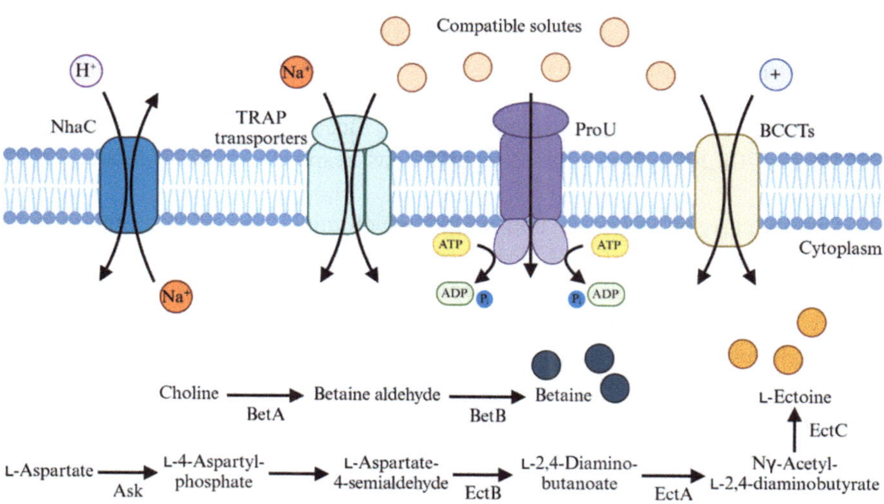

Fig. 1 Schematic overview of prominent high salt stress responses in *V. natriegens*. Abbreviations: *ADP*, adenosine diphosphate; *Ask*, aspartate kinase; *ATP*, adenosine triphosphate; *BCCTs*, betaine/carnitine/choline transporters; *BetA*, choline dehydrogenase; *BetB*, betaine-aldehyde dehydrogenase; *EctA*, L-2,4-diaminobutyrate acyltransferase; *EctB*, L-2,4-diaminobutyrate aminotransferase; *EctC*, ectoine synthase; *NhaC*, family of Na^+/H^+ antiporters; P_i, inorganic phosphate; *ProU*, proline/glycine betaine ABC transporters; *TRAP*, tripartite ATP-independent periplasmic. Figure is based on the data of Huang et al. and Coppens et al. [39, 42]

[37, 40]. Comparative transcriptomics and proteomics showed a significant increase in *ectABC-ask* and *proWXV* expression as response to an osmotic upshift. In the transcriptomic data, seven of the eight genes with the highest fold change belong to the two mentioned operons, highlighting ectoine synthesis as most relevant to counteract increasing osmotic pressure in *V. natriegens* [39, 42].

Another widespread compatible solute is glycine betaine (betaine), predominantly produced by a two-step oxidation of the precursor choline by the choline dehydrogenase BetA and betaine-aldehyde dehydrogenase BetB (Fig. 1). In the genome of *V. natriegens*, the corresponding genes are organized together with the *betI* gene encoding the repressor BetI as *betIBA* operon which is located next to the operon *proXWV* encoding a second proline/glycine betaine ABC transporter ProU2 [35–38]. *betIBA* transcription during high salt conditions was shown to be significantly higher compared to low salt conditions in the transcriptomic analysis [39]. Proteomics found a higher abundance of ProX for higher salt conditions [42]. Inactivation of ectoine biosynthesis in *V. parahaemolyticus* resulted in growth deficiencies in high salt conditions which is not observed in betaine biosynthesis knock-out mutants [38, 41]. For *V. natriegens*, no experiments regarding essentiality for both compatible solutes were conducted so far.

Aside from the already mentioned ABC transporters ProU1 and ProU2, *V. natriegens* is also equipped with six putative betaine/carnitine/choline transporters (BCCTs, Fig. 1). BCCTs can transport other compatible solutes beyond the name giving betaine, carnitine, and choline. In *V. parahaemolyticus*, the four BCCTs each accept a different substrate variety, including proline, dimethylglycine, dimethylsulfoniopropionate, and ectoine [43–45]. The BCCT with the broadest substrate specificity has a homologue in *V. natriegens* with 90% protein sequence identity (PN96_06025) and shows increased transcription levels under high salt conditions. Additionally, a homologue to BccT3 of *V. parahaemolyticus* (PN96_04170) is significantly upregulated as response to an osmotic upshift [39, 44].

Another class of secondary transporters which might participate in the osmotic stress response is tripartite ATP-independent periplasmic (TRAP) transporters (Fig. 1). TRAP transporters are Na^+ dependent and are highly diverse with a great variety of substrates, such as ectoine and 5-hydroxyectoine [46–50]. *V. natriegens* has 16 putative TRAP transporters annotated with one transporter (PN96_18175-80) being transcriptionally upregulated as response to high salt concentrations and one subunit of a second TRAP transporter (PN96_23190) being more abundant in proteomic analysis of high salt conditions [39, 42].

Moreover, the transcriptomic analysis revealed an upregulation of genes encoding one Na^+/H^+ antiporter of the NhaC family (PN96_18400, Fig. 1) [39]. *V. natriegens* has each one NhaA-, NhaB-, and NhaD-type antiporter annotated and four NhaC antiporters, similar to *V. parahaemolyticus* and *V. cholerae* [51]. While NhaC antiporters are not thoroughly investigated in *Vibrio* species yet, NhaA can be essential to *V. cholerae's* growth when the Na^+-NQR is inactivated [52]. Indeed, the Na^+-NQR might be part of *V. natriegens*' sodium balancing system as the proteome at higher salt concentrations showed an increase in Na^+-NQR abundance [42].

These findings provide the first foundation to understand the adaptation of *V. natriegens* to osmotic changes; however, further investigations are essential to unravel the osmoregulation. Notably, *V. natriegens* possesses a homologue of CosR, a global regulator for osmotic stress in *V. cholerae* [53, 54].

2.4 Carbon Metabolism

Based on the genetic information, the first publicly available genome-scale metabolic model (GEM) for *V. natriegens* was created, encompassing 1,096 metabolites involved in 982 cytoplasmic reactions [42]. By comparing predicted phenotypes such as growth rates, fluxes through the central metabolism, and predicted essential genes to experimental data, the model was validated and further applied to investigate 173 compounds for their potential as carbon source. Through experimental assessment, it was determined that 64 compounds led to in vivo respiratory activity, which was used as representative for growth. The model predicted 48 of the 64 positively tested compounds and 97 of the 109 negative compounds correctly, leaving the overall prediction accuracy of the model at 83% [42]. The 64 compounds showing respiratory activity coincide with the 66 substrates already reported to show biomass formation (reviewed in [11]). Notably exceptions are δ-aminovalerate, caprate, D-glucuronate, *myo*-inositol, α-ketoglutarate, and salicin. These compounds do not exhibit respiratory activity but are reported as substrate for biomass formation, whereas formate shows respiratory activity but no biomass formation [6, 42]. This broad range of substrates enables a variety of possible applications. *V. natriegens* was even found to be the most versatile member of all tested strains from the back then clustered *Beneckea* genus [6]. Advancing GEMs of *V. natriegens* can be a powerful tool to understand large data sets created by –omics technologies and can speed up metabolic engineering approaches by appropriate target selection [55, 56]. A general outline for GEM generation was published with the application case of 1,3-propanediol production [57].

The fluxes of the central carbon metabolism of *V. natriegens* for growth on glucose were charted by Long et al. using ^{13}C analysis [58]. The Embden-Meyerhof-Parnas (EMP) pathway and to a small degree the pentose phosphate pathway (PPP) are the main routes for aerobic glucose catabolism with 80–92% and 8–18% flux, respectively [58, 59]. Flux over the Entner-Doudoroff pathway (ED) is virtually absent [58], and the involved enzymes are not detected in extracts [59]. This aligns with findings of other *Vibrio* species using EMP over ED during growth on glucose [60]. Using gluconate as substrate, around 80% of carbon is channeled through ED in *V. natriegens* [59]. The carbon flux distribution was also simulated by the above-introduced GEM, which not only follows the experimental data of a general EMP preference but also models minor ED involvement for glucose as substrate [42].

C1 compounds are attractive substrates for biotechnological research [61–63]. *V. natriegens*' potential for formate utilization was assessed by Tian et al. by transcriptomic comparison of cells growing with and without additional formate in the medium [64]. Several genes of the reductive glycine pathway, serine cycle and TCA cycle, were found to be upregulated in the presence of formate. By introducing several genomic mutations, the authors fused the serine and TCA cycle. Additional disruption of the *serA* gene encoding 3-phosphoglycerate dehydrogenase forced the formation of serine from formate via the THF cycle rather than via glycolysis. These combined modifications enabled formate utilization via THF, serine, and TCA cycle. Using ALE, the formate consumption rate and the tolerance to formate were further improved yielding a strain which was able to grow in the presence of 140 g sodium formate L^{-1} showing a consumption rate of up to 1.42 g L^{-1} h^{-1}. The majority of formate is presumed to be oxidized to CO_2 generating reducing equivalents and energy as ^{13}C analysis showed only 12.1% of the consumed formate to be assimilated in biomass [64].

3 Biotechnological Application of *V. natriegens*

3.1 First Chassis Strains Are Available

To date, numerous metabolic engineering approaches have been carried out to exploit *V. natriegens*' potential as a production host for several biotechnological products (Table 1). To obtain a more robust platform strain, the two prophage regions VNP1 and VNP2, which are activated under stress and also spontaneously under standard conditions to induce the lytic phage cycle, were identified and consequently removed from the genome [65]. Studies on the ATCC14048 strain revealed the deletion of the *dns* gene, encoding an extracellular nuclease, increases plasmid transformation efficiency via heat shock transformation [66]. Commercially available strains such as Vmax™ Express and the next-generation strain Vmax™ X2 (Synthetic Genomics Inc., La Jolla, CA, USA) are already marketed and feature an adjusted genetic background for improved protein expression and incorporation of noncanonical α-amino acids [67, 68]. Both strains possess a reduced genome and provide a genomically integrated and inducible T7-RNA polymerase system, enabling the usage of pET and pET-like vectors [5, 69].

3.2 Fermentative Production

The suitability of *V. natriegens* as a production host has already been demonstrated for multiple products. An overview of the KPIs of engineered *V. natriegens* strains is given in Table 1. In 2017, Dalia et al. used Multiplex Genome Editing by Natural Transformation (MuGENT) to enhance the production of the biopolymer

Table 1 Key performance indicators of engineered *V. natriegens* strains

Organism	Substrate	Product	Complex/ organic supplements	Volumetric productivity ($g\ L^{-1}\ h^{-1}$)	Titer ($g\ L^{-1}$)	Yield ($g\ g^{-1}$)	Duration (h)	References	Process
V. natriegens Δ*lldh* Δ*dldh* Δ*pfl* Δ*ald* Δ*dns*::pyc_{Cg}	Glucose $KHCO_3$	Succinate	–	8.6	60	0.74	7	[78]	Anaerobic resting cell approach
V. natriegens Δ*lldh* Δ*dldh* Δ*pfl* Δ*ald* Δ*dns*::pyc_{Cg} Δ*cpsR*	Glucose $KHCO_3$	Succinate	–	8.5	51	0.90	6	[77]	Anaerobic resting cell approach
V. natriegens p455_Ediss	Glucose Citric acid Fe(III)citrate	2,3-Butanediol Acetoin	Yeast extract	3.9	50	0.22	12	[81]	Microaerobic pulsed fed-batch process
V. natriegens Δ*frdA* Δ*ldhA*-pETRABC	Glucose	2,3-Butanediol	–	3.4	41	0.39	12	[82]	Non-sterilized fed-batch process
V. natriegens Δ*adhE* Δ*ldhA* Δ*pfl* Δ*pta-ackA* Δ*arcA* Δ*glpR* Δ*sthA* Δ*frdABCD* Δ*aldA* Δ*aldB* Δ*glpD* pTrc99a-J23106-doy-phaP-pntAB-glpD3	Glycerol	1,3-Propanediol	Yeast extract Vitamin B12	2.9	70	0.51	24	[79]	Fed-batch process
V. natriegens Δ*adhE* Δ*ldhA* Δ*pfl* Δ*pta-ackA* Δ*frdABCD* Δ*glpD* pTrc99a-dhaBCE-aldH-glpD3	Glycerol	3-Hydroxy-propionate	Yeast extract Vitamin B12	2.7	65	0.65	24	[79]	Fed-batch process
V. natriegens Δ*VNP1* Δ*VNP2* Δ*pflB* Δ*lldh*	Glucose Citric acid	Pyruvate	–	3.4	54	0.57	16	[90]	Fed-batch process

(continued)

Table 1 (continued)

Organism	Substrate	Product	Complex/ organic supplements	Volumetric productivity (g L^{-1} h^{-1})	Titer (g L^{-1})	Yield (g g^{-1})	Duration (h)	References	Process
Δ*dldh* Δ*pps1* Δ*pps2* P2-*aceE*^{TTG} P2-*ppc*^{ATG}									
V. natriegens Δ*dldh* Δ*lldh* Δ*pfl* Δ*mdh*	Glucose $NaHCO_3$	Alanine	–	34	17	0.73	0.5	[3]	Anaerobic resting cell approach
V. natriegens Vmax pET-28a(+)-Fn-TPL	Catechol Sodium pyruvate Ammonium acetate	3,4-Dihydroxy-phenyl-L-alanine	Pyridoxal 5'phosphate EDTA	5.4	54	N/A	10	[84]	Fed-batch process
V. natriegens S-TCA-2.0 IE (Δ*PN96_00930* Δ*PN96_07295* Δ*PN96_10585* Δ*PN96_14755* Δ*PN96_06470* Δ*PN96_19465* Δ*PN96_11695*::*KmR* Δ*dns*::*CmR* pColE1-idgs-sfp + ALE mutations)	Sodium formate	Indigoidine	Yeast extract Peptone	2.3	29	N/A	72	[64]	Pulsed fed-batch process
V. natriegens p15A-*phaBAC*	Crude glycerol	Poly-β-hydroxybutyrate	Tryptone Yeast extract	0.05	2.4	N/A	48	[70]	Non-sterile batch process

V. natriegens pJV-Tyr1	L-Tyrosine	Melanin	–	0.32	0.5	N/A	2	[75]	N/A
V. natriegens pJV-Tyr1	Disodium tyrosine Glucose Glycerol	Melanin	Casamino acids Aspartate Thiamine	0.5	7.6	N/A	O/N	[74]	Shaking flask
V. natriegens pJV-ß-carotene-MVA	–	ß-Carotene	Tryptone Yeast extract	N/A	0.013	N/A	O/N	[73]	Shaking flask
V. natriegens pVio	–	Violacein	Tryptone Yeast extract	N/A	0.003	N/A	44	[73]	Shaking flask

Abbreviations: O/N, overnight; N/A, not applicable; ALE, adaptive laboratory evolution

poly-β-hydroxybutyrate (PHB) by targeting nine genes involved in PHB synthesis [28]. The resulting strain formed around 100-fold more PHB in comparison with the parental strain. Another optimized strain for PHB production, *V. natriegens* p15A-*phaBAC*, outperformed *E. coli* MG1655 p15A-*phaBAC* in a non-sterile system in rich medium supplemented with crude glycerol as additional carbon source. A titer of 2.39 g PHB L^{-1} was achieved [70] which is, however, still significantly lower compared to other microbial systems such as *Cupriavidus necator* [71, 72]. Ellis et al. reported the overproduction of β-carotene (3 mg L^{-1}) and violacein (13 mg L^{-1}) in a rich medium [73]. For the former product, genes from the β-carotene biosynthetic pathway of *Vibrio campbellii* as well as the mevalonate biosynthetic pathway from *Lactobacillus acidophilus* were expressed. For the latter product, a gene cluster from *Chromobacterium violaceum* was introduced into *V. natriegens*. The pigment melanin was produced with *V. natriegens* as an alternative to the conventionally used chemical synthesis or isolation from natural sources like sepia ink [74]. For this purpose, the tyrosinase gene (*tyr1*) from *Priestia megaterium* (formerly *Bacillus megaterium*) was overexpressed. The resulting strain *V. natriegens* pJV-Tyr1 formed melanin more rapidly from L-tyrosine compared to previously reported heterologous systems with a titer of around 0.45 g L^{-1} within 2 h [75]. Smith et al. optimized the minimal medium (M9v2) for melanin production and achieved a titer of 7.6 g L^{-1} from overnight cultures with *V. natriegens* pJV-Tyr1 [74]. This corresponds to one of the highest reported volumetric productivities of 473 mg L^{-1} h^{-1} to date. Recently, Li et al. presented a strategy for carbon-negative biosynthesis with a compatible phototrophic community [76]. In this approach, a sucrose-producing CO_2 sequestration module (*Synechococcus elongatus*) was coupled with a sucrose utilization module (*V. natriegens*) that harbors the respective production pathway. In this symbiotic community, the cyanobacterium produced sucrose from CO_2, and the engineered *V. natriegens* strains were applied for the production of different chemicals (313 mg lactic acid L^{-1}, 137 mg 2,3-butanediol L^{-1}, 25 mg *p*-coumaric acid L^{-1}, and 9 mg melanin L^{-1}) [76].

High maximal volumetric productivities of around 21 g L^{-1} h^{-1} and 34 g L^{-1} h^{-1} have been reported for resting cell approaches for anaerobic succinate [77, 78] and alanine [3] production, respectively. For the latter, Hoffart et al. engineered *V. natriegens* to abolish or reduce the formation of lactate, acetate, formate, succinate, and ethanol by deletion of genes encoding two lactate dehydrogenases, pyruvate formate lyase and malate dehydrogenase, respectively. The resulting strain *V. natriegens* Δ*dldh* Δ*lldh* Δ*pfl* Δ*mdh* produced 17 g alanine L^{-1} from glucose within 0.5 h using an anaerobic resting cell approach in a minimal medium [3]. Based on *V. natriegens* Δ*dldh* Δ*lldh* Δ*pfl*, Thoma et al. optimized the strain for the anaerobic production of succinate by deletion of the *ald* gene, encoding alanine dehydrogenase. To enhance the anaplerotic flux, the *pyc* gene from *C. glutamicum*, encoding a pyruvate carboxylase, was additionally integrated into the genome. Using a zero-growth process with minimal medium, *V. natriegens* Δ*lldh* Δ*dldh* Δ*pfl* Δ*ald* Δ*dns*::pyc_{Cg} (Succ1) formed 60 g succinate L^{-1} from glucose which corresponds to a yield of 1.14 mol succinate mol^{-1} glucose

[78]. Schulze et al. observed high viscosities, likely caused by exopolysaccharide formation, during succinate production which negatively impacted the process and consequently inactivated the transcriptional biofilm regulator CpsR in the parental strain Succ1. Thereby, no viscosity changes occurred anymore during the fermentation, and the succinate yield was improved by 22% to 1.39 mol succinate mol^{-1} glucose which correspond to 81% of the theoretical maximum [77]. Fed-batch processes from glycerol were conducted for the production of 1,3-propanediol and 3-hydroxypropionate. The production strains were derived from *V. natriegens* by deleting genes related to by-product formation and metabolic regulation and by introducing a high-copy plasmid pTrc99a for gene overexpression. A plasmid maintenance system was established by shifting the essential gene *glpD* for glycerol degradation from the chromosome onto the plasmid. The resulting strains *V. natriegens* Δ*adhE* Δ*ldhA* Δ*pfl* Δ*pta-ackA* Δ*frdABCD* Δ*glpD* pTrc99a-dhaBCE-aldH-glpD3 and *V. natriegens* Δ*adhE* Δ*ldhA* Δ*pfl* Δ*pta-ackA* Δ*arcA* Δ*glpR* Δ*sthA* Δ*frdABCD* Δ*aldA* Δ*aldB* Δ*glpD* pTrc99a-J23106-doy-phaP-pntAB-glpD3 achieved titers of 65 g 3-hydroxypropionate L^{-1} and 70 g 1,3-propanediol L^{-1}, respectively [79]. Erian et al. used a pulsed microaerobic fed-batch to produce 2,3-butanediol and acetoin with *V. natriegens* p455_Ediss. The plasmid harbors the 2,3-butanediol biosynthesis genes *budA*, *budB*, and *budC* from *Enterobacter cloacae* subsp. *dissolvens* which were already successfully applied in *E. coli* W [80, 81]. A diol titer of 50 g L^{-1} was achieved, which corresponds to an overall volumetric productivity of 3.9 g L^{-1} h^{-1} [81]. Meng et al. used a seawater-based medium for the production of 2,3-butanediol and achieved a titer of 41 g L^{-1} with *V. natriegens* Δ*frdA* Δ*ldhA*-pETRABC using a non-sterilized fed-batch fermentation. A volumetric productivity of 3.4 g L^{-1} h^{-1} was obtained [82]. To enable pyruvate production with *V. natriegens*, Wu et al. eliminated two prophage regions and genes for the biosynthesis of byproducts. Furthermore, translation of the *aceE* gene, coding for the E1 component of the pyruvate dehydrogenase complex, was attenuated, and expression of the *ppc* gene, coding for phosphoenolpyruvate carboxylase, was fine-tuned. The resulting strain *V. natriegens* Δ*VNP1* Δ*VNP2* Δ*pflB* Δ*lldh* Δ*dldh* Δ*pps1* Δ*pps2* P2-$aceE^{TTG}$ P2-ppc^{ATG} produced 54 g pyruvate L^{-1} from glucose within 16 h using a fed-batch process [83]. The plasmid-based expression of the *tpl* gene from *Fusobacterium nucleatum*, which codes for tyrosine phenol lyase, led to the production of 3,4-dihydroxyphenyl-L-alanine (L-DOPA) in Vmax. The final strain secreted 54 g L-DOPA L^{-1} with an average volumetric productivity of 5.4 g L^{-1} h^{-1} during a fed-batch process [84]. Most recently, it was shown that *V. natriegens* possesses a high tolerance against formate and the capability to efficiently utilize this promising C1 feedstock [64]. To improve formate assimilation, the serine and TCA cycle were rewired to eventually couple serine synthesis and formate utilization. The resulting strain S-TCA.1.0 exhibited an increased formate consumption per biomass. However, total formate utilization was lower, and growth was impaired compared to the wild type. ALE was applied to improve the formate tolerance and the growth of the mutant. The resulting strain S-TCA-2.0 was able to grow in the presence of 140 g sodium formate L^{-1}. Sequencing of three isolates revealed nine mutated genes, two of which (*fumA*

and *sdhC*) were associated with formate consumption. Lastly, an indigoidine-production pathway was introduced by integration of *idgs* from *Streptomyces lividans* and *sfp* from *B. subtilis* to convert formate into the valuable pigment. The final strain S-TCA-2.0-IE consumed 165 g formate L^{-1} and formed 29 g indigoidine L^{-1} within 72 h [64]. Other examples for the successful application of *V. natriegens* are, e.g., the fast production of selenium and selenium nanoparticles [85] or the formation of a novel antibiotic against Gram-negative pathogens, darobactin A, that could be heterologous produced by an engineered Vmax strain [86].

3.3 Biotransformation

Peng et al. applied whole-cell biotransformation at high temperatures with *V. natriegens* Vmax pETDuet-*Txage-EcnanA* to convert *N*-acetylglucosamine and pyruvate into *N*-acetyl-D-neuraminic acid. At 60°C, a titer of 126 g L^{-1} was obtained within 3 h, corresponding to an average volumetric productivity of 42 g *N*-acetyl-D-neuraminic acid L^{-1} h^{-1} [87]. Recently, Bahlawan et al. tested different levansucrases to catalyze the transfructosylation of lactose/sucrose into lactosucrose and fructooligosaccharides. In addition, whey permeate and milk permeate were used as a potential lactose source. Compared to the enzymes LS1 from *Gluconobacter oxydans*, LS3 from *Novosphingobium aromaticivorans*, and LS4 from *Burkholderia graminis*, the expression of levansucrase L2 from *V. natriegens* led to a higher concentration of lactosucrose of 328 g L^{-1} and 251 g L^{-1} from lactose/sucrose and whey permeate/sucrose, respectively [88]. Chen et al. applied in silico exploration to identify new levanases for the synthesis of levan-type fructooligosaccharides. LEV79 derived from *V. natriegens* was chosen as one of five promising new levanases screened from 113 potential candidates [89].

3.4 Recombinant Protein Production

V. natriegens has been shown to be amenable for recombinant protein production, and examples of reporter (e.g., GFP [5, 15, 83, 91, 92], superfolder GFP (sfGFP) [92], mCherry [65, 92], mSCFP3 [92], and LacZ [92]) and isotopically labeled proteins (e.g., FK506 binding protein, enhanced yellow fluorescent protein, and tobacco etch virus protease [93]) as well as membrane protein complexes (e.g., transport systems Na^{+}-NQR and Mrp [94]) have been reported. Lately, González et al. showed that Vmax X2 is capable of producing sfGFP containing up to five noncanonical α-amino acids with approximately 25-fold higher yields compared to commercial expression strains and more than tenfold higher yields than the optimized *E. coli* strains C321.ΔA.opt and C321.ΔA.exp [67]. Kormanová et al. succeeded in overexpressing the membrane-associated enzymes D, D-carboxypeptidases PBP4 and PBP5/6, resulting in a leakage of intracellular proteins [95]. Without the usage of secretion signals, fusion tags, or additional permeabilization methods, the authors were able to purify a manganese-dependent bacterial peroxidase, a thermostable catalase-peroxidase from *Archaeoglobus fulgidus*, or Taq polymerase in a two-to-three-digit milligram range directly from the growth medium. This approach could be a promising option for the production of recombinant proteins, as it simplifies the downstream processing and lowers overall production costs [95]. Xu et al. presented a forward genetic method to improve recombinant protein production with *V. natriegens* VnDX [13]. VnDX carries an integrated T7 RNAP expression and spectinomycin resistance cassette in the *dns* locus of *V. natriegens* ATCC14048. Genes of interest (GOIs) encoding biocatalytic enzymes such as an ω-transaminase derived from *Ochrobactrum anthropi* were selected as targets since these genes were only weakly expressed in *V. natriegens* but strongly expressed in *E. coli* using a pET expression system. Fusion of the C-terminal domain of the GOIs with GFP followed by ultraviolet mutagenesis and fluorescence-activated cell sorting was used to successfully screen for mutants with an enhanced protein yield and enzyme activity [13]. Aminoacylases are a green alternative for the formation of acyl-amino acids to replace the environmentally harmful chemical synthesis. Haeger et al. identified a novel aminoacylase gene from *Mycolicibacterium smegmatis* MKD 8 and used a T7-based expression system in *E. coli* BL21 (DE3) and *V. natriegens* Vmax [96]. *V. natriegens* achieved better results compared to *E. coli* when isopropyl ß-d-1-thiogalactopyranoside (IPTG) induction was used. Since the recombinant enzyme tended to aggregate to inclusion bodies, co-expression of the chaperones GroEL/S was used to improve the solubility and hence activity. As a result, the aminoacylase activity in the cell-free extract of the recombinant *V. natriegens* was increased by a factor of 1.8 [96].

3.5 Cell-Free Protein Synthesis

The fast growth of *V. natriegens* demands for high protein synthesis and thus the bacterium is equipped with 11–13 rRNA operons, 129 genes for tRNAs, and around 115,000 ribosomes per cell (at $\mu = 2.8\ h^{-1}$) [15, 97]. This high translational capacity makes *V. natriegens* attractive for cell-free protein synthesis (CFPS), and consequently some groups have already approached this method based on plasmid DNA as template [98–100]. The usage of linear DNA is desirable for high-throughput approaches since the synthesis is less laborious and expensive, but to date only low product yields and rapid DNA template degradation have been reported [100]. To overcome this issue, Zhu et al. added a double-stranded DNA-binding protein (single-chain Cro protein, scCro) to the CFPS reaction. The addition of 2 μM of scCro was sufficient to protect about 64 nM of linear expression template which improved the sfGFP yield in the *V. natriegens* CFPS platform by a factor of 18 [101]. Zhang et al. investigated four CFPS systems based on *E. coli*, *V. natriegens*, *B. subtilis*, and *C. glutamicum* and systematically improved the performance by optimizing the reaction parameters. For *V. natriegens*, the key factors for improving protein expression were the plasmid concentration (600 ng plasmid DNA μL^{-1}) and the addition of PEG8000 (2%), Mg^{2+} (10 mM), and PEP (20 mM). Furthermore, the correct choice of the host cell as a plasmid source is crucial. The highest expression of sfGFP was achieved when *E. coli* XL1-Blue cells were used as the plasmid source. Optimization of the aforementioned reaction parameters led to an increase from 570 mg sfGFP L^{-1} to 725 mg sfGFP L^{-1} [102].

3.6 V. natriegens *Is Electroactive*

The high growth rate of *V. natriegens* also requires a high oxygen uptake, and thus the oxygen transfer rate in a bioreactor might be limited at a larger scale. However, the flexibility of *V. natriegens* in using different electron acceptors makes it theoretically possible to address this issue in a bioelectrochemical system with an anode as electron acceptor. *V. natriegens* is capable of extracellular electron transfer [103], and recently it was demonstrated that this bacterium can transfer electrons to the anode either by direct contact or via an artificial redox mediator generating currents of 6.6 μA cm^{-2} and 196 μA cm^{-2}, respectively [104, 105]. Nevertheless, the fermentative pathway is still the preferred route to balance the redox state of the cell, and formate, acetate, and succinate are produced from glucose in the mediated system [104]. Microbial fuel cells (MFC) convert chemical into electrical energy and have the potential to treat industrial wastewater streams, which might contain not only organic matter but also high levels of salt [106]. Therefore, the application of the halophilic and electroactive *V. natriegens* with nutritional flexibility seems to be promising. In this context, Gong et al. tested several media and carbon sources in a MFC setup and achieved the highest power density of around 112 mW cm^{-2} with 1% glycerol as sole carbon source [105].

3.7 Bioprocess Engineering

To reach industrial relevance, KPIs such as titer, rate, and yield must be high. High titers of around 100 g L^{-1} and yields of 80–90% as well as productivities of at least 2.5 g L^{-1} h^{-1} are desirable [11, 107]. Therefore, high biomass concentrations are required, and fed-batch processes are usually applied to achieve high space time yields. Schulze et al. established an aerobic glucose-limited fed-batch process with *V. natriegens* WT at 37°C, which yielded a final biomass concentration of up to 28 g_X L^{-1}. An 800-fold increase in viscosity caused by exopolysaccharide formation was reported, which might negatively impact heat transfer, oxygen supply, and downstream processing. To tackle this problem, it was shown that the inactivation of the transcriptional biofilm regulator CpsR or deletion of the capsular polysaccharide cluster *cps* resulted in a consistently low viscosity during the overall process and improved the mass transfer. This resulted in a 23% higher oxygen transfer rate compared to cultivations of the wild-type strain. However, no further biomass formation was observed, indicating other limitations besides oxygen. Besides the reduction of viscosity, inactivation of CpsR or the *cps* cluster might be a useful strategy to channel the carbon more efficiently toward the desired product as already shown for anaerobic succinate production [77]. Thiele et al. used a combination of an exponential and a constant glucose feeding strategy during a fed-batch process together with a reduced temperature to achieve high biomass concentrations. To maintain the dissolved oxygen above 40% for high biomass concentrations, an automated cascade consisting of an increased stirrer speed (200–1,500 rpm), increased air flow (0.5–2 vvm), and increased oxygen in the gas mix (0–70%) was applied. The authors described a beneficial effect of lowering the cultivation temperature from 37 to 30°C on cell growth and were thus able to enhance the biomass concentration of up to 55 g_X L^{-1} [108]. In general, the high growth rate of *V. natriegens* results in a high oxygen uptake rate which might exceed the mass transfer capacity of large-scale bioreactors hindering unlimited aerobic growth to high cell densities. Therefore, substrate-limited fed-batch processes have to be applied which might be operated with an optimized stirrer and aeration setup at elevated pressure or oxygen-enriched aeration [109]. *V. natriegens* possesses also under anaerobic conditions very high substrate consumption rates [3] which might be utilized in zero-growth processes [110] to circumvent mass transfer issues.

V. natriegens high demand for sodium, which is typically satisfied with sodium chloride, can pose a challenge for process development since chloride is known to cause corrosion on metal surfaces in bioreactors. Further, the high sodium chloride content can have a negative impact on downstream processing particularly on product-broth separation [3]. To overcome this issue, Hoffart et al. developed a medium that only contains traces of chloride but still provides sufficient sodium for growth. This allowed the growth rate of *V. natriegens* to be maintained at 91% [3]. More recently, Biener et al. completely replaced sodium chloride with disodium hydrogen phosphate, disodium sulfate, and sodium citrate while maintaining the same sodium concentration. With this medium, the authors reported the highest biomass concentration achieved to date of around 60 g_X L^{-1} at 30°C using a fed-batch process [111]. Although high salt loads may negatively impact the bioprocess, the halotolerance of *V. natriegens* can also be used to advantage. Sterilization of media and feedstocks significantly contributes to the overall production costs, and fermentation processes consume large quantities of fresh water. The halotolerance of *V. natriegens* enables the utilization of high salinity water such as sea water or even wastewater streams. In addition, the use of high salinity water allows to avoid media sterilization since growth of a vast number of microorganisms is inhibited or abolished under such conditions. Some authors already took advantage of this approach and used non-sterile process conditions for the production of 2,3-butanediol [82] and PHB [70] with *V. natriegens*. In general, PHB production is limited by high costs for sterilization and substrates such as glucose, which contributes up to 50% of the overall production costs [70]. Li et al. used crude glycerol as substrate and achieved a titer of 2.39 g PHB L^{-1} with *V. natriegens* p15A-*phaBAC* under non-sterile conditions [70]. For 2,3-butanediol production, Meng et al. applied a non-sterile fermentation process with seawater-based minimal medium. The engineered strain *V. natriegens* Δ*frdA* Δ*ldhA*-pETRABC achieved 41 g 2,3-butanediol L^{-1} within 12 h [82]. Recently, Zhou et al. expressed a PET hydrolase in Vmax using unsterilized seawater-based media, which resulted in great energy and cost reductions per gram of protein compared to *E. coli* in sterile freshwater medium [112].

4 Synthetic Biology Tools for *V. natriegens*

4.1 Natural Transformation-Based Engineering

Starting with the first published genomes of *V. natriegens* [5, 15–17], over the last decade, several tools for genome modification in *V. natriegens* were described. While the first strains have been modified using conjugation [5], today's most commonly used genome editing tools for *V. natriegens* involve the natural competence system (Fig. 2) [28, 30]. Natural competence is spread throughout different clades of bacteria and facilitates the uptake of environmental DNA as nutrient source and for potential horizontal gene transfer [113–115]. In Gram-negative bacteria, the DNA is bound to a pilus of the type IV family and pulled through an outer membrane pore into the periplasm. For further transport into the cytoplasm, one strand of a linear DNA is degraded, and the ssDNA is translocated. If the ssDNA contains sufficient homologous regions, it can be genomically integrated by homologous recombination. This process has previously been described in detail in different reviews and articles [114–116]. While the regulation of this mechanism appears to be broadly conserved in *Vibrio* species, it seems that different strains of the same species can lose the functionality or strongly differ in natural transformation frequencies. In some cases, the strain-to-strain variation in terms of transformation efficiency under the same conditions can exceed the variation in-between different species [117]. Commonly, in *Vibrio* species, the mechanism is controlled by the presence of chitin and quorum sensing [116–120]. The native competence regulation can be overcome by heterologous expression of the master regulator TfoX [28, 117]. In *V. natriegens*, this leads to the activation and repression of different sets of genes involved in natural competence [31] and enables the transformation and homologous recombination of linear transfer DNA (tDNA) (Fig. 2a). While the original protocol already enabled multiplexing by MuGENT, it required an additional selection marker [28] (Fig. 2b). Similar to *V. cholerae*, by using FRT recombination, the selection marker can be removed from the genome of *V. natriegens*, leaving a short remaining scar while requiring an additional working step [66, 121]. The curability of MuGENT plasmid was enhanced by an additional *sacB* for sucrose counterselection [27]. More recent approaches developed scarless genome editing in *V. natriegens* based on natural transformation (NT) combined with CRISPR counter selection (see below). Next to electroporation and chemical transformation, natural competence also enables the uptake of plasmids into

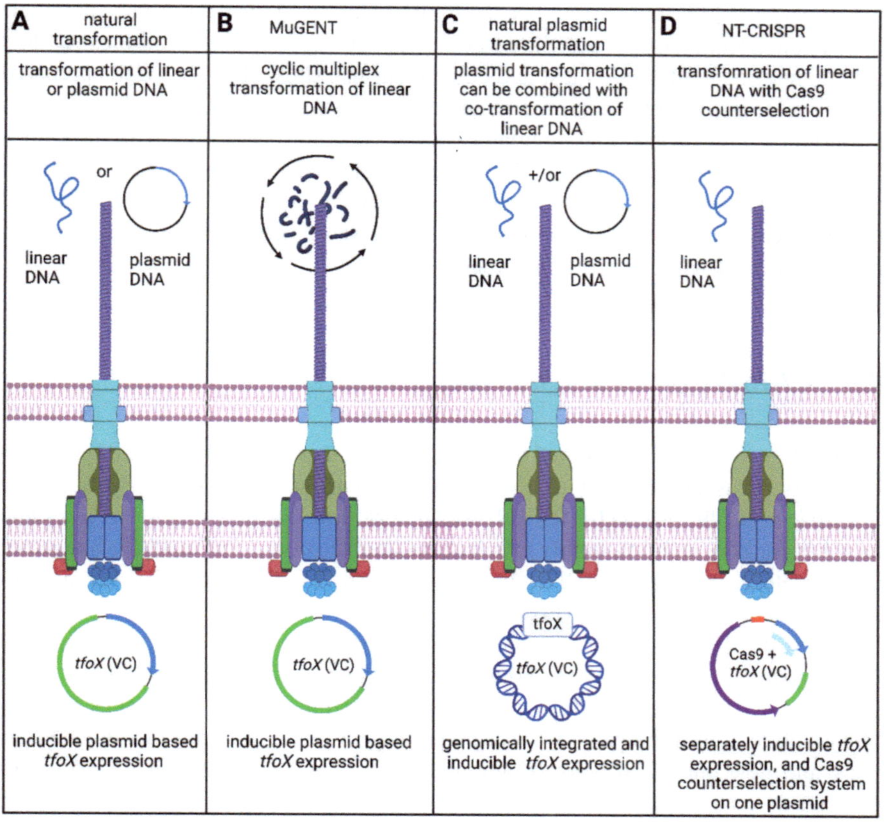

Fig. 2 Comparing the amino acid sequences of the proteins required for the formation of the DNA uptake system in *V. cholerae* O1 biovar El Tor N16961 (viz., PilA, PilB, PilD, PilM, PilN, PilO, PilP, PilQ, PilT, ComEA, and ComEC) with equivalents in *V. natriegens* using BLASTP [130], we detected high similarities (query cover >97%; Per. Ident. 39–87%). Therefore, we assumed the model for the natural competence system to be comparable with the one in *V. cholerae* and based the figure on the model of Matthey and Blokesch [131]. (a) Heterologous expression of the regulator TfoX enables the uptake of linear tDNA and plasmid DNA [28, 117]. (b) The same system facilitates multiplex genome editing and can be used for repeated library transformation by switching the selection marker in each cycle of transformation enabling the increase of unselected gene deletions in every round of natural transformation [28]. (c) The genomic integration of an inducible heterologous TfoX expression system yields a strain for fast and easy plasmid transformation as an alternative to classic chemically and electrocompetent cells. Due to the high transformation efficiency of the natural competence, plasmid transformation can be used alongside linear DNA as a potential selection marker to facilitate scarless deletions after PCR screening [28, 117]. (d) The additional expression of *cas9* on a *tfoX*-carrying vector enables highly efficient counter selection and simplifies the procedure of creating deletions and integrations [30]

V. natriegens (Fig. 2a) [29, 117]. Although, as previously described, the uptake mechanism is expected only to translocate linear ssDNA through the inner membrane [114], plasmid transformation by natural competence is shown for several

Vibrio species [29, 117, 122]. The co-transformation of plasmid DNA alongside tDNA can circumvent the utilization of a selection marker. This combines scarless genetic modifications with a fast and convenient workflow (Fig. 2c). Cells prepared for natural transformation can even be frozen and stored at −80°C [29]. A strain with a genomically integrated *tfoX*-regulator gene leaves more flexibility with the choice of antibiotic resistance marker [29]. While this method requires later screening for the unselected genome edit, no additional gRNA vector has to be constructed and transformed prior to making a modification. The strain promotes low-cost transformation protocol without additional vector design [31].

4.2 CRISPR-Based Engineering

Natural transformation combined with CRISPR counter selection (NT-CRISPR) has been proven as an efficient tool to introduce scarless genomic modifications in *V. natriegens* [29, 30]. NT-CRISPR uses natural transformation prior to *cas9* and *gRNA* induction enabling highly efficient counter selection. Successfully integrated DNA protects the genomic DNA from cleavage by Cas9 without the use of additional selection markers [30] (Fig. 2d).

CRISPR SWAPnDROP combines CRISPR-Cas9 counter selection with the λRED system to mediate homologous recombination without the direct use of linear DNA [123]. This method is fully plasmid based and uses a complex set of vectors. The combination of the pSWAP-vector and the cr3 helper plasmid facilitates insertions and deletions. The deleted genomic region can be loaded on the pSWAP-plasmid by swapping with predesigned homologous regions. A third vector, pDROP, enables the transfer and integration of the loaded region on pSWAP from one organism to another by conjugation. Since both organisms require the helper plasmid, cr3 is compatible with different Golden Gate-based modular cloning systems [66, 124] and customizable for interspecies gene transfer. The success of the CRISPR SWAPnDROP transformation procedure is indicated at each step by an incorporated multi-color co-selection system [123]. As proof of concept, the CRISPR SWAPnDROP system was applied to transfer the *lacZω* operon from *E. coli* DH5α to *V. natriegens* enabling functional alpha complementation and the possibility of blue-white screening in this strain [123].

CRISPR interference (CRISPRi) is a valuable tool enabling repression of gene expression with a guided but catalytically dead Cas9 (dCas9) lacking endonuclease activity [125]. This approach was established for *V. natriegens* and successfully applied to target over 99% of genes on both chromosomes to identify genes being essential or growth related under specific conditions [126]. In a recent preprint, this tool was further combined with a modular cloning approach enabling induced expression of up to four gRNAs and the dCas9 from a single vector [127]. Combining gRNA modification including variations of defined mismatches [128] and gRNA truncation [129] enables gradual knockdown regulation in *V. natriegens* parallel for multiple genomic targets [127].

4.3 Genetic Parts

Efficient host engineering requires an adjusted synthetic biology tool box and therefore the design and characterization of (regulatory) genetic parts, enabling the construction of vectors and transcription units (TUs) à la carte. Several methods and protocols for plasmid transformation and even cloning by transformation of Golden Gate assembly reactions in *V. natriegens* are reported [5, 29, 30, 66, 92]. Notably, part characterization has to be interpreted in the context of the tested environment. Cellular ribosomal protein and rRNA content is proportional to the growth rate [132, 133]. During exponential growth, the growth rate should correlate if not influenced by other parameters with constitutive heterologous gene expression [134–137]. This correlation is represented and taken into account by different calculations and predictions for gene expression [137–139]. Therefore, factors affecting the growth rate like carbon source and growth media [137], oxygen availability [140], temperature [141], metabolic burden [142], or the time points of measurement and the growth stage of the cells [135] are relevant for genetic part characterization. Furthermore, consistent reference standards can increase the inter-comparability of genetic characterizations regarding gene expression in between labs and publications [143] and may accelerate establishing *V. natriegens*, putting the different efforts made in part characterization in a uniform context.

The combination of different factors and the composition of a vector are paramount for gene expression and well characterized for *E. coli* [144]. Genetic parts and vectors in *V. natriegens* follow the same principles and were summarized recently [14]. Hong et al. give a broad overview about characterized and tested constitutive and inducible promoters, RBSs, degradation tags, reporters, and origins of replication [14, 66, 92]. While it could be shown that genetic parts from *E. coli* appear to be generally functional in *V. natriegens* [5, 66, 92], also new genetic parts in the form of artificial regulatory sequences from randomized libraries have been designed for *V. natriegens* and were tested and compared in different media [83, 145]. A stronger reliance on a Shine-Dalgarno (SD) motif in *V. natriegens* indicates that unlike *E. coli*, *V. natriegens* might not have a SD-independent translation mechanism [145].

Considering a gap of one nucleotide in between the organisms −10 and −35 consensus sequences, rRNA promoters are conserved comparing *V. natriegens* and *E. coli* [97, 145]. Additionally, the binding sides of the sigma factors, which are present in both organisms, share high similarity [145]. In regard of translation, *V. natriegens* and *E. coli* share a common anti-Shine-Dalgarno sequence for the 16S rRNA. Prediction tools for promoters and RBSs from *E. coli* have been applied to *V. natriegens* 5'UTR libraries to estimate among others transcription and translation rates with mixed results working partially but unreliable at least for translation [145]. More of such comparisons have to be made in the future to adjust predictive tools enabling reliable estimations.

4.4 Codon Usage and Optimization

After the publication of first codon usage pattern for *V. natriegens* [15], approaches have been made to apply codon optimization to improve the expression of different reporter genes [66]. The overexpression of the Lux operon influences the growth rate, thereby implicating infliction of metabolic burden in *V. natriegens*. The toleration of this burden could be improved by codon optimization of the antibiotic resistance cassette [66]. On average, it is expected that organism-specific codon usage can decrease the metabolic burden on a cell posed by heterologous gene expression [142, 146]. Despite this, the mechanism is barely understood with multiple effects intertwining. Apart from global translation regulation and RBS binding affinity, local translation appears also to be affected by properties of the coding sequence in the form of the secondary mRNA structure [146] and the codon usage [147], especially close to the 5'UTR [146–148]. At high folding-free energies, resulting in weak mRNA structures, the codon bias becomes the determining factor for translation efficiency [147]. Comparing 20 genomes of all three domains of life, the average of endogenous genes appears to have a universally conserved low translation efficiency ramp domain (ramp) of 30–50 weak codons after the start codon. The ramp slows down translation directly after translation initiation and thereby locally increases the ribosome density on the mRNA. This region, followed by codons promoting faster translation, is expected to reduce following events of ribosome traffic jams. This is expected to reduce translation abortion at states with higher energy cost and increase the overall efficiency of translation [148]. It surely will not solely explain *V. natriegens* fast growth, but it would be interesting to see if *V. natriegens*, having a high ribosomes content [97], has a different average translation efficiency profile as estimated universally for other organisms [147, 148].

5 Conclusion and Outlook

In recent years, there has been an astonishing increase in publications with *V. natriegens*. Researchers have taken note of the organism's inherit advantages, including its fast growth, high substrate uptake rate, salt tolerance, and nutritional versatility. Its potential as an emerging chassis organism for the production of a variety of different products as well as its potential for CFPS has already been proven. However, although numerous encouraging KPIs have been reported, further research is needed to achieve economic viable processes, particularly by investigating *V. natriegens*' performance in scale-up scenarios. Metabolic engineering approaches will benefit from now available metabolic models [42], which can facilitate the identification of metabolic bottlenecks and side products to more efficiently engineer production strains.

Strain development requires functional genetic parts and engineering methods, establishing a growing synthetic biology field around *V. natriegens* [14]. While it is feasible to engineer single or a small number of targets with high efficiency, the application of multiplex approaches in *V. natriegens* is only beginning to gain traction [145]. The transformation efficiency is lower compared to the laboratory workhorse *E. coli*, making it currently inferior as cloning host. For protein expression, the commercially available strain Vmax™ X2 has been used successfully, including proteins which are difficult to express in *E. coli* [68]. This shows that *V. natriegens* could be used as an alternative host for protein production in the future.

While *V. natriegens* was seen simply as a faster *E. coli* in the beginning, it gets recognized for more of its outstanding qualities. Especially, the halotolerance is not only utilized with its electroactivity for microbial fuel cells but also in unsterilized fermentations with sea water, which lowers the overall production costs [82, 105]. This trait also opens up the utilization of so far untapped waste streams as substrate sources in a circular economy.

A significant progress has been made in understanding *V. natriegens* at a systems level, particularly regarding halotolerance and ROS response [31, 39, 42], both important fields for further strain optimization. However, still poorly understood are the VBNC state and CILC in *V. natriegens*. Improving the bacteria's vulnerability would improve the basic laboratory handling and the performance in the industrial environment. Furthermore, *V. natriegens* possesses yet unexplored quorum sensing capabilities. Investigating its regulatory mechanisms and effectors might prove vital for scaling up processes. Closely linked to quorum sensing is the EPS formation that *V. natriegens* exhibits [77], and a reduction or removal of EPS would benefit production strains.

In conclusion, while *V. natriegens* showcases a remarkable potential across various research fields, continued research efforts are essential to harness its full capabilities and address the remaining challenges for widespread industrial applications.

Additional Information

All figures were created using BioRender.

Author Contributions MH, CS, JH, and BB conceived and wrote the paper. All authors read and approved the manuscript.

Funding This work was supported by the BL1408/2-1 grant (Deutsche Forschungsgemeinschaft) and by the TUM WACKER Institute for Industrial Biotechnology.

Conflict of Interests The authors declare no competing interests.

References

1. Blombach B, Grünberger A, Centler F et al (2022) Exploiting unconventional prokaryotic hosts for industrial biotechnology. Trends Biotechnol 40:385–397. https://doi.org/10.1016/j.tibtech.2021.08.003
2. Eagon RG (1962) *Pseudomonas natriegens*, a marine bacterium with a generation time of less than 10 minutes. J Bacteriol 83:736–737. https://doi.org/10.1128/jb.83.4.736-737.1962
3. Hoffart E, Grenz S, Lange J et al (2017) High substrate uptake rates empower *Vibrio natriegens* as production host for industrial biotechnology. Appl Environ Microbiol 83: e01614–e01617. https://doi.org/10.1128/AEM.01614-17
4. Payne WJ (1958) Studies on bacterial utilization of uronic acids. III. Induction of oxidative enzymes in a marine isolate. J Bacteriol 76:301–307. https://doi.org/10.1128/jb.76.3.301-307.1958
5. Weinstock MT, Hesek ED, Wilson CM et al (2016) *Vibrio natriegens* as a fast-growing host for molecular biology. Nat Methods 13:849–851. https://doi.org/10.1038/nmeth.3970
6. Baumann P, Baumann L, Mandel M (1971) Taxonomy of marine bacteria: the genus *Beneckea*. J Bacteriol 107:268–294. https://doi.org/10.1128/jb.107.1.268-294.1971
7. Payne WJ (1960) Effects of sodium and potassium ions on growth and substrate penetration of a marine pseudomonad. J Bacteriol 80:696–700. https://doi.org/10.1128/jb.80.5.696-700.1960
8. Varma A, Palsson BO (1994) Stoichiometric flux balance models quantitatively predict growth and metabolic by-product secretion in wild-type *Escherichia coli* W3110. Appl Environ Microbiol 60:3724–3731. https://doi.org/10.1128/aem.60.10.3724-3731.1994
9. Sauer U, Hatzimanikatis V, Hohmann HP et al (1996) Physiology and metabolic fluxes of wild-type and riboflavin-producing *Bacillus subtilis*. Appl Environ Microbiol 62:3687–3696. https://doi.org/10.1128/aem.62.10.3687-3696.1996
10. Radoš D, Turner DL, Fonseca LL et al (2014) Carbon flux analysis by ^{13}C nuclear magnetic resonance to determine the effect of CO_2 on anaerobic succinate production by *Corynebacterium glutamicum*. Appl Environ Microbiol 80:3015–3024. https://doi.org/10.1128/AEM.04189-13
11. Thoma F, Blombach B (2021) Metabolic engineering of *Vibrio natriegens*. Essays Biochem 65:381–392. https://doi.org/10.1042/EBC20200135
12. Hoff J, Daniel B, Stukenberg D et al (2020) *Vibrio natriegens*: an ultrafast-growing marine bacterium as emerging synthetic biology chassis. Environ Microbiol 22:4394–4408. https://doi.org/10.1111/1462-2920.15128
13. Xu J, Yang S, Yang L (2022) *Vibrio natriegens* as a host for rapid biotechnology. Trends Biotechnol 40:381–384. https://doi.org/10.1016/j.tibtech.2021.10.007
14. Hong C, Kim Y, Lee H et al (2024) *Vibrio* species as next-generation chassis for accelerated synthetic biology. Biotechnol Bioproc E. https://doi.org/10.1007/s12257-024-00023-7
15. Lee HH, Ostrov N, Wong BG et al (2016) *Vibrio natriegens*, a new genomic powerhouse. bioRxiv:58487. https://doi.org/10.1101/058487
16. Maida I, Bosi E, Perrin E et al (2013) Draft genome sequence of the fast-growing bacterium *Vibrio natriegens* strain DSMZ 759. Genome Announc 1. https://doi.org/10.1128/genomea.00648-13
17. Wang Z, Lin B, Hervey WJ et al (2013) Draft genome sequence of the fast-growing marine bacterium *Vibrio natriegens* strain ATCC 14048. Genome Announc 1. https://doi.org/10.1128/genomea.00589-13
18. Okada K, Iida T, Kita-Tsukamoto K et al (2005) *Vibrios* commonly possess two chromosomes. J Bacteriol 187:752–757. https://doi.org/10.1128/JB.187.2.752-757.2005
19. Oliver JD (2000) The public health significance of viable but nonculturable bacteria. In: Colwell RR, Grimes DJ (eds) Nonculturable microorganisms in the environment. Springer US, Boston, pp 277–299. https://doi.org/10.1007/978-1-4757-0271-2_16

20. Oliver JD (2010) Recent findings on the viable but nonculturable state in pathogenic bacteria. FEMS Microbiol Rev 34:415–425. https://doi.org/10.1111/j.1574-6976.2009.00200.x
21. Wang S, Wang B, You X et al (2023) Transcriptomic responses of the fast-growing bacterium *Vibrio natriegens* during cold-induced loss of culturability. Appl Microbiol Biotechnol 107: 3009–3019. https://doi.org/10.1007/s00253-023-12487-3
22. Kong I-S, Bates TC, Hülsmann A et al (2004) Role of catalase and *oxyR* in the viable but nonculturable state of *Vibrio vulnificus*. FEMS Microbiol Ecol 50:133–142. https://doi.org/10.1016/j.femsec.2004.06.004
23. Morgan RW, Christman MF, Jacobson FS et al (1986) Hydrogen peroxide-inducible proteins in *Salmonella typhimurium* overlap with heat shock and other stress proteins. Proc Natl Acad Sci U S A 83:8059–8063. https://doi.org/10.1073/pnas.83.21.8059
24. Loprasert S, Fuangthong M, Whangsuk W et al (2000) Molecular and physiological analysis of an OxyR-regulated *ahpC* promoter in *Xanthomonas campestris* pv. *phaseoli*. Mol Microbiol 37:1504–1514. https://doi.org/10.1046/j.1365-2958.2000.02107.x
25. Chiang SM, Schellhorn HE (2012) Regulators of oxidative stress response genes in *Escherichia coli* and their functional conservation in bacteria. Arch Biochem Biophys 525: 161–169. https://doi.org/10.1016/j.abb.2012.02.007
26. Anand A, Chen K, Catoiu E et al (2020) OxyR is a convergent target for mutations acquired during adaptation to oxidative stress-prone metabolic states. Mol Biol Evol 37:660–667. https://doi.org/10.1093/molbev/msz251
27. Glasgo LD, Lukasiak KL, Zinser ER (2024) Expanding the capabilities of MuGENT for large-scale genetic engineering of the fastest-replicating species, *Vibrio natriegens*. Microbiol Spectr:e0396423. https://doi.org/10.1128/spectrum.03964-23
28. Dalia TN, Hayes CA, Stolyar S et al (2017) Multiplex genome editing by natural transformation (MuGENT) for synthetic biology in *Vibrio natriegens*. ACS Synth Biol 6:1650–1655. https://doi.org/10.1021/acssynbio.7b00116
29. Specht DA, Sheppard TJ, Kennedy F et al (2024) Efficient natural plasmid transformation of *Vibrio natriegens* enables zero-capital molecular biology. PNAS Nexus 3:pgad444. https://doi.org/10.1093/pnasnexus/pgad444
30. Stukenberg D, Hoff J, Faber A et al (2022) NT-CRISPR, combining natural transformation and CRISPR-Cas9 counterselection for markerless and scarless genome editing in *Vibrio natriegens*. Commun Biol 5:265. https://doi.org/10.1038/s42003-022-03150-0
31. Shin J, Rychel K, Palsson BO (2023) Systems biology of competency in *Vibrio natriegens* is revealed by applying novel data analytics to the transcriptome. Cell Rep 42:112619. https://doi.org/10.1016/j.celrep.2023.112619
32. Chattopadhyay MK, Tabor CW, Tabor H (2003) Polyamines protect *Escherichia coli* cells from the toxic effect of oxygen. Proc Natl Acad Sci U S A 100:2261–2265. https://doi.org/10.1073/pnas.2627990100
33. Jung IL, Kim IG (2003) Transcription of *ahpC*, *katG*, and *katE* genes in *Escherichia coli* is regulated by polyamines: polyamine-deficient mutant sensitive to H_2O_2-induced oxidative damage. Biochem Biophys Res Commun 301:915–922. https://doi.org/10.1016/S0006-291X(03)00064-0
34. Kang I-H, Kim J-S, Kim E-J et al (2007) Cadaverine protects *Vibrio vulnificus* from superoxide stress. J Microbiol Biotechnol 17:176–179
35. da Costa MS, Santos H, Galinski EA (1998) An overview of the role and diversity of compatible solutes in bacteria and archaea. Adv Biochem Eng Biotechnol 61:117–153. https://doi.org/10.1007/BFb0102291
36. Oren A (2008) Microbial life at high salt concentrations: phylogenetic and metabolic diversity. Saline Syst 4:2. https://doi.org/10.1186/1746-1448-4-2
37. Gregory GJ, Boyd EF (2021) Stressed out: bacterial response to high salinity using compatible solute biosynthesis and uptake systems, lessons from *Vibrionaceae*. Comput Struct Biotechnol J 19:1014–1027. https://doi.org/10.1016/j.csbj.2021.01.030

38. Naughton LM, Blumerman SL, Carlberg M et al (2009) Osmoadaptation among *Vibrio* species and unique genomic features and physiological responses of *Vibrio parahaemolyticus*. Appl Environ Microbiol 75:2802–2810. https://doi.org/10.1128/AEM.01698-08
39. Huang L, Ni J, Zhong C et al (2022) Establishment of a salt-induced bioremediation platform from marine *Vibrio natriegens*. Commun Biol 5:1352. https://doi.org/10.1038/s42003-022-04319-3
40. Widderich N, Höppner A, Pittelkow M et al (2014) Biochemical properties of ectoine hydroxylases from extremophiles and their wider taxonomic distribution among microorganisms. PLoS One 9:e93809. https://doi.org/10.1371/journal.pone.0093809
41. Ongagna-Yhombi SY, Boyd EF (2013) Biosynthesis of the osmoprotectant ectoine, but not glycine betaine, is critical for survival of osmotically stressed *Vibrio parahaemolyticus* cells. Appl Environ Microbiol 79:5038–5049. https://doi.org/10.1128/AEM.01008-13
42. Coppens L, Tschirhart T, Leary DH et al (2023) *Vibrio natriegens* genome-scale modeling reveals insights into halophilic adaptations and resource allocation. Mol Syst Biol 19:e10523. https://doi.org/10.15252/msb.202110523
43. Ongagna-Yhombi SY, McDonald ND, Boyd EF (2015) Deciphering the role of multiple betaine-carnitine-choline transporters in the halophile *Vibrio parahaemolyticus*. Appl Environ Microbiol 81:351–363. https://doi.org/10.1128/aem.02402-14
44. Gregory GJ, Dutta A, Parashar V et al (2020) Investigations of dimethylglycine, glycine betaine, and ectoine uptake by a betaine-carnitine-choline transporter family transporter with diverse substrate specificity in *Vibrio* species. J Bacteriol 202:e00314–e00320. https://doi.org/10.1128/jb.00314-20
45. Gregory GJ, Boas KE, Boyd EF (2021) The organosulfur compound dimethylsulfoniopropionate (DMSP) is utilized as an osmoprotectant by *Vibrio* species. Appl Environ Microbiol 87:e02235–e02220. https://doi.org/10.1128/AEM.02235-20
46. Grammann K, Volke A, Kunte HJ (2002) New type of osmoregulated solute transporter identified in halophilic members of the bacteria domain: TRAP transporter TeaABC mediates uptake of ectoine and hydroxyectoine in *Halomonas elongata* DSM 2581(T). J Bacteriol 184:3078–3085. https://doi.org/10.1128/JB.184.11.3078-3085.2002
47. Lecher J, Pittelkow M, Zobel S et al (2009) The crystal structure of UehA in complex with ectoine-A comparison with other TRAP-T binding proteins. J Mol Biol 389:58–73. https://doi.org/10.1016/j.jmb.2009.03.077
48. Mulligan C, Fischer M, Thomas GH (2011) Tripartite ATP-independent periplasmic (TRAP) transporters in bacteria and archaea. FEMS Microbiol Rev 35:68–86. https://doi.org/10.1111/j.1574-6976.2010.00236.x
49. Rosa LT, Bianconi ME, Thomas GH et al (2018) Tripartite ATP-independent periplasmic (TRAP) transporters and tripartite tricarboxylate transporters (TTT): from uptake to pathogenicity. Front Cell Infect Microbiol 8:33. https://doi.org/10.3389/fcimb.2018.00033
50. Vetting MW, Al-Obaidi N, Zhao S et al (2015) Experimental strategies for functional annotation and metabolism discovery: targeted screening of solute binding proteins and unbiased panning of metabolomes. Biochemistry 54:909–931. https://doi.org/10.1021/bi501388y
51. Krulwich TA, Hicks DB, Ito M (2009) Cation/proton antiporter complements of bacteria: why so large and diverse? Mol Microbiol 74:257–260. https://doi.org/10.1111/j.1365-2958.2009.06842.x
52. Herz K, Vimont S, Padan E et al (2003) Roles of NhaA, NhaB, and NhaD Na^+/H^+ antiporters in survival of *Vibrio cholerae* in a saline environment. J Bacteriol 185:1236–1244. https://doi.org/10.1128/jb.185.4.1236-1244.2003
53. Shikuma NJ, Davis KR, Fong JNC et al (2013) The transcriptional regulator, CosR, controls compatible solute biosynthesis and transport, motility and biofilm formation in *Vibrio cholerae*. Environ Microbiol 15:1387–1399. https://doi.org/10.1111/j.1462-2920.2012.02805.x

54. Gregory GJ, Morreale DP, Boyd EF (2020) CosR is a global regulator of the osmotic stress response with widespread distribution among bacteria. Appl Environ Microbiol 86:e00120–e00120. https://doi.org/10.1128/AEM.00120-20
55. Gu C, Kim GB, Kim WJ et al (2019) Current status and applications of genome-scale metabolic models. Genome Biol 20:121. https://doi.org/10.1186/s13059-019-1730-3
56. Passi A, Tibocha-Bonilla JD, Kumar M et al (2021) Genome-scale metabolic modeling enables in-depth understanding of big data. Metabolites 12:14. https://doi.org/10.3390/metabo12010014
57. Zhang Y, Liu D, Chen Z (2023) Genome-scale modeling and systems metabolic engineering of *Vibrio natriegens* for the production of 1,3-propanediol. Methods Mol Biol 2553:209–220. https://doi.org/10.1007/978-1-0716-2617-7_11
58. Long CP, Gonzalez JE, Cipolla RM et al (2017) Metabolism of the fast-growing bacterium *Vibrio natriegens* elucidated by ^{13}C metabolic flux analysis. Metab Eng 44:191–197. https://doi.org/10.1016/j.ymben.2017.10.008
59. Eagon RG, Wang CH (1962) Dissimilation of glucose and gluconic acid by *Pseudomonas natriegens*. J Bacteriol 83:879–886. https://doi.org/10.1128/JB.83.4.879-886.1962
60. Klingner A, Bartsch A, Dogs M et al (2015) Large-scale ^{13}C flux profiling reveals conservation of the Entner-Doudoroff pathway as a glycolytic strategy among marine bacteria that use glucose. Appl Environ Microbiol 81:2408–2422. https://doi.org/10.1128/AEM.03157-14
61. Dürre P, Eikmanns BJ (2015) C1-carbon sources for chemical and fuel production by microbial gas fermentation. Curr Opin Biotechnol 35:63–72. https://doi.org/10.1016/j.copbio.2015.03.008
62. Yishai O, Lindner SN, Gonzalez L, de Cruz J et al (2016) The formate bio-economy. Curr Opin Chem Biol 35:1–9. https://doi.org/10.1016/j.cbpa.2016.07.005
63. Jiang W, Hernández Villamor D, Peng H et al (2021) Metabolic engineering strategies to enable microbial utilization of C1 feedstocks. Nat Chem Biol 17:845–855. https://doi.org/10.1038/s41589-021-00836-0
64. Tian J, Deng W, Zhang Z et al (2023) Discovery and remodeling of *Vibrio natriegens* as a microbial platform for efficient formic acid biorefinery. Nat Commun 14:7758. https://doi.org/10.1038/s41467-023-43631-2
65. Pfeifer E, Michniewski S, Gätgens C et al (2019) Generation of a prophage-free variant of the fast-growing bacterium *Vibrio natriegens*. Appl Environ Microbiol 85:e00853–e00819. https://doi.org/10.1128/AEM.00853-19
66. Stukenberg D, Hensel T, Hoff J et al (2021) The Marburg collection: a golden gate DNA assembly framework for synthetic biology applications in *Vibrio natriegens*. ACS Synth Biol 10:1904–1919. https://doi.org/10.1021/acssynbio.1c00126
67. González SS, Ad O, Shah B et al (2021) Genetic code expansion in the engineered organism Vmax X2: high yield and exceptional fidelity. ACS Cent Sci 7:1500–1507. https://doi.org/10.1021/acscentsci.1c00499
68. Mojica N, Kersten F, Montserrat-Canals M et al (2024) Using *Vibrio natriegens* for high-yield production of challenging expression targets and for protein perdeuteration. Biochemistry 63:587–598. https://doi.org/10.1021/acs.biochem.3c00612
69. Xu J, Dong F, Wu M et al (2021) *Vibrio natriegens* as a pET-compatible expression host complementary to *Escherichia coli*. Front Microbiol 12:627181. https://doi.org/10.3389/fmicb.2021.627181
70. Li H-H, Wu J, Liu J-Q et al (2023) Nonsterilized fermentation of crude glycerol for polyhydroxybutyrate production by metabolically engineered *Vibrio natriegens*. ACS Synth Biol. https://doi.org/10.1021/acssynbio.3c00498
71. Cavalheiro JM, de Almeida MCM, Grandfils C et al (2009) Poly(3-hydroxybutyrate) production by *Cupriavidus necator* using waste glycerol. Process Biochem 44:509–515. https://doi.org/10.1016/j.procbio.2009.01.008

72. Zhang L, Jiang Z, Tsui T-H et al (2022) A review on enhancing *Cupriavidus necator* fermentation for poly(3-hydroxybutyrate) (PHB) production from low-cost carbon sources. Front Bioeng Biotechnol 10:946085. https://doi.org/10.3389/fbioe.2022.946085
73. Ellis GA, Tschirhart T, Spangler J et al (2019) Exploiting the feedstock flexibility of the emergent synthetic biology chassis *Vibrio natriegens* for engineered natural product production. Mar Drugs 17:679. https://doi.org/10.3390/md17120679
74. Smith AD, Tschirhart T, Compton J et al (2023) Rapid, high-titer biosynthesis of melanin using the marine bacterium *Vibrio natriegens*. Front Bioeng Biotechnol 11:1239756. https://doi.org/10.3389/fbioe.2023.1239756
75. Wang Z, Tschirhart T, Schultzhaus Z et al (2020) Melanin produced by the fast-growing marine bacterium *Vibrio natriegens* through heterologous biosynthesis: characterization and application. Appl Environ Microbiol 86:e02749–e02719. https://doi.org/10.1128/AEM.02749-19
76. Li C, Wang R, Wang J et al (2023) A highly compatible phototrophic community for carbon-negative biosynthesis. Angew Chem Int Ed Engl 62:e202215013. https://doi.org/10.1002/anie.202215013
77. Schulze C, Hädrich M, Borger J et al (2024) Investigation of exopolysaccharide formation and its impact on anaerobic succinate production with *Vibrio natriegens*. J Microbial Biotechnol 17:e14277. https://doi.org/10.1111/1751-7915.14277
78. Thoma F, Schulze C, Gutierrez-Coto C et al (2022) Metabolic engineering of *Vibrio natriegens* for anaerobic succinate production. J Microbial Biotechnol 15:1671–1684. https://doi.org/10.1111/1751-7915.13983
79. Zhang Y, Sun Q, Liu Y et al (2021) Development of a plasmid stabilization system in *Vibrio natriegens* for the high production of 1,3-propanediol and 3-hydroxypropionate. Bioresour Bioprocess 8:125. https://doi.org/10.1186/s40643-021-00485-0
80. Erian AM, Gibisch M, Pflügl S (2018) Engineered *E. coli* W enables efficient 2,3-butanediol production from glucose and sugar beet molasses using defined minimal medium as economic basis. Microb Cell Fact 17:190. https://doi.org/10.1186/s12934-018-1038-0
81. Erian AM, Freitag P, Gibisch M et al (2020) High rate 2,3-butanediol production with *Vibrio natriegens*. Bioresour Technol Rep 10:100408. https://doi.org/10.1016/j.biteb.2020.100408
82. Meng W, Zhang Y, Ma L et al (2022) Non-sterilized fermentation of 2,3-butanediol with seawater by metabolic engineered fast-growing *Vibrio natriegens*. Front Bioeng Biotechnol 10:955097. https://doi.org/10.3389/fbioe.2022.955097
83. Wu F, Chen W, Peng Y et al (2020) Design and reconstruction of regulatory parts for fast-frowing *Vibrio natriegens* synthetic biology. ACS Synth Biol 9:2399–2409. https://doi.org/10.1021/acssynbio.0c00158
84. Liu X, Han X, Peng Y et al (2022) Rapid production of l-DOPA by *Vibrio natriegens*, an emerging next-generation whole-cell catalysis chassis. J Microbial Biotechnol 15:1610–1621. https://doi.org/10.1111/1751-7915.14001
85. Fernández-Llamosas H, Castro L, Blázquez ML et al (2017) Speeding up bioproduction of selenium nanoparticles by using *Vibrio natriegens* as microbial factory. Sci Rep 7:16046. https://doi.org/10.1038/s41598-017-16252-1
86. Wuisan ZG, Kresna IDM, Böhringer N et al (2021) Optimization of heterologous Darobactin A expression and identification of the minimal biosynthetic gene cluster. Metab Eng 66:123–136. https://doi.org/10.1016/j.ymben.2021.04.007
87. Peng Y, Ma L, Xu P et al (2023) High-performance production of *N*-acetyl-D-neuraminic acid with whole cells of fast-growing *Vibrio natriegens* via a thermal strategy. J Agric Food Chem 71:20198–20209. https://doi.org/10.1021/acs.jafc.3c07259
88. Bahlawan R, Karboune S, Liu L et al (2023) Investigation of biocatalytic production of lactosucrose and fructooligosaccharides using levansucrases and dairy by-products as starting materials. Enzyme Microb Technol 169:110279. https://doi.org/10.1016/j.enzmictec.2023.110279

89. Chen L, Hill A, Petit J-L et al (2023) Discovery and enzymatic screening of genome-mined microbial levanases to produce second-generation β-(2,6)-fructooligosaccharides: catalytic properties. ACS Chem Biol 18:465–475. https://doi.org/10.1021/acschembio.2c00728
90. Wu F, Wang S, Peng Y et al (2023) Metabolic engineering of fast-growing *Vibrio natriegens* for efficient pyruvate production. Microb Cell Fact 22:172. https://doi.org/10.1186/s12934-023-02185-0
91. Eichmann J, Oberpaul M, Weidner T et al (2019) Selection of high producers from combinatorial libraries for the production of recombinant proteins in *Escherichia coli* and *Vibrio natriegens*. Front Bioeng Biotechnol 7:254. https://doi.org/10.3389/fbioe.2019.00254
92. Tschirhart T, Shukla V, Kelly EE et al (2019) Synthetic biology tools for the fast-growing marine bacterium *Vibrio natriegens*. ACS Synth Biol 8:2069–2079. https://doi.org/10.1021/acssynbio.9b00176
93. Becker W, Wimberger F, Zangger K (2019) *Vibrio natriegens*: an alternative expression system for the high-yield production of isotopically labeled proteins. Biochemistry 58:2799–2803. https://doi.org/10.1021/acs.biochem.9b00403
94. Schleicher L, Muras V, Claussen B et al (2018) *Vibrio natriegens* as host for expression of multisubunit membrane protein complexes. Front Microbiol 9:2537. https://doi.org/10.3389/fmicb.2018.02537
95. Kormanová Ľ, Levarski Z, Minich A et al (2023) Novel expression system based on enhanced permeability of *Vibrio natriegens* cells induced by D,D-carboxypeptidase overexpression. World J Microbiol Biotechnol 39:277. https://doi.org/10.1007/s11274-023-03723-z
96. Haeger G, Wirges J, Tanzmann N et al (2023) Chaperone assisted recombinant expression of a mycobacterial aminoacylase in *Vibrio natriegens* and *Escherichia coli* capable of N-lauroyl-L-amino acid synthesis. Microb Cell Fact 22:77. https://doi.org/10.1186/s12934-023-02079-1
97. Aiyar SE, Gaal T, Gourse RL (2002) rRNA promoter activity in the fast-growing bacterium *Vibrio natriegens*. J Bacteriol 184:1349–1358. https://doi.org/10.1128/JB.184.5.1349-1358.2002
98. Des Soye BJ, Davidson SR, Weinstock MT et al (2018) Establishing a high-yielding cell-free protein synthesis platform derived from *Vibrio natriegens*. ACS Synth Biol 7:2245–2255. https://doi.org/10.1021/acssynbio.8b00252
99. Failmezger J, Scholz S, Blombach B et al (2018) Cell-free protein synthesis from fast-growing *Vibrio natriegens*. Front Microbiol 9:1146. https://doi.org/10.3389/fmicb.2018.01146
100. Wiegand DJ, Lee HH, Ostrov N et al (2018) Establishing a cell-free *Vibrio natriegens* expression system. ACS Synth Biol 7:2475–2479. https://doi.org/10.1021/acssynbio.8b00222
101. Zhu B, Gan R, Cabezas MD et al (2020) Increasing cell-free gene expression yields from linear templates in *Escherichia coli* and *Vibrio natriegens* extracts by using DNA-binding proteins. Biotechnol Bioeng 117:3849–3857. https://doi.org/10.1002/bit.27538
102. Zhang L, Lin X, Wang T et al (2021) Development and comparison of cell-free protein synthesis systems derived from typical bacterial chassis. Bioresour Bioprocess 8:58. https://doi.org/10.1186/s40643-021-00413-2
103. Conley BE, Weinstock MT, Bond DR et al (2020) A hybrid extracellular electron transfer pathway enhances the survival of *Vibrio natriegens*. Appl Environ Microbiol 86:e01253–e01220. https://doi.org/10.1128/AEM.01253-20
104. Gemünde A, Gail J, Holtmann D (2023) Anodic respiration of *Vibrio natriegens* in a bioelectrochemical system. ChemSusChem 16:e202300181. https://doi.org/10.1002/cssc.202300181
105. Gong Z, Xie R, Zhang Y et al (2023) Identification of emerging industrial biotechnology chassis *Vibrio natriegens* as a novel high salt-tolerant and feedstock flexibility electroactive microorganism for microbial fuel cell. Microorganisms 11:490. https://doi.org/10.3390/microorganisms11020490
106. Lefebvre O, Habouzit F, Bru V et al (2004) Treatment of hypersaline industrial wastewater by a microbial consortium in a sequencing batch reactor. Environ Technol 25:543–553. https://doi.org/10.1080/09593330.2004.9619345

107. Werpy T, Petersen G (2004) Top value added chemicals from biomass: volume I – results of screening for potential candidates from sugars and synthesis gas. National Renewable Energy Lab (NREL), Golden. https://doi.org/10.2172/15008859
108. Thiele I, Gutschmann B, Aulich L et al (2021) High-cell-density fed-batch cultivations of *Vibrio natriegens*. Biotechnol Lett 43:1723–1733. https://doi.org/10.1007/s10529-021-03147-5
109. Garcia-Ochoa F, Gomez E (2009) Bioreactor scale-up and oxygen transfer rate in microbial processes: an overview. Biotechnol Adv 27:153–176. https://doi.org/10.1016/j.biotechadv.2008.10.006
110. Lange J, Takors R, Blombach B (2017) Zero-growth bioprocesses: a challenge for microbial production strains and bioprocess engineering. Eng Life Sci 17:27–35. https://doi.org/10.1002/elsc.201600108
111. Biener R, Horn T, Komitakis A et al (2023) High-cell-density cultivation of *Vibrio natriegens* in a low-chloride chemically defined medium. Appl Microbiol Biotechnol 107:7043–7054. https://doi.org/10.1007/s00253-023-12799-4
112. Zhou Y, Shen B, You S et al (2024) Development of a novel "4E" polyethylene terephthalate bio-recycling process with the potential for industrial application: efficient, economical, energy-saving, and eco-friendly. Bioresour Technol 391:129913. https://doi.org/10.1016/j.biortech.2023.129913
113. Redfield RJ (1993) Genes for breakfast: the have-your-cake-and-eat-it-too of bacterial transformation. J Hered 84:400–404. https://doi.org/10.1093/oxfordjournals.jhered.a111361
114. Mell JC, Redfield RJ (2014) Natural competence and the evolution of DNA uptake specificity. J Bacteriol 196:1471–1483. https://doi.org/10.1128/jb.01293-13
115. Chen I, Dubnau D (2004) DNA uptake during bacterial transformation. Nat Rev Microbiol 2:241–249. https://doi.org/10.1038/nrmicro844
116. Sun Y, Bernardy EE, Hammer BK et al (2013) Competence and natural transformation in *vibrios*. Mol Microbiol 89:583–595. https://doi.org/10.1111/mmi.12307
117. Simpson CA, Podicheti R, Rusch DB et al (2019) Diversity in natural transformation frequencies and regulation across *Vibrio* species. MBio 10:e02788–e02719. https://doi.org/10.1128/mbio.02788-19
118. Meibom KL, Blokesch M, Dolganov NA et al (2005) Chitin induces natural competence in *Vibrio cholerae*. Science 310:1824–1827. https://doi.org/10.1126/science.1120096
119. Yamamoto S, Izumiya H, Mitobe J et al (2011) Identification of a chitin-induced small RNA that regulates translation of the *tfoX* gene, encoding a positive regulator of natural competence in *Vibrio cholerae*. J Bacteriol 193:1953–1965. https://doi.org/10.1128/jb.01340-10
120. Metzger LC, Blokesch M (2016) Regulation of competence-mediated horizontal gene transfer in the natural habitat of *Vibrio cholerae*. Curr Opin Microbiol 30:1–7. https://doi.org/10.1016/j.mib.2015.10.007
121. De Souza SO, Blokesch M (2010) Genetic manipulation of *Vibrio cholerae* by combining natural transformation with FLP recombination. Plasmid 64:186–195. https://doi.org/10.1016/j.plasmid.2010.08.001
122. Jeffrey WH, Paul JH, Stewart GJ (1990) Natural transformation of a marine *Vibrio* species by plasmid DNA. Microb Ecol 19:259–268. https://doi.org/10.1007/BF02017170
123. Teufel M, Klein CA, Mager M et al (2022) A multifunctional system for genome editing and large-scale interspecies gene transfer. Nat Commun 13:3430. https://doi.org/10.1038/s41467-022-30843-1
124. Klein CA, Emde L, Kuijpers A et al (2019) MoCloFlex: a modular yet flexible cloning system. Front Bioeng Biotechnol 7:271. https://doi.org/10.3389/fbioe.2019.00271
125. Qi LS, Larson MH, Gilbert LA et al (2013) Repurposing CRISPR as an RNA-guided platform for sequence-specific control of gene expression. Cell 152:1173–1183. https://doi.org/10.1016/j.cell.2013.02.022

126. Lee HH, Ostrov N, Wong BG et al (2019) Functional genomics of the rapidly replicating bacterium *Vibrio natriegens* by CRISPRi. Nat Microbiol 4:1105–1113. https://doi.org/10.1038/s41564-019-0423-8
127. Stukenberg D, Faber A, Becker A (2024) Graded-CRISPRi, a novel tool for tuning the strengths of CRISPRi-mediated knockdowns in *Vibrio natriegens* using gRNA libraries. bioRxiv. https://doi.org/10.1101/2024.01.29.577714
128. Hawkins JS, Silvis MR, Koo B-M et al (2020) Mismatch-CRISPRi reveals the co-varying expression-fitness relationships of essential genes in *Escherichia coli* and *Bacillus subtilis*. Cell Syst 11:523–535.e9. https://doi.org/10.1016/j.cels.2020.09.009
129. Fu Y, Sander JD, Reyon D et al (2014) Improving CRISPR-Cas nuclease specificity using truncated guide RNAs. Nat Biotechnol 32:279–284. https://doi.org/10.1038/nbt.2808
130. Camacho C, Coulouris G, Avagyan V et al (2009) BLAST+: architecture and applications. BMC Bioinformatics 10:421. https://doi.org/10.1186/1471-2105-10-421
131. Matthey N, Blokesch M (2016) The DNA-uptake process of naturally competent *Vibrio cholerae*. Trends Microbiol 24:98–110. https://doi.org/10.1016/j.tim.2015.10.008
132. Miura A, Krueger JH, Itoh S et al (1981) Growth-rate-dependent regulation of ribosome synthesis in *E. coli*: expression of the *lacZ* and *galK* genes fused to ribosomal promoters. Cell 25:773–782. https://doi.org/10.1016/0092-8674(81)90185-9
133. Scott M, Klumpp S, Mateescu EM et al (2014) Emergence of robust growth laws from optimal regulation of ribosome synthesis. Mol Syst Biol 10:747. https://doi.org/10.15252/msb.20145379
134. Klumpp S, Zhang Z, Hwa T (2009) Growth rate-dependent global effects on gene expression in bacteria. Cell 139:1366–1375. https://doi.org/10.1016/j.cell.2009.12.001
135. Keren L, Zackay O, Lotan-Pompan M et al (2013) Promoters maintain their relative activity levels under different growth conditions. Mol Syst Biol 9:701. https://doi.org/10.1038/msb.2013.59
136. Liang S, Bipatnath M, Xu Y et al (1999) Activities of constitutive promoters in *Escherichia coli*. J Mol Biol 292:19–37. https://doi.org/10.1006/jmbi.1999.3056
137. Scott M, Gunderson CW, Mateescu EM et al (2010) Interdependence of cell growth and gene expression: origins and consequences. Science 330:1099–1102. https://doi.org/10.1126/science.1192588
138. Bienick MS, Young KW, Klesmith JR et al (2014) The interrelationship between promoter strength, gene expression, and growth rate. PLoS One 9:e109105. https://doi.org/10.1371/journal.pone.0109105
139. Leveau JH, Lindow SE (2001) Predictive and interpretive simulation of green fluorescent protein expression in reporter bacteria. J Bacteriol 183:6752–6762. https://doi.org/10.1128/jb.183.23.6752-6762.2001
140. Losen M, Frölich B, Pohl M et al (2004) Effect of oxygen limitation and medium composition on *Escherichia coli* fermentation in shake-flask cultures. Biotechnol Prog 20:1062–1068. https://doi.org/10.1021/bp034282t
141. Ratkowsky DA, Lowry RK, McMeekin TA et al (1983) Model for bacterial culture growth rate throughout the entire biokinetic temperature range. J Bacteriol 154:1222–1226. https://doi.org/10.1128/jb.154.3.1222-1226.1983
142. Glick BR (1995) Metabolic load and heterologous gene expression. Biotechnol Adv 13:247–261. https://doi.org/10.1016/0734-9750(95)00004-A
143. Kelly JR, Rubin AJ, Davis JH et al (2009) Measuring the activity of BioBrick promoters using an *in vivo* reference standard. J Biol Eng 3:4. https://doi.org/10.1186/1754-1611-3-4
144. Pouresmaeil M, Azizi-Dargahlou S (2023) Factors involved in heterologous expression of proteins in *E. coli* host. Arch Microbiol 205:212. https://doi.org/10.1007/s00203-023-03541-9
145. Tietze L, Mangold A, Hoff MW et al (2022) Identification and cross-characterisation of artificial promoters and 5′ untranslated regions in *Vibrio natriegens*. Front Bioeng Biotechnol 10:826142. https://doi.org/10.3389/fbioe.2022.826142

146. Kudla G, Murray AW, Tollervey D et al (2009) Coding-sequence determinants of gene expression in *Escherichia coli*. Science 324:255–258. https://doi.org/10.1126/science.1170160
147. Tuller T, Waldman YY, Kupiec M et al (2010) Translation efficiency is determined by both codon bias and folding energy. Proc Natl Acad Sci U S A 107:3645–3650. https://doi.org/10.1073/pnas.0909910107
148. Tuller T, Carmi A, Vestsigian K et al (2010) An evolutionarily conserved mechanism for controlling the efficiency of protein translation. Cell 141:344–354. https://doi.org/10.1016/j.cell.2010.03.031

Adv Biochem Eng Biotechnol (2025) 192: 117–154
https://doi.org/10.1007/10_2024_272

Published online: 24 November 2024

Shewanella oneidensis: Biotechnological Application of Metal-Reducing Bacteria

Lukas Kneuer, René Wurst, and Johannes Gescher

Contents

Abstract What is an unconventional organism in biotechnology? The γ-proteobacterium *Shewanella oneidensis* might fall into this category as it was initially established as a laboratory model organism for a process that was not seen as potentially interesting for biotechnology. The reduction of solid-state extracellular electron acceptors such as iron and manganese oxides is highly relevant for many biogeochemical cycles, although it turned out in recent years to be quite relevant for many potential biotechnological applications as well. Applications started with the production of nanoparticles and dramatically increased after understanding that electrodes in bioelectrochemical systems can also be used by these organisms. From the potential production of current and hydrogen in these systems and the development of biosensors, the field expanded to anode-assisted fermentations

Lukas Kneuer and René Wurst contributed equally to this work.

L. Kneuer, R. Wurst, and J. Gescher (✉)
Institute of Technical Microbiology, University of Technology Hamburg, Hamburg, Germany
e-mail: Johannes.gescher@tuhh.de

enabling fermentation reactions that were – so far – dependent on oxygen as an electron acceptor. Now the field expands further to cathode-dependent production routines. As a side product to all these application endeavors, *S. oneidensis* was understood more and more, and our understanding and genetic repertoire is at eye level to *E. coli*. Corresponding to this line of thought, this chapter will first summarize the available arsenal of tools in molecular biology that was established for working with the organism and thereafter describe so far established directions of application. Last but not least, we will highlight potential future directions of work with the unconventional model organism *S. oneidensis*.

Keywords Bioelectrochemistry, Biotechnology, Exoelectrogens, Shewanella, Sustainability

1 Introduction

Biotechnology needs more model organisms because individual organisms come with a specific spectrum of possible applications. In this chapter we focus on *S. oneidensis* as a potential organism for future biotechnological application.

In 1988, Ken Nealson and Charles Myers published a seminal report detailing bacterial manganese reduction and growth using manganese as the sole electron acceptor [1]. They studied a bacterium initially known as *Alteromonas putrefaciens* MR-1. Since then, this strain has been reclassified as *Shewanella oneidensis* to honor the Scottish microbiologist James M. Shewan [2]. The genus *Shewanella* is known for its exceptional respiratory versatility, with many species capable of reducing various inorganic and organic electron acceptors, whether soluble (such as dimethylsulfoxide, fumarate, nitrate, nitrite, trimethylamine-N-oxide, oxygen, humic acids) or in solid forms (like ferrihydrite, hematite, birnessite, or electrodes) [3]. These bacteria thrive in redox-stratified environments where electron acceptor availability, rather than the electron donor, dictates their ecological niche. *S. oneidensis* MR-1 remains the most extensively studied model organism for understanding extracellular electron transfer (EET) processes to date, and most applications studied focus on this ability of transferring respiratory electrons to solid-state electron acceptors (see below).

As shown in Fig. 1, electron transport from the cytoplasm via the periplasm to the cell surface takes place primarily via *c*-type cytochromes [4]. Cytochromes are proteins that carry heme groups as cofactors. The central iron atom in the heme complex changes its oxidation state when it receives and releases an electron. The genome of *S. oneidensis* contains 42 genes for *c*-type cytochromes [5].

The oxidoreductase is anchored in the cytoplasmic membrane and transfers electrons to the menaquinol pool. From there, the electrons are transferred to CymA, which plays an important role as a distributor of electrons to various periplasmic *c*-type cytochromes [7]. The fumarate reductase FccA and the small

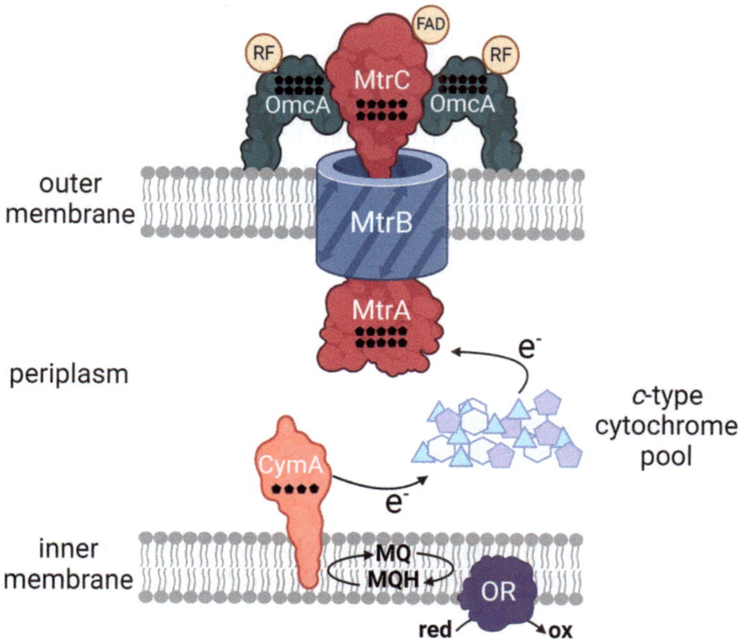

Fig. 1 Scheme of the extracellular electron transfer (EET) pathway in *S. oneidensis*. The number of heme cofactors in *c*-type cytochromes is indicated by black pentagons. Electrons are distributed from the cytoplasmic membrane protein A (CymA) to a *c*-type cytochrome pool in the periplasm and finally to the outer membrane module consisting of the metal-reducing protein A (MtrA), which spans the outer membrane through the pore-forming metal-reducing protein B (MtrB) and transfers electrons either to the metal-reducing protein C (MtrC) or to the outer membrane cytochrome A (OmcA). OR stands for oxidoreductase, and RF and FAD represent riboflavin and flavin adenine dinucleotide. MQ and MQH represent the oxidized and reduced states of the menaquinol pool, respectively (Figure is modified from Klein et al. [6])

tetraheme cytochrome (STC) should also be emphasized here. Double deletion mutants of these two proteins are no longer capable of electron transfer to terminal reductases [8]. The periplasmic cytochromes transfer the electrons to MtrA, which together with MtrB and MtrC forms a protein complex that spans the outer membrane. Electron transfer then takes place from MtrA to MtrC, with MtrB acting as a pore to ensure spatial proximity between the two. Finally, the terminal reductase MtrC transfers electrons to the extracellular electron acceptors [9]. In addition to MtrC, OmcA also plays a role in the reduction of extracellular acceptors [10].

S. oneidensis belongs to the γ-proteobacteria, is consequently closely related to *E. coli*, and shares some characteristics such as the facultative anaerobic metabolism and the ability to grow aerobically on agar plates overnight. Initially spurred by the motivation to gain a fundamental understanding of the biochemical underpinnings of extracellular electron transfer, there has been a steady development of new genetic tools or the adaptation of the existing tools toward application in *S. oneidensis*. It seems now fair to say that genetic engineering in *S. oneidensis* can be conducted with a similar speed compared to *Escherichia*, *Pseudomonas*, or *Bacillus* strains that are widely applied in biotechnology. Moreover, we gained a systems biology understanding that is also similar to other existing model organisms due to a multitude of omics studies under a wide variety of conditions.

We understand this chapter as a starting point to think about new ways to valorize *S. oneidensis*, which is why we provide an in-depth overview of the existing genetic tools, before we focus on so far established directions for its application.

2 Genetic Tools

For use in industrial applications, a fundamental understanding of the processes in the microorganisms used is essential, which in turn can foremost be achieved using genetic tools. *E. coli* is currently one of the most important production organisms due to its broad substrate spectrum and the numerous possibilities for genetic manipulation. *S. oneidensis* is also remarkable for the variety of genetic tools available and its ease of use compared to other electroactive bacteria. The following section takes a closer look at the established tools, highlighting their advantages and limitations.

2.1 Transposon Libraries

Since transposons are universal tools that do not depend on host organism proteins or genomic sequences, they are usually the first tools used in genetic engineering. Transposons are mobile genetic elements flanked by inverted repeats (IRs). A transposase recognizes and binds to these inverted repeats and facilitates the movement of the transposable element into any given DNA by a cut or copy and paste mechanism. A transposon library is generated using hyperactive transposases that facilitate gene disruption by randomly inserting transposons into the genome at high frequencies, followed by the collection of a sufficient number of colonies so that most non-essential genes are disrupted. However, the goal is to disrupt only one gene in each colony to avoid overlapping phenotypes. Production of a transposon library is followed by a screening to identify cells with impaired growth under certain conditions, such as iron respiration. The disrupted gene responsible for the phenotype can then be identified by PCR followed by sequencing. In addition to randomly integrating transposases, the directed transposase Tn7, which facilitates insertions into the *attTn7*, has also been used in *S. oneidensis* [11].

Even before the whole genome of *S. oneidensis* was sequenced, a basic understanding of the minimal electron transport chain was achieved mainly by using transposon libraries. Myers and colleagues [12, 13] identified CymA as the main entry point for electron transport from the menaquinone pool to the periplasm. In addition, they were able to elucidate the role of the corresponding protein FccA as a fumarate reductase [14, 15]. While Beliaev and colleagues [16, 17] elucidated the role of the MtrCAB complex in the outer membrane and provided a strikingly accurate sketch of the minimal electron transport chain in 2001, Newman and colleagues first proposed that a soluble, nonprotein compound is involved in EET [18]. Several years later, flavins were found to be responsible for 70% of the electron transfer to extracellular electron acceptors under batch conditions [19].

Even with the development of new genetic tools, transposon libraries followed by directed mutagenesis are still useful to elucidate the involvement of previously overlooked genes [20–24]. Conte et al. established the MiniISY100 transposase in *S. oneidensis*, providing a system with a preference for TA integration [25]. Bouhenni and colleagues used a modified *mariner* transposon, mini*Himar* RB1, to generate a more stable genetic background of the transposon mutants. The trick here is that only the transposon, encoding a kanamycin resistance, is inserted into the genome and not the transposase. The plasmid carries the R6K origin of replication, which requires the π protein for stable replication [26]. Since *S. oneidensis* does not encode the corresponding *pir* gene, the plasmid and thus the transposase will not be present in all daughter cells. Finally, transposase-mediated genetic drift cannot occur during high-throughput screens, making them more reliable [27]. In the following years, this system was widely used for recently developed methods like signature-tagged transposon mutagenesis, Tn-Seq, or Knockout Sudoku.

These techniques were all enabled through the invention of fast and cheap high-throughput sequencing methods. In signature-tagged mutagenesis, unique DNA barcodes are integrated in the transposons. After genome sequencing, these barcodes, and therefore the transposon insertion location, can be easily identified in the DNA sequence [28–30]. Tn-Seq is an evolution of this method. Pooled cells of a barcoded transposon library are cultured under the conditions of interest and then sequenced. The abundance of reads before and after cultivation provides information about which genes are essential under these conditions [31–36].

Baym and colleagues used so-called Knockout Sudoku to generate a curated whole-genome knockout collection. Therefore, they created a barcoded transposon library with over 39,000 clones, followed by pooling and massive parallel Illumina sequencing. Using bioinformatic and combinatorial methods to assign single colonies to certain gene knockouts resulted in the isolation of 3,667 single knockout which is in line with the estimated number of non-essential genes (3,635 to 3,814) [37].

2.2 *Homologous Recombination*

To gain a fundamental understanding of the complex metabolic and respiratory pathways in microorganisms, directed mutagenesis tools are of great importance. The most common approach, homologous recombination with suicide plasmids, is well established in *S. oneidensis* for more than two decades [38–42] and can be used similarly to *E. coli*. In general, sequences flanking the region of interest are amplified, fused together, and introduced into a vector carrying an antibiotic resistance cassette, a counter-selectable marker, an origin of transfer to facilitate conjugation, and an origin of replication which is only functional in the donor strain but not in the recipient strain (Fig. 2a). This vector is then transformed into the target strain and streaked onto selection plates. Note that only cells with the plasmid stably integrated into the genome will retain the antibiotic resistance cassette, since the origin of replication is not functional in the recipient strain. This integration is facilitated through the host's own homologous recombination system and is determined by the homologous sequences provided by the plasmid DNA. An outgrow followed by selection on plates containing the counter-selection marker, for instance, sucrose, if *sacB* is used for negative selection, allows selection toward colonies that have excised the plasmid DNA in a second crossover step. This can result in either a markerless mutant or the revertant and can be rapidly distinguished using a colony test PCR. This method can be used to either delete, insert, or modify a gene at any position in the genome without polar effects.

The importance of this high-precision tool is illustrated by the sheer number of publications using this technique to gain a fundamental understanding of extracellular electron transport [8, 10, 40, 42–53], biofilm formation [41, 54], or metabolic pathways [55].

Various functions of genes and the interaction of proteins could only be investigated with multiple knockout mutants.

Gao and colleagues [52] generated 36 out of 41 predicted *c*-type cytochrome single knockout mutants using the cre-lox system and homologous recombination. The Cre recombinase recognizes the originally introduced loxP sites flanking the gene of interest and excises the enclosed DNA sequence, resulting in a mutant with a loxP element remaining in the genome. They tested the mutants for their fitness during anoxic and oxic growth and showed that the proteins for extracellular electron transport to Cr(VI) are significantly different from those used for EET to Mn(IV) or Fe(III) species. However, they also observed that multiple cytochromes contribute to electron transport to the same electron acceptor, suggesting that a complex cytochrome network is used rather than a single chain.

This finding was indeed supported by subsequent studies using multiple knockout mutants. Several studies have been performed to unravel the complexity of extracellular electron transport using the allele replacement system to generate strains lacking a number of essential genes for EET, encoding proteins that facilitate electron transport through the periplasm [8, 42] or across the outer membrane to an extracellular electron acceptor [9, 10, 48, 50].

In addition, not only knockout mutants but also knock-in or gene-exchange strains have been used to optimize electron flux across the outer membrane to an electrode [56].

Li and colleagues [57, 58] introduced a quorum-sensing promoter system, *luxR* and *luxI*, to control genes for extracellular electron transport in order to establish a self-regulating system with improved biomass formation in the first phase and

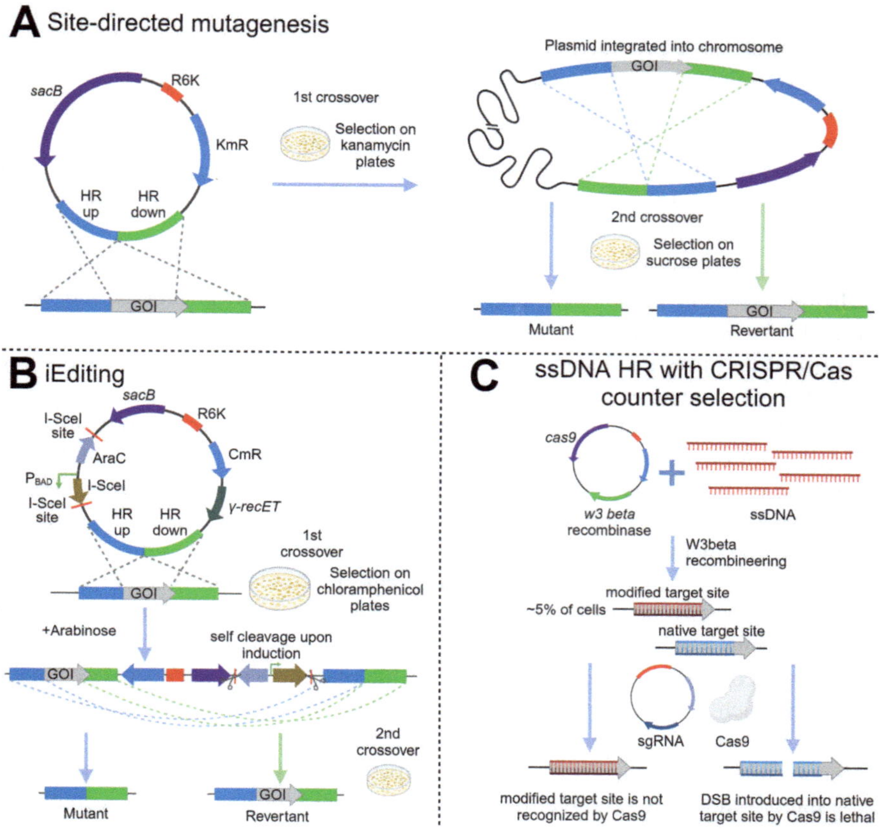

Fig. 2 Homologous recombination techniques used in *S. oneidensis*. (a) Double crossover site-directed mutagenesis with a suicide plasmid. The plasmid containing large homology regions to the up- and downstream areas of the target sequence is introduced via conjugation into *S. oneidensis*. Only cells having the plasmid integrated in the chromosome can survive antibiotic selection (first crossover). Outgrow with LB medium is followed by plating the cells on LB-sucrose plates. Counter-selection is facilitated by sacB expression encoding a levansucrase. Sucrose is converted into a toxic metabolite by cells still containing the plasmid, and therefore only cells with a second crossover can survive. The possible results are the mutant or the revertant. (b) The iEditing approach uses a self-cleavage system instead of sucrose counter-selection. I-SceI is induced after the first crossover and will introduce a double-strand break into chromosomally inserted DNA which triggers homologous recombination with the introduced homology regions [61]. (c) A plasmid carrying cas9 and a w3 beta recombinase is transformed into the target strain. First, the recombinase is induced, and the ssDNA oligos along with a plasmid expressing sgRNA constitutively are introduced via electroporation. The recombinase facilitates introduction of the ssDNA oligos with short homology arms to the target region. Afterward, cas9 is induced, and the nuclease is guided to targets that were not edited by the recombinase. The introduced DSBs only occur in wild-type cells and are lethal, thus facilitating an efficient counter-selection [63]. KmR, kanamycin resistance; CmR, chloramphenicol resistance; HR, homologous region; GOI, gene of interest

enhanced electron transport in the late phase. They could show that this system outcompetes classic promoter approaches in the late growth phase.

Li and colleagues [59] could reach a superior performance in MFCs (almost 40-fold higher compared to the wild type) through systematic assessment and improvement of the whole biofilm formation cycle and overexpression of critical genes for EET. They combined the genetic engineering approach with technical engineering and added reduced graphene oxide carbon nanotubes to provide a conductive matrix [62].

However, due to the complex metabolic versatility of *S. oneidensis*, it is still a challenge to unravel certain gene functions since their possible phenotypes are often masked by redundant proteins. In the last decade, three new approaches have been established that may significantly improve the genetic tractability or accelerate the generation of mutants in *S. oneidensis*.

Enyeart et al. [60] combined a targetron with a Cre-lox approach. A targetron can be used to introduce an intron site-specifically into a target sequence, while the Cre recombinase recognizes lox sites (introduced with the targetron) and deletes or inverts sequences between two lox sites without the need for counter-selection. However, this method always leaves a scar in the genome and is time-consuming compared to deletions or insertions with suicide plasmids and is therefore rarely used. Nevertheless, a major advantage of this system is that large genomic rearrangements can be achieved.

Fan and colleagues designed a self-cleavable plasmid to avoid sucrose counter-selection which is often a rate-limiting step since spontaneous mutations in *sacB* can provide a resistance strategy to the cells. Therefore, they introduced a homing endonuclease, I-SceI, flanked by the respective recognition sites and regulated by an arabinose-inducible promoter to a standard plasmid for homologous recombination. Induction of the endonuclease will lead to a double-strand break (DSB) and cleavage of the endonuclease encoding gene. Consequently, the DSB is repaired by homologous recombination with the homologous sequences provided. Additionally, they introduced the RecET system to overcome limitations of the unfavorable *S. oneidensis* HR system (Fig. 2b). Taken together, they were able to generate the mutant strains slightly faster or, if combined with sucrose counter-selection, at very high editing efficiencies of up to 100% [61]. They called their method "iEditing".

In another approach, a plasmid harboring the efficient w3 beta recombinase from a close relative, *Shewanella* sp. W3-18-1, was used to facilitate homologous recombination with ssDNA. The DNA can be provided via oligos which would omit cloning steps and therefore heavily increase the speed of mutant generation (Fig. 2c). First, the strain is prepared via the transformation of the plasmid harboring the recombinase. Second, synthesized oligos as short as 40 nucleotides can be transformed as single-stranded DNA followed by the last step, screening for recombination-positive transformants. The advantage is that the DNA can be delivered as a string, eliminating the need for cloning and potentially allowing rapid, high-throughput editing. The authors stated that 5% of the colonies were edited, thus providing a tool for homologous recombination that is comparably efficient to those used in *E. coli*. To provide sufficient amounts of ssDNA, the authors established a highly efficient electroporation protocol. Until then, transformation in *S. oneidensis* was mainly done via conjugation which takes several days, while electroporation can be carried out in 1 day [62]. In a second publication, the authors combined the ssDNA recombination system with CRISPR-Cas (CRISPR, clustered regularly interspaced short palindromic repeats; Cas, CRISPR-associated protein) counter-selection to simplify screening. A Cas9 protein is directed to the target sequence after ssDNA insertion, causing a DSB only in wild-type cells because the corresponding sequence is altered in the mutants. This DSB then leads to cell death since there is no template for the homology-directed repair (HDR). This additional step not only leads to a longer cloning time but also makes mutants with a strong growth-deficit accessible, since no revertant can be generated as it is the case for the suicide plasmid-based double crossover or the iEditing method [63]. CRISPR-Cas is an efficient tool for straightforward cloning and will be reviewed in the next section.

2.3 CRISPR Editing

While random mutagenesis approaches are excellent for high-throughput screening to identify the function of specific genes, they suffer from the necessity to discover insertion sites in subsequent steps. In contrast, classical targeted approaches are superior for studying single or multiple gene knockouts but are limited by the time-consuming cloning procedure. Both methods do not allow the modification of (conditionally) essential genes. Briefly, a CRISPR-associated protein (Cas9 or variants) is guided to a target region nearby a protospacer adjacent motif (PAM) by the single-guide RNA (sgRNA), consisting of the CRISPR RNA (crRNA) and trans-activating CRISPR RNA (tracrRNA), and introduces a DSB. This DSB can then be cured with HDR with a provided template containing homologous regions.

However, very few publications have shown that DSBs can be efficiently repaired in *S. oneidensis* [63–66]. Chen and colleagues found that Cas9 appears to be toxic to *S. oneidensis* even without a sgRNA [64]. Therefore, they tested Cas9 variants to reduce the toxicity of the gene editing system and showed that AsCpf1 and BhCas12b can be used in *S. oneidensis* with low toxicity and high DSB efficiency. They provide an all-in-one approach using a single plasmid for the entire editing process. As proof of principle, they have exchanged a promoter using relatively long homology arms of 1,000 bps. Another advantage is that these variants require different PAM sites (TTTV and ATTN vs. NGG for Cas9), thus expanding the editable sequences (Fig. 3a). In a recent study, Cas9 and sgRNA were regulated by an IPTG-inducible lac promoter to prevent cell toxicity. Then, similar to the iEditing approach, all modules, including 1,000 bp homology arms and an arabinose-inducible homing endonuclease I-SceI, were fused onto a plasmid. After successful editing, the plasmid could be cured upon arabinose induction of I-SceI, enabling rapid iterative multiplexed gene exchange [65].

Besides conventional CRISPR-Cas gene replacement, the ShCAST system (CRISPR-associated transposase from cyanobacteria *Scytonema hofmanni* [67])

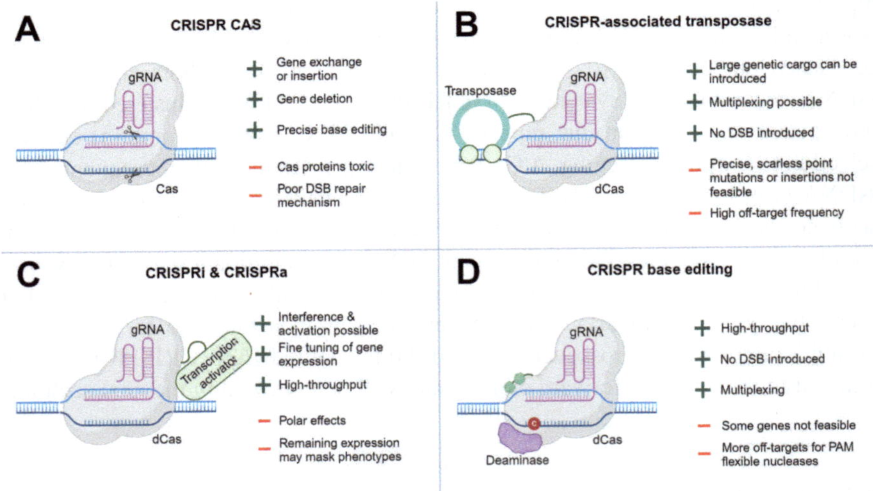

Fig. 3 Overview of CRISPR-related methods established in *S. oneidensis*

has been established in *S. oneidensis* (Fig. 3b). Here, a naturally nuclease-deficient Cas12k protein is fused together with a Tn7-like transposase facilitating transposon integration directed by the crRNA. Cheng and colleagues [68] integrated up to 30 kbp sequences in different genomic target regions. Moreover, they were able to integrate the DNA simultaneously in different loci. This could contribute to a significant improvement in the gain of function screens.

In addition to gene exchange, insertion or deletion, there are several other approaches that use a deactivated nuclease (dCas) that is still directed to the target sequence but does not cause a DSB. This could then be used to either interfere with transcription by simply blocking the sequence (CRISPRi) or to activate transcription (CRISPRa). Several strategies for CRISPR interference have been implemented, each with its own advantages (Fig. 3c). In one of the first studies using CRISPR in *Shewanella oneidensis*, Cao et al. [69] designed a simple system in which a dCAS9 protein was expressed from an IPTG-inducible Tac promoter, while the sgRNA was regulated by a constitutive pCI promoter on the same plasmid. A BsaI overlapping restriction site for Golden Gate cloning was introduced for easy and fast target exchange. With this plasmid, a 180-fold suppression of *gfp* encoded on the same plasmid was achieved in *Shewanella oneidensis*, showing that the CRISPRi approach works. They then showed that up to three genes could be silenced simultaneously. To demonstrate this, they silenced the *mtrCAB* complex, which forms the electron channel in the outer membrane, either simultaneously or one gene at a time. They were able to show nicely that multiple gene silencing did not change the level of silencing compared to single gene silencing. However, knocking down *mtrA* alone only reduced power density by 1.9-fold (fivefold higher than the Δ*mtrA* control), while knocking down the entire *mtrCAB* complex resulted in a more than

13-fold reduction. Then, they combined the CRISPRi system with the chaperone Hfq, which can block expression at the translational level guided by an sRNA and showed that this led to a further reduction in the current generation when *mtrA* is blocked at the transcriptional and translational level. Such CRISPRi silencing systems have been used in the following years, for example, to improve the production of the biotechnologically relevant 5-aminolevulinic acid [70] or to study the role of outer membrane vesicles and the influence of multiple genes on outer membrane composition using a modular screening approach [71].

Another group used a CAS variant from *Acidaminococcus* spp. BV3L6, ddAsCpf1, for CRISPRi as it is known for low off-target recognition and recognizes T-rich PAM sites, which could be particularly useful for manipulating the usually AT-rich promoter regions [72]. They designed a two-plasmid system with ddAsCpf1 encoded on the first plasmid and regulated under an IPTG-inducible lac promoter, while the crRNA was constitutively expressed from a second plasmid. In contrast to dCas9, ddAsCpf1 showed no significant growth defects when expressed without crRNA, indicating good host compatibility. In addition to demonstrating good gene silencing of up to 100% in a *gfp* and *lacZ* reporter strain, the authors were able to show in a proof of principle that the system can be used to screen for target genes that would, for example, improve EET by redirecting electron flow.

In addition to gene silencing, CRISPR can also be used to activate transcription. A transcriptional regulator is fused to the dCAS protein, which then recruits RNA polymerase to the target site. In two studies by Chen and colleagues, the best transcriptional regulators and the most suitable binding position of the sgRNA in *S. oneidensis* were carefully evaluated [73, 74]. First, the most promising candidate, *rpoD*, was used to upregulate genes responsible for cell division *(minCDE)* while downregulating SO3166, a gene involved in cell motility. Therefore, the transcriptional activator *rpoD* was fused to dCas9, allowing the sgRNA to be placed in front of a transcriptional start site, ultimately leading to the recruitment of RNA polymerase and increased transcription levels. Of note, the same protein delivered to a region within a coding DNA sequence results in steric hindrance of transcription and thus downregulation of gene expression levels allowing up- and downregulation with the Cas9-RpoD protein [73]. In a second study, the authors showed that using the type I-F CAS system from *Pseudomonas aeruginosa* is even better for gene upregulation. Multiple Cas proteins are involved, with Cas7 forming a stable complex with six subunits. Fusing the transcriptional regulator RpoD to Cas7 and thus providing multiple activators further improved gene activation. Finally, it has been shown that simultaneous gene activation and repression can be achieved with this system, although the effects were lower compared to activation or repression alone, possibly due to competition for the Cas system by the guide RNAs [74]. Gene activation offers a great potential to rapidly study complex reaction cascades such as electron transport chains.

A smart combination of CRISPRi, flux balance analysis, and homologous recombination was provided by Ford and colleagues [75]. They noticed that the deletion of *gpmA* was not feasible, thus indicating that it is an essential gene. However, they used flux balance analysis to identify possible supplements that could rescue the growth-deficient phenotype, repressed the gene expression to a certain extent via CRISPRi, and conducted screenings with the identified supplements added. The supplement facilitating less growth deficiencies was chosen to be added during the homologous recombination which enabled the gene deletion.

CRISPR interference and activation can be a useful tool to quickly and easily identify proteins that might enhance or reduce extracellular electron transfer. However, CRISPRi can only repress gene expression to a limited extent, which is also highly dependent on the available recognition sites. This means that many phenotypes can only be achieved by completely knocking out certain genes.

In addition to the conventional CRISPR-Cas gene replacement method, a second approach for gene knockdown, CRISPR base editing, has been established in *S. oneidensis*. For CRISPR base editing, a cytidine deaminase is fused to a nuclease-deficient Cas protein. The dCAS protein would be directed to a specific genetic target with a sgRNA, but would not cause a DSB, similar to the process of CRISPR interference or activation. The fused deaminase would then facilitate C to T conversion in a highly sequence-specific manner. This makes it possible to introduce premature stop codons when targeting either CAG, CAA, CGA, or TGG (Fig. 3d). However, this method is limited by the requirement of a specific PAM site within a narrow editing window of a suitable triplet that can be modified into a stop codon. In addition, fusion of the cytidine deaminase rAPOBEC1 to nCas9 suffered from poor editing efficiency when GC was the target motif [76]. Thus, several studies have focused on expanding the editing window to enable whole-genome editing [77–79]. Chen and colleagues carefully evaluated the influence of different Cas proteins (dCas9, nCas9(H840A), and nCas9(D10A)), the sgRNA length, and the architecture of the Cas-AID (activation-induced deaminase) fusion protein and were able to extend the editing window to a region between 22 and 9 bp upstream of the PAM site. However, this is only possible when all approaches are taken together and is not a one-size-fits-all method. The authors also introduced PAM-flexible Cas9 variants, which opened the possibility of introducing a stop codon in any single gene with a suitable codon. Of course, this has the disadvantage of potentially more off-targets, but it is still a considerable tool for genes that would otherwise not be editable [77]. Expression of the dCas9-AID fusion protein together with multiple gRNAs from the same plasmid allowed the simultaneous knockdown of up to eight genes [78]. This multiplexed gene knockdown approach clearly represents a powerful tool for rapid cloning. In another approach, a cytidine deaminase and an adenosine deaminase, which facilitates A to G conversion, have been fused to a PAM-flexible Cas variant. This system is designed for high-throughput screening, as precise base editing is not possible with two deaminases. Nevertheless, it has been nicely demonstrated how multiple changes in either the *ribA/B* promoter region or *cymA*, rapidly introduced by the dual deaminase CAS system and followed by a high-throughput screening, led to improved performance under the conditions tested [79].

2.4 Expression Systems

In the last decade, more and more expression systems have been developed to fine-tune strains for biotechnological applications.

It has been shown that *Shewanella oneidensis* strains can be used as biosensors using a promoter that responds to environmental triggers. Golitsch and colleagues [80] were able to measure a current response to elevated arabinose concentrations in such a strain. Based on the previous approach, Webster et al. [81] have developed a strain that can detect the toxic metalloid arsenic. Several studies have been conducted to expand the substrate spectrum or to improve the reduction capabilities of *S. oneidensis* using modular plasmid systems [82–85].

However, to enable even more complex systems, it was necessary to test the applicability and combination possibilities of different replication origins, promoters, and resistance cassettes. Hajimorad et al. [86] demonstrated superior conjugation efficiencies using the RK2 ori of transfer (*oriT*) compared to the mob-based pBBR1 *oriT* and that multiple plasmid expression is possible in *S. oneidensis*.

Ng et al. [88] improved the transformation efficiency by introducing the ori *repB* and showed that a LacI promoter can be used for constitutive gene expression in *S. oneidensis*.

Cao et al. [87] designed and characterized a complete plasmid toolkit with easily interchangeable modules using the BioBrick standard. They investigated the influence of different origins of replication (pBBR1, ColE, p15A, pSC101), inducible (pBAD, pLlacO1, placUV5, pTac, pTet, pXyl), and constitutive (pCI, pJ23119 and pTrc*) promoters as well as kanamycin and chloramphenicol resistance cassettes. The work greatly increases the number of plasmid systems that can be used and simplifies construction by using interchangeable modules.

In the following years, Yi et al. [90] integrated the T7 RNA polymerase into the genome and used the T7 promoter for protein overexpression, while Cheng and colleagues [91] introduced the rhamnose-inducible promoter to *S. oneidensis* to further broaden the possibilities. Fan et al. [89] were even able to establish antibiotic-free plasmid maintenance using auxotrophic strains. An overview of established genetic elements for expression systems used in *S. oneidensis* is given in Table 1.

2.5 Phages

Whole-genome sequencing of *S. oneidensis* revealed that it harbors at least three prophages, namely, MuSo1, MuSo2, and λSo [92]. Recently, a fourth cryptic prophage, CP4So, which is induced at low temperatures, has been elucidated [93]. Prophages are integrated into the host genome and remain there in a dormant state until they are induced under stressful conditions, such as oxidative stress or by an SOS response, which initiates the phage lytic cycle and ultimately lyses the host cells. Strains lacking MuSo2 and the λSo phages do not form infectious particles,

Table 1 Selection of genetic elements for expression plasmids used in *S. oneidensis*. Incompatibility groups are defined arbitrarily and arc only compared with each other. Plasmids with the same incompatibility group should not be transformed together. Categories are not related to each other. The list is not intended to be comprehensive but only represents a selection of frequently used elements

Promoter			Origin of replication			
Strength	Inducible	Constitutive	Name	Copy number	Incompatibility group	Counter-selection
High	pTac [87] pBAD [87] pXyl [87]	pLacI [88] pCI [87] pJ23119 [87]	pBBR1 [87]	23	D	Kanamycin Chloramphenicol Tetracycline Gentamycin Streptomycin Auxotrophy [89]
Medium	pTet [87]	pJ23100 [90] pTrc* [87]	ColE1 [87]	54	A	
			p15A [87, 90]	33	B	
Low	pLacO1 [87] pLacUV5 [87]	pJ23105 [90] pJ23109 [90] pCopA [90] pYeaR [90]	pSC101 [87]	40	C	
			pBR322 [90]	72	A	

suggesting that both MuSo1 and CP4So lack essential proteins to exit cells [94, 95]. However, CP4So is still a prominent biofilm regulator at low temperatures. During a temperature downshift, histone-like nucleoid structuring protein (H-NS) levels are reduced, leading to de-repression of the putative excisionase *alpA*. The corresponding enzyme excises CP4So from the genome in a small subpopulation, leading to a change in the sequence of the tmRNA and thus to enhanced biofilm formation. A toxin/antitoxin system ensures that cells reintegrate the prophage into the genome at elevated temperatures [96]. This shows how even a cryptic prophage can help the community cope with stress conditions.

In contrast, the temperate prophages MuSo2 and λSo encode all the proteins necessary for plaque formation and thus induce the lytic cycle [94, 95]. Gödeke and colleagues have shown that these two prophages play an important role in normal biofilm formation. Again, prophage gene expression is induced in a subpopulation of the community, resulting in the lytic cycle and cell lysis. Along with the phages, cellular material such as DNA and other cytoplasmic substances are released. The released extracellular DNA (eDNA) can then serve as a matrix for biofilm formation by the remaining cells. While the prophage and the host organism live in a synergistic manner, the prophages themselves are in a competitive relationship for the resources of the host organism [95]. Recently, it has been shown that λSo encodes lytic pathway proteins that repress MuSo2 prophage transcription [95]. Thus, each prophage contributes to biofilm formation in different growth phases. In the early developmental phase, MuSo2 is predominant, whereas λSo enters the lytic cycle several hours later, suggesting that MuSo2 responds even faster to the induction trigger or that its lytic cycle is induced by different factors. It has been shown that iron induces λSo in a highly RecA-dependent manner, whereas the induction factors during biofilm growth remain unclear for MuSo2 [95, 97]. However, the release of eDNA during the lytic cycle of λSo has been shown to be one of the major components necessary for three-dimensional structures in a more mature biofilm. A λSo deletion mutant, in contrast to a ΔMuSo2 mutant, is not affected in early biofilm formation but is no longer able to cover the whole surface [94]. With regard to biotechnological applications, which are often highly dependent on stable biofilm formation, further investigation of prophages is still important.

Recently some phages have been studied that may be of interest for biotechnological or biomedical applications as well as genetic tools. Interestingly, the Thanatos phages have been shown to promote biofilm formation when added to planktonic cells at an early growth stage while negatively affecting a mature biofilm [98]. The Dolos phage itself is hijacked by a phage satellite, pDOLOS, which is an autonomously replicating plasmid and could potentially be used as a genetic tool [99]. Finally, it was shown that the phage FishSpeaker recognizes its prey by the flagellum and, by OmcA, a key protein in the extracellular electron transport chain.

3 Applications

3.1 *Bioelectrochemical Systems*

3.1.1 Production of Electricity and Hydrogen

Bacteria in microbial fuel cells (MFCs) can oxidize organic and inorganic (waste) substrates at an anode surface as part of their respiratory chain and thereby convert chemical energy into electrical current. Electricity is produced by electron transfer to the anode, which is the result of biochemical reactions catalyzed by the microorganisms. Microbial electrolysis cells (MECs) on the other hand need an additional energy input, facilitating the generation of hydrogen at the cathode. MFC and MEC performance is influenced by many factors as cell architecture, electrode material, anode potential, electrolyte concentration, presence of mediators, bacteria used, and many more [100]. This makes the comparison of different setups rather difficult, since these parameters often vary from one experiment to another. Therefore, the following section will focus on a general overview, outlining the capabilities and drawbacks of *S. oneidensis* in bioelectrochemical systems for electricity and hydrogen production.

Watson and Logan [101] compared the performance of pure cultures of *S. oneidensis* with mixed cultures derived from wastewater in different MFC configurations. The mixed culture produced 68–480% more power than pure *S. oneidensis*, depending on reactor configuration. Maximum power densities were 858 ± 9 mW m^{-2} for the mixed culture and 148 ± 20 mW m^{-2} for *S. oneidensis*, respectively. Lower internal resistance was identified as the main source for the higher power produced by the mixed culture. Additionally, mixed cultures produced more power, even though their redox potential was less negative. In an approach that was focusing on the fabrication of an ultra-low cost, disposable all-paper MFC, anode and cathode were deposited on paper, while hydrophobic parchment paper was used as proton exchange membrane (PEM). Inoculation with *S. oneidensis* led to a maximum power generation of 8.33 mW m^{-2}, demonstrating that even very simple and affordable setups can be exploited for energy production. Possible application could be in the field of disposable diagnostic devices [102].

Bacterial hydrogen production from organic matter in MECs can be achieved with comparatively low energy consumption. A theoretical potential of 0.2 V can already sustain hydrogen formation in MECs, whereas for conventional water hydrolysis, at least 1.23 V is necessary. Additionally, the organic matter can be derived from waste streams, coupling waste treatment with simultaneous hydrogen production. Hu et al. [103] developed a membrane-free MEC which resulted in stable hydrogen production from lactate, while mixed cultures suffered from methane production.

Proper combination of species in a BES can lead to higher energy production, making the setup more resilient and broadening the substrate scope. For example, a coculture of *S. oneidensis* and *E. coli* significantly enhanced MFC energy generation from sarcocarp harvested from coconuts as substrate, compared to single cultures [104]. Coculturing with *Geobacter sulfurreducens* aids in the formation of stable biofilms with comparatively high current production [105]. To expand the carbon spectrum, *S. oneidensis* was cocultured with the glucose-fed fungus *Saccharomyces cerevisiae* in which the ethanol pathway was knocked out and a lactic acid biosynthesis pathway was introduced to facilitate a continuous lactate supply for the bacterium. This metabolic coordination led to a stable bioelectricity production in a glucose-fed MFC and shows the possibility of coculturing *S. oneidensis* with other synergistic organisms, giving access to a wide range of substrates [106].

3.1.2 Biosensors

Microbial biosensors are analytical devices that combine biological detection of a target analyte with physical transduction to produce a proportional, measurable signal. This signal may, for example, be of an electrochemical, optical, and mechanical nature. Biosensors exploit the biological capabilities of microorganisms to provide sensitive and selective measurements. They are well suited for applications in environmental monitoring due to the microbial ability to detect the bioavailability and toxicity of chemical substances [107]. Regarding *S. oneidensis* sensing is mostly achieved via electrochemical measurement in BES. Proof of principle for a highly specific biosensor based on a genetically engineered *S. oneidensis* strain was presented by Golitsch et al. [80]. Its ability to produce electrical current was set under control of a heterologous L-arabinose promotor which resulted in a linear correlation and allowed sensing of arabinose in a range from 0.1 to 1 mM with high accuracy ($R^2 = 0.92$) and a response time of 2 h.

This work in developing a BES-based biosensor using an engineered strain of *S. oneidensis* laid the groundwork for other similar biosensors. Webster et al. [81] developed an arsenic sensing strain driven by an arsenic-inducible promotor. Upon inoculation into a BES, current production increases in response to arsenic. Arsenite was found to linearly increase mean peak current over a range from 0–100 μM ($R^2 = 0.97$) with a lower detection limit of 40 μM. Although this value is still well above the maximum chronic exposure limit set by the WHO (~133 nM), this approach indicates that *S. oneidensis* is suitable for remote monitoring of environmental contamination.

Whole-cell biosensing with *S. oneidensis* is not limited to the abovementioned anodic processes with genetically engineered strains. For instance, it was possible to detect and quantify fumarate with wild-type *S. oneidensis* MR-1 through inward electron flow at the cathode of a BES. The range of fumarate sensing was 2 μM–10 mM ($R^2 = 0.9997$) with a response time of several minutes and stable measurement for at least 30 successive runs. To test the applicability for possible uses in food industry and disease diagnosis, additional measurements with fumarate-spiked apple juice or mice kidney were conducted, which also resulted in reliable fumarate quantification [108].

Besides whole-cell applications, it is also possible to facilitate proteins expressed by *S. oneidensis* to create biosensors. For instance, MtrC was harnessed to investigate its peroxidase activity. Therefore, it was immobilized onto an ITO electrode, and protein film voltammetry (PFV) scans were conducted and compared to horseradish peroxidase (HRP), the state-of-the-art biocatalyst for hydrogen peroxide reduction. Immobilized MtrC showed a comparable activity as HRP, which demonstrates the untapped potential of multiheme cytochromes, found in exoelectrogenic bacteria for biosensors [109].

Furthermore, there are also advances in materials science, mainly focused on the cell-electrode interface [110–112] and artificial neural networks [113] that improve and broaden the scope of possible biosensor application, paving the way for future applications of *S. oneidensis* as model organism for bioelectrochemical sensing.

3.1.3 Electrofermentation

Due to the capability of electroactive bacteria (EAB) like *S. oneidensis* to exchange electrons with electrodes, it is also possible to steer so-called electrofermentations (EF) or unbalanced fermentations. In conventional fermentation processes, the total redox equivalent needs to be balanced between substrates and products, whereas in EF oxidized or reduced products can be obtained utilizing the electrodes as additional inexhaustible electron donor or acceptor, which broadens the product scope and might increase product yields as well as specificity of metabolic routes [114, 115].

Extracellular electron transfer (EET) is crucial for EF. Regarding *S. oneidensis*, the Mtr pathway is most important and well-characterized. To enable unbalanced fermentation, Flynn et al. [116] employed genetic engineering to facilitate the electrode-dependent conversion of glycerol to ethanol and CO_2, a reaction where the products are more oxidized than the substrate. Ethanol production was in a comparable range with engineered *E. coli* production strains and could be further improved by increasing the electrode surface area. This work proved that *S. oneidensis* with its native EET mechanism is an apt organism for production of chemicals via EF.

For production applications it is also necessary to further improve genetic robustness. Bursac et al. [117] developed a λ-prophage deletion mutant of *S. oneidensis* which shows 2.3-fold higher cell density on an anode surface compared to wild type. This strain was further modified to enable production of acetoin, an industrially important platform chemical [118]. The yield of 78% is in the range of other production strains, like *Bacillus subtilis*, but energy-intensive aeration would be obsolete with EF.

In a more recent endeavor, Wohlers et al. [119] metabolically engineered *S. oneidensis* to optimize production of glutamate 72-fold and to enable production of itaconic acid via redirection of carbon flux from acetate toward the tricarboxylic acid (TCA) cycle. This study is highlighting how redirection of carbon flux can broaden the product spectrum of *S. oneidensis*; nonetheless its application in EF is yet to be shown.

Delgado et al. [56] improved EET in *S. oneidensis* via genetic engineering of the periplasmic protein content. Overproduction of STC and higher flavin concentrations might be beneficial for the realization of EF.

In order to expand the rather limited substrate scope of *S. oneidensis*, it is possible to coculture it with other organisms. Uno et al. [120] report coculturing with *Streptococcus bovis* 148, which metabolizes starch to lactic acid, making it available for *S. oneidensis*.

Overall, EF promises an elegant solution to overcome boundaries of conventional fermentation, and *S. oneidensis* is a well-suited model organism for this approach, which is in theory applicable to all EAB. Although EF is still in its developing phase, further research must be done to enable industrial applications.

3.1.4 Microbial Electrosynthesis

Under native conditions the Mtr respiratory pathway of *S. oneidensis* catalyzes electron flow from the cytoplasm to extracellular acceptors such as anodes. Reversing this process, i.e., pumping electrons into microorganisms, by poising the working electrode with a negative potential would enable new routes for microbial electrosynthesis, such as the reduction of CO_2 to produce valuable products [121]. Reversing the Mtr pathway in *S. oneidensis* was examined by Ross et al. [122] to determine feasibility of the process and experimental parameters. In a BES a biofilm of *S. oneidensis* poised with a potential of -0.36 V (vs. SHE) showed an immediate electron uptake upon fumarate addition. Therefore, the Mtr pathway is fully reversible and can drive reductive reactions. In addition to the MtrCAB complex, CymA and FccA are important factors in this cascade. This finding also contrasts with data for *Geobacter sulfurreducens*, which uses two separate pathways for in- and outward electron flow [123].

Conventional biocatalytical CO_2 reduction usually requires highly purified and relatively expensive hydrogen gas, so alternative approaches to harness the greenhouse gas are needed. Microbial electrosynthesis is well suited to circumvent these obstacles because of its ability to replace gaseous hydrogen [124]. Besides aiding in the production of isobutanol [125], *S. oneidensis* was found to be able to catalyze the synthesis of formate from CO_2 using electrons supplied by a cathode as the sole energy source [126, 127]. Reduction kinetics seem to be strongly influenced by proton transfer [127].

For effective electrosynthesis, it is preferable that electrons are transferred to an intracellular electron carrier with a lower potential. Since electron transfer from reduced quinones to NAD^+ is thermodynamically unfavorable, it is necessary to generate proton motif force to drive NADH dehydrogenases in reverse. Therefore, Tefft et al. [128] integrated proteorhodopsin, a heterologous light-driven proton-pump into a *S. oneidensis* strain to facilitate NADH-dependent cathodic reduction of acetoin to generate 2,3-butanediol.

Another approach of high relevance is presented by Li et al. [129]. Instead of oxidized carbon species, nitrate is utilized as a substrate for the electrosynthesis of ammonia. This method could potentially help address two challenges at once. On the one hand, it could assist in wastewater treatment, as rising nitrate pollution poses a threat to public health. On the other hand, it could serve as an alternative to the traditional synthesis of ammonia via the Haber-Bosch process, which demands immense energy and releases vast amounts of CO_2.

Several studies have also shown that riboflavin accelerates the electron uptake rate [122, 127, 130], though flavin photochemistry can vice versa lead to generation of radicals and therefore to decreased electron transfer rates on the other hand [128, 131].

Further endeavors in the field of microbial electrosynthesis include the improvement of electron uptake rates. This can be achieved, e.g., via addition of electron carriers like soluble iron [132], cell membrane intercalation of a redox-active conjugated oligoelectrolyte [133], or creation of conductive 3D matrices that enable higher cell densities per electrode area and enhanced bio-abiotic interfacing [111, 134, 135].

S. oneidensis already serves as a model organism for EET in anodic setups and is proving to be a promising candidate for investigating the mechanism of CO_2 reduction at the cathode and electrosynthesis in general due to its well-described electron transport chain and carbon metabolism. Microbial electrosynthesis represents a means of energy storage, which is becoming increasingly important as intermittent regenerative energy sources become more prevalent.

3.2 Biodegradation, Bioremediation, and Wastewater Treatment

3.2.1 Metal Reduction

Shewanella's capability of performing EET to a broad range of substrates enables it to reduce several soluble and insoluble metal species, including Mn(IV), Fe(III), Cr (VI), U(VI), Au(III), Pd(II), Co(III), Ag(I), Pt(II), and Cu(II), which makes it well-suited for bioremediation applications due to the influence of the redox state of metals on their solubility and toxicity [136–143]. Complexation of heavy metals by ligands makes real-world applications more difficult, since they enhance their stability, change reduction mechanisms, and might be cytotoxic themselves [140, 144]. Among these, especially hexavalent chromium and uranium are of particular interest due to the threat they pose on human health and the environment. Cr(VI) is hemotoxic, genotoxic, and carcinogenic. It is mainly released into the environment through chromate production for dyes, pigments, coatings, or paints. Trivalent chromium is comparatively insoluble, has low toxicity, and is an essential nutrient [145]. Therefore, the reduction of Cr(VI) to Cr(III) by *S. oneidensis* may be a valuable strategy for mitigating its toxicity. Miller et al. [144] found that

S. oneidensis is not capable of growing under anoxic conditions with Cr(VI) as the sole terminal electron acceptor. This might have to do with Cr(VI) toxicity and insufficient amounts of the electron acceptor. Aerobic growth is observed with media containing 0.01 mM and 0.1 mM Cr(VI), but not with 1 mM. Under conditions not linked with growth, the organism was capable of reducing Cr(VI) at concentrations of ≤1 mM under anoxic conditions.

Hexavalent uranium is highly toxic and occurs in nature mainly in the form UO_2^{2+}, which has good solubility in water [146]. *S. oneidensis*-mediated reduction yields low-soluble and therefore less harmful U(IV) species which makes it a candidate for bioremediation of U(VI) contaminated soils [147, 148].

As in other fields of application, electron transfer rate is crucial for efficient removal of toxic heavy metal species. The addition of advanced materials can promote metal reduction. These materials facilitate degradation of heavy metals through *S. oneidensis* metabolism by various mechanisms that include electron shuttles for MET and conductivity promotion of DET, serving as catalysts, adsorbents, and reducing agents [149–152]. Some materials serve multiple functions at once, e.g., metal-organic frameworks (MOFs) are good absorbents, while also acting as growth substrate and facilitating Cr(VI) reduction [152]. Additionally, MOFs were harnessed for improved Au(III) precipitation with *S. oneidensis*. Since the global demand for gold is still rising, this could become a useful strategy for gold extraction from waste effluents or environmental water sources and supplement the existing mining infrastructure [153].

Synthetic biology is also capable of improving metal reduction. Zhang et al. [154] developed synergistic bacterial interaction by asymmetric modification and covalent linking of *S. oneidensis* with *Lactobacillus rhamnosus* which can metabolize glucose to lactate. This metabolic synergy leads to increased substrate availability, exhibiting superior metal reduction rates compared to conventional cocultures of these microorganisms without the necessity for genetic engineering and enabled reduction of heavy metals like Pb(II) in the kidney and liver tissue of living mice.

3.2.2 Nanoparticle Formation

The ability to reduce various metals also enables *S. oneidensis* to biosynthesize metal nanoparticles (NPs) including precious metals like Pd, Ag, Cu, Au, and Pt [141–143, 155, 156]. These NPs pose a high surface-area-to-volume ratio and are demanded for their outstanding catalytic activities. The very limited availability of precious metals and increasing demand make them very valuable goods. While chemical synthesis is currently the standard route for metal NP synthesis, it has several disadvantages, like the requirement of toxic and expensive reagents for reduction and stabilization [157]. Biological synthesis is less harmful to the environment because it does not produce a secondary toxic waste stream [158]. In addition, the microbial surface acts as a stabilizer. This eliminates the need for expensive chemicals, making the process energy efficient and cost-effective [159]. Bio-NPs are generally synthesized on cell walls, on cell membranes, or within the cytoplasm, and their synthesis mostly depends on OMCs and hydrogenases [3, 158, 160].

Biosynthesis of Pd-NPs by *S. oneidensis* is comparatively well described and can occur inside and outside of the cell [155]. Outside the cell, biosorption of Pd (II) occurs by ion exchange, electrostatic interaction, complexation, or physical adsorption to the cell wall and is thereafter converted to zerovalent Pd by chemical and autocatalytic conversion, as well as through bioreduction by OMCs as OmcA and MtrC. Inside the cell, NADH dehydrogenase, [FeFe]-hydrogenase, and [NiFe]-hydrogenase seem to be essential for bio-Pd NP synthesis [157, 161]. Bio-Pd NPs can be exploited for bioremediation as reusable and highly active catalysts. Additionally, they may serve in improved MES as electrode coating to increase in- and outward electron flux [155, 162, 163].

Intracellular bioreduction of Cu(II) yields NPs with sizes around 20–40 nm that are capable of catalyzing click-reactions [143]. De Corte et al. [138] demonstrated the formation of gold NPs from Au(III), which was interestingly also possible with heat-killed cells, indicating a non-enzymatic reduction mechanism.

The potential to use bioreduction to selectively target and remove (precious) metal ions from wastewater presents an opportunity to reclaim and add value to various industrial effluents, including those generated by mining operations. This additional financial incentive underscores the need to further evaluate and explore this emerging technology.

Besides synthesis of metal NPs, *S. oneidensis* is also able to produce a range of other nanomaterials. Lehner et al. [164] showed that the microorganism can produce graphene through reduction of graphene oxide (GO) flakes. This resulted in promising properties, e.g., more than two orders of magnitude increased conductance. Compared to chemically reduced GO, it exhibited a higher surface-area-to-thickness ratio, accompanied by improved storability. This makes it a potential candidate for conductive inks, biosensors, and improved BESs. Upscaling of this process would be more sustainable and eco-friendlier than conventional chemical synthesis due to the mild reaction conditions [134, 164].

The bioreduction of toxic and bioavailable Se(IV) yielding in nanostructured selenium species is also within the scope of *S. oneidensis*. These structures pose unique properties with possible applications in the fields of catalysis, semiconducting, and nanophotonics. This provides a potential for combining bioremediation with sustainable production of nanomaterials [165].

Synthesis of bio-FeS NPs by *S. oneidensis* is another interesting approach. This material shows an excellent performance in environmental remediation and energy recovery applications and indicates a promising potential for use in cancer treatment. Fu et al. [166] created a bio-hybrid consisting of *S. oneidensis* and self-assembled bio-FeS, which not only covered the cell surface but also constituted an artificial transmembrane electron channel, resulting in improved Cr(VI) removal rates. Beyond that, Dang et al. [167] were able to regulate the synthesis rate and yield by altering the initial iron precursor concentration and were able to highly efficiently kill cancer cells at extremely low concentrations of less than 12 $\mu g\ ml^{-1}$ in vivo.

These examples showcase the high versatility and value of *S. oneidensis* in the biogenic production of a vast number of diverse nanomaterials.

3.2.3 Inorganic Non-metallic Pollutants

S. oneidensis is capable of respiring non-metallic inorganic ions as terminal electron acceptors. Among them are common environmental pollutants like nitrite, nitrate, sulfite, thiosulfate, or iodate [140]. Nitrate is extensively used in fertilizers, while only about half of the nitrogen ends up in crops. The other half ends up in ground and surface waters and soils, making nitrate pollution of freshwaters an increasing global problem [168]. In the gastrointestinal tract, nitrate can be reduced to nitrite, which in turn can react with dietary amines and amides to form highly carcinogenic N-nitroso compounds [169]. Through its periplasmic nitrate reduction system (NAP) and nitrite reduction system (NRF), *S. oneidensis* achieves ammonification and can aid in detoxification of freshwaters and soil. Coculturing with denitrification microorganisms allows interspecies electron transfer facilitated by *S. oneidensis*, leading to production of molecular nitrogen. Additionally, *S. oneidensis* can stimulate biofilm formation of other microorganisms, improving electron transfer and nitrate removal [140]. Microbial electrosynthetic nitrate reduction to ammonia might also play a role in future applications [129]. Sulfur-containing ions like sulfites or thiosulfates are also suitable as potential terminal electron acceptors [170], which in the case of thiosulfate is not only possible under anoxic but also under oxic conditions. It is noteworthy that *S. oneidensis* is the only microorganism known so far to possess both thiosulfate reductase and thiosulfate dehydrogenase [171].

Additionally, it is able to reduce iodate to iodide extracellularly with a so far unknown reductase which relies on the Mtr pathway [172]. Oxidized antimony species, mainly Sb(III) and Sb(V), are components of smelting slag and possess high toxicity and carcinogenicity. *S. oneidensis* can immobilize these ions and mitigate their toxicity. This provides the possibility to treat antimony-polluted sites with *S. oneidensis* [173].

3.2.4 Organic Pollutants

Organic Pollutants as Electron Acceptors

Synthetic azo dyes are designed to have high chemical stability in order to prolong their lifetime. This property entails an accumulation in nature since the compounds are poorly degradable via photolysis or by microorganisms under oxic conditions. Since some azo dyes are toxic, mutagenic, or carcinogenic, this poses a threat to human and environmental health. Under anoxic conditions though, biodegradation is possible, and *S. oneidensis* is among the organisms that are capable of azo dye discoloration [140, 174]. Usually, azo dyes consist of an azo group, which is linking two (potentially substituted) aryl groups. By azo bond reduction, dye discoloration can be achieved. Reduction takes place extracellularly via the MtrCAB complex. Azo dyes with higher hydrophobicity are membrane permeable and can also be reduced by CymA in the periplasm [175].

Aromatic nitro compounds as the highly toxic nitrobenzene or the explosive 2,4,6-trinitrotoluene (TNT) are widely used in industry and contaminate soils and groundwater [176]. *S. oneidensis* can reduce these compounds via a similar mechanism as for azo dyes. Additionally, small and low polar aromatic nitro compounds as the two aforementioned can enter the cytoplasm, where they are reduced by the intracellular NAD(P)H nitroreductase NfnB [177]. Besides, *S. oneidensis* enhances dechlorination of pentachlorophenol (PCP) and reduction of the organoarsenic pollutant roxarsone [178, 179].

Organic Pollutants as Electron Donors

While reduction of pollutants that act as electron acceptors for *S. oneidensis* relies on the addition of electron donors, e.g., lactate, the oxidation of organic pollutants does not necessitate further electron acceptors, if an anode is present.

S. oneidensis can oxidize the polycyclic aromatic hydrocarbon phenanthrene in a MFC setup. Due to π-π interaction between the compound and graphite, adsorption of the pollutant on the anode is facilitated, improving availability for the microorganism as substrate and therefore degradation rates [180]. Phenanthrene degradation was also shown by Yeferni et al. [181], which enhanced bioremediation of contaminated sediments in a coculture with meiobenthic nematodes. The mechanism of this process is yet still unclear and needs further clarification.

It has also been shown that addition of *S. oneidensis* to a mixed species MFC improves the degradation of the aromatic hydrocarbon o-xylene by 35–76% [182]. These results indicate that *S. oneidensis* is capable of using organic pollutants as electron donors, even though its substrate scope is rather small. Future work in this field is needed in order to either elucidate and broaden the available carbon sources or explore the ability of improving turnover rates in cocultures with synergistic organisms.

Fenton Reaction

The Fenton reaction, which is an advanced oxidation process, is an effective method to degrade many types of organic pollutants. However, a continuous supply of Fe (II) and H_2O_2 is required for the reaction, leading to high costs [183]. *S. oneidensis* is capable of respiring Fe(III) under anoxic and oxygen under oxic conditions, yielding in Fe(II) and H_2O_2, respectively. It can drive the Fenton reaction under alternating respiration conditions, regenerating Fe(II) and H_2O_2 without the need of a continuous supply, therefore lowering process costs. This approach has been tested to degrade several organic pollutants as 1,4-dioxane, tetrachloroethene (PCE), trichloroethene (TCE), anthracene, and pyrene [184–187]. Additionally,

S. oneidensis-driven Fenton reaction can be utilized to degrade cellulose and hemicellulose into short-chain oligosaccharides and monomeric sugars [188].

Other Applications

Besides the abovementioned applications, *S. oneidensis* can be utilized for further uses. For instance, it can biosynthesize a head-to-head long-chain olefinic hydrocarbon, which supposedly aids the cells in sudden temperature drops. As fossil fuels are exhaustible resources and become more expensive, the industrial need for renewable replacement is growing, to sustain production of specialty chemicals and to produce high-energy biofuels. Harnessing *S. oneidensis* for effective high-yield synthesis of such hydrocarbons might be part of the transition to a more sustainable petrochemical industry [189]. *S. oneidensis* can be used to produce caustic soda (NaOH), a widespread strong base, in a MEC setup. Its conventional synthesis relies on the chlor-alkali electrolysis process, which has a high energy demand and forms chlorine gas as a byproduct. In a MEC, with a Nafion membrane that separates anodic and cathodic compartment, while being permeable for Na^+ ions, biosynthesis of NaOH is possible on the cathode side. Since electrons are withdrawn from the anodic chamber, flowing toward the cathode, cations are permeating the membrane in the same direction, which allows NaOH formation [190].

As another example, *S. oneidensis* is used as a model organism to investigate a potential environmental impact of hazardous substances like the corrosion inhibitor benzothiazole or nanoscale lithium nickel manganese cobalt oxide, which is recently used as a cathode material for lithium-ion electrodes, by monitoring population growth and respiration [191, 192].

4 Future Directions and Outlook

Microbial whole-cell biocatalysis will become more and more driven toward continuous long-term processes in the future. These processes benefit from evolution, if they are accompanied with a potential benefit for the organism, but will have to cope with genetic drift, if the organism is negatively impacted by the process. Hence, genetic stabilization has to be a focus for future research to enable a predictable production. Some studies in this direction were already conducted for other model organisms, and it will be interesting to see whether they can be adapted to *S. oneidensis* as well.

S. oneidensis could be well suited for continuous processes as it is a well-researched biofilm forming organism. In these productive biofilm-driven processes, the biocatalyst will remain attached to the biofilm substratum allowing for controlled plug flow reactor operation [193]. The organism will benefit from its ability to respire with anodes in anoxic anode-driven processes, but this ability would have to be intensified in order to allow for competitive space-time yields in anode-assisted fermentations [194]. Although many studies reported on potential ways to highly increase electron transfer rates with *S. oneidensis*, it will be relevant to study how sustainable these technologies are for long-term usage [6]. Additionally, cocultivation with synergistic organisms ensures a promising path to more resilient and versatile bioprocesses [104–106, 120, 181, 182].

Several studies also pointed toward the possibility to develop cathode-driven processes using *S. oneidensis* [195, 196]. This could be an interesting field for future research. Although most cathodic bioelectrosynthesis processes are currently conducted using autotrophic organisms, these organisms usually have the disadvantage that electron transfer from the electrode to the organisms has to be conducted via the production of hydrogen on the cathode. Using organisms with developed electron transfer chains for the interaction with solid electron donors and acceptors could allow for more efficient electron import and less costs for electrical energy. Using this technology would allow to develop processes in which the desired end product is more reduced than the substrate or to decouple energy production from central metabolism [128].

References

1. Myers CR, Nealson KH (1988) Bacterial manganese reduction and growth with manganese oxide as the sole electron acceptor. Science 240:1319–1321. https://doi.org/10.1126/SCIENCE.240.4857.1319
2. Venkateswaran K, Moser DP, Dollhopf ME, Lies DP, Saffarini DA, MacGregor BJ, Ringelberg DB, White DC, Nishijima M, Sano H, Burghardt J, Stackebrandt E, Nealson KH (1999) Polyphasic taxonomy of the genus *Shewanella* and description of *Shewanella oneidensis* sp. nov. Int J Syst Bacteriol 49(Pt 2):705–724. https://doi.org/10.1099/00207713-49-2-705
3. Fredrickson JK, Romine MF, Beliaev AS, Auchtung JM, Driscoll ME, Gardner TS, Nealson KH, Osterman AL, Pinchuk G, Reed JL, Rodionov DA, Rodrigues JLM, Saffarini DA, Serres MH, Spormann AM, Zhulin IB, Tiedje JM (2008) Towards environmental systems biology of *Shewanella*. Nat Rev Microbiol 6:592–603. https://doi.org/10.1038/nrmicro1947
4. Fonseca BM, Paquete CM, Neto SE, Pacheco I, Soares CM, Louro RO (2013) Mind the gap: cytochrome interactions reveal electron pathways across the periplasm of *Shewanella oneidensis* MR-1. Biochem J 449:101–108. https://doi.org/10.1042/BJ20121467
5. Meyer TE, Tsapin AI, Vandenberghe I, De Smet L, Frishman D, Nealson KH, Cusanovich MA, Van Beeumen JJ (2004) Identification of 42 possible cytochrome C genes in the *Shewanella oneidensis* genome and characterization of six soluble cytochromes. OMICS 8:57–77. https://doi.org/10.1089/153623104773547499
6. Klein EM, Knoll MT, Gescher J (2023) Microbe–anode interactions: comparing the impact of genetic and material engineering approaches to improve the performance of microbial electrochemical systems (MES). J Microbial Biotechnol 16:1179–1202. https://doi.org/10.1111/1751-7915.14236
7. Marritt SJ, McMillan DGG, Shi L, Fredrickson JK, Zachara JM, Richardson DJ, Jeuken LJC, Butt JN (2012) The roles of CymA in support of the respiratory flexibility of *Shewanella oneidensis* MR-1. Biochem Soc Trans 40:1217–1221. https://doi.org/10.1042/BST20120150

8. Sturm G, Richter K, Doetsch A, Heide H, Louro RO, Gescher J (2015) A dynamic periplasmic electron transfer network enables respiratory flexibility beyond a thermodynamic regulatory regime. ISME J 9:1802–1811. https://doi.org/10.1038/ismej.2014.264
9. Coursolle D, Gralnick JA (2010) Modularity of the Mtr respiratory pathway of *Shewanella oneidensis* strain MR-1. Mol Microbiol 77:995–1008. https://doi.org/10.1111/j.1365-2958.2010.07266.x
10. Bücking C, Popp F, Kerzenmacher S, Gescher J (2010) Involvement and specificity *of Shewanella oneidensis* outer membrane cytochromes in the reduction of soluble and solid-phase terminal electron acceptors. FEMS Microbiol Lett 306:144–151. https://doi.org/10.1111/j.1574-6968.2010.01949.x
11. Thormann KM, Saville RM, Shukla S, Pelletier DA, Spormann AM (2004) Initial phases of biofilm formation in *Shewanella oneidensis* MR-1. J Bacteriol 186:8096–8104. https://doi.org/10.1128/JB.186.23.8096-8104.2004
12. Myers CR, Myers JM (1993) Role of menaquinone in the reduction of fumarate, nitrate, iron (III) and manganese(IV) by *Shewanella putrefaciens* MR-1. FEMS Microbiol Lett 114:378–1097. https://doi.org/10.1111/j.1574-6968.1993.tb06576.x
13. Myers CR, Myers JM (1997) Cloning and sequence of *cymA*, a gene encoding a tetraheme cytochrome *c* required for reduction of iron(III), fumarate, and nitrate by *Shewanella putrefaciens* MR-1. J Bacteriol 179:1143–1152. https://doi.org/10.1128/JB.179.4.1143-1152.1997
14. Myers CR, Myers JM (1997) Isolation and characterization of a transposon mutant of *Shewanella putrefaciens* MR-1 deficient in fumarate reductase. Lett Appl Microbiol 25:162–168. https://doi.org/10.1046/j.1472-765X.1997.00196.x
15. Maier TM, Myers JM, Myers CR (2003) Identification of the gene encoding the sole physiological fumarate reductase in *Shewanella oneidensis* MR-1. J Basic Microbiol 43:312–327. https://doi.org/10.1002/jobm.200390034
16. Beliaev AS, Saffarini DA (1998) Shewanella putrefaciens mtrB encodes an outer membrane protein required for Fe(III) and Mn(IV) reduction. J Bacteriol 180:6292. https://doi.org/10.1128/JB.180.23.6292-6297.1998
17. Beliaev AS, Saffarini DA, McLaughlin JL, Hunnicutt D (2001) MtrC, an outer membrane decahaem *c* cytochrome required for metal reduction in *Shewanella putrefaciens* MR-1. Mol Microbiol 39:722–730. https://doi.org/10.1046/J.1365-2958.2001.02257.X
18. Newman DK, Kolter R (2000) A role for excreted quinones in extracellular electron transfer. Nature 405:94–97. https://doi.org/10.1038/35011098
19. Kotloski NJ, Gralnick JA (2013) Flavin electron shuttles dominate extracellular electron transfer by *Shewanella oneidensis*. MBio 4. https://doi.org/10.1128/mbio.00553-12
20. Theunissen S, De Smet L, Dansercoer A, Motte B, Coenye T, Van Beeumen JJ, Devreese B, Savvides SN, Vergauwen B (2010) The 285 kDa Bap/RTX hybrid cell surface protein (SO4317) of *Shewanella oneidensis* MR-1 is a key mediator of biofilm formation. Res Microbiol 161:144–152. https://doi.org/10.1016/J.RESMIC.2009.12.002
21. Covington ED, Gelbmann CB, Kotloski NJ, Gralnick JA (2010) An essential role for UshA in processing of extracellular flavin electron shuttles by *Shewanella oneidensis*. Mol Microbiol 78:519–532. https://doi.org/10.1111/J.1365-2958.2010.07353.X
22. Mueller J, Shukla S, Jost KA, Spormann AM (2013) The mxd operon in *Shewanella oneidensis* MR-1 is induced in response to starvation and regulated by ArcS/ArcA and BarA/UvrY. BMC Microbiol 13:119. https://doi.org/10.1186/1471-2180-13-119
23. Fu H, Chen H, Wang J, Zhou G, Zhang H, Zhang L, Gao H (2013) Crp-dependent cytochrome *bd* oxidase confers nitrite resistance to *Shewanella oneidensis*. Environ Microbiol 15:2198–2212. https://doi.org/10.1111/1462-2920.12091
24. Fu H, Jin M, Ju L, Mao Y, Gao H (2014) Evidence for function overlapping of CymA and the cytochrome bc_1 complex in the *Shewanella oneidensis* nitrate and nitrite respiration. Environ Microbiol 16:3181–3195. https://doi.org/10.1111/1462-2920.12457

25. Conte E, Mende L, Grainge I, Colloms SD (2019) A mini-ISY100 transposon delivery system effective in γ proteobacteria. Front Microbiol 10(280). https://doi.org/10.3389/FMICB.2019.00280
26. Rakowski SA, Filutowicz M (2013) Plasmid R6K replication control. Plasmid 69:231–242. https://doi.org/10.1016/j.plasmid.2013.02.003
27. Bouhenni R, Gehrke A, Saffarini D (2005) Identification of genes involved in cytochrome *c* biogenesis in *Shewanella oneidensis*, using a modified mariner transposon. Appl Environ Microbiol 71:4935–4937. https://doi.org/10.1128/AEM.71.8.4935-4937.2005
28. Oh J, Fung E, Price MN, Dehal PS, Davis RW, Giaever G, Nislow C, Arkin AP, Deutschbauer A (2010) A universal TagModule collection for parallel genetic analysis of microorganisms. Nucleic Acids Res 38:e146. https://doi.org/10.1093/nar/gkq419
29. Deutschbauer A, Price MN, Wetmore KM, Shao WJ, Baumohl JK, Xu ZC, Nguyen M, Tamse R, Davis RW, Arkin AP (2011) Evidence-based annotation of gene function in *Shewanella oneidensis* MR-1 using genome-wide fitness profiling across 121 conditions. PLoS Genet 7:e1002385. https://doi.org/10.1371/journal.pgen.1002385
30. Groh JL, Luo QW, Ballard JD, Krumholz LR (2005) A method adapting microarray technology for signature-tagged mutagenesis of *Desulfovibrio desulfuricans* G20 and *Shewanella oneidensis* MR-1 in anaerobic sediment survival experiments. Appl Environ Microbiol 71: 7064–7074. https://doi.org/10.1128/AEM.71.11.7064-7074.2005
31. Brutinel ED, Gralnick JA (2012) Anomalies of the anaerobic tricarboxylic acid cycle in *Shewanella oneidensis* revealed by Tn-seq. Mol Microbiol 86:273–283. https://doi.org/10.1111/J.1365-2958.2012.08196.X
32. Wetmore KM, Price MN, Waters RJ, Lamson JS, He J, Hoover CA, Blow MJ, Bristow J, Butland G, Arkin AP, Deutschbauer A (2015) Rapid quantification of mutant fitness in diverse bacteria by sequencing randomly bar-coded transposons. MBio 6. https://doi.org/10.1128/MBio00306-15
33. Yang H, Krumholz EW, Brutinel ED, Palani NP, Sadowsky MJ, Odlyzko AM, Gralnick JA, Libourel IGL (2014) Genome-scale metabolic network validation of *Shewanella oneidensis* using transposon insertion frequency analysis. PLoS Comput Biol 10:e1003848. https://doi.org/10.1371/JOURNAL.PCBI.1003848
34. Phan J, Macwan S, Gralnick JA, Yee N (2024) Extracellular organic disulfide reduction by *Shewanella oneidensis* MR-1. Microbiol Spectr 12:e04081–e04023. https://doi.org/10.1128/spectrum.04081-23
35. Bennett BD, Redford KE, Gralnick JA (2018) MgtE homolog FicI acts as a secondary ferrous iron importer in *Shewanella oneidensis* strain MR-1. Appl Environ Microbiol 84. https://doi.org/10.1128/AEM.01245-17
36. van Opijnen T, Bodi KL, Camilli A (2009) Tn-seq: high-throughput parallel sequencing for fitness and genetic interaction studies in microorganisms. Nat Methods 6:767–772. https://doi.org/10.1038/nmeth.1377
37. Baym M, Shaket L, Anzai IA, Adesina O, Barstow B (2016) Rapid construction of a whole-genome transposon insertion collection for *Shewanella oneidensis* by Knockout Sudoku. Nat Commun 7:13270. https://doi.org/10.1038/ncomms13270
38. Beliaev AS, Thompson DK, Fields MW, Wu L, Lies DP, Nealson KH, Zhou J (2002) Microarray transcription profiling of a *Shewanella oneidensis etrA* mutant. J Bacteriol 184: 4612–4616. https://doi.org/10.1128/JB.184.16.4612-4616.2002
39. Schwalb C, Chapman SK, Reid GA (2003) The Tetraheme cytochrome CymA is required for anaerobic respiration with dimethyl sulfoxide and nitrite in *Shewanella oneidensis*. Biochemistry 42:9491–9497. https://doi.org/10.1021/bi034456f
40. Lies DP, Hernandez ME, Kappler A, Mielke RE, Gralnick JA, Newman DK (2005) *Shewanella oneidensis* MR-1 uses overlapping pathways for iron reduction at a distance and by direct contact under conditions relevant for biofilms. Appl Environ Microbiol 71:4414–4426. https://doi.org/10.1128/AEM.71.8.4414-4426.2005

41. Thormann KM, Saville RM, Shukla S, Spormann AM (2005) Induction of rapid detachment in *Shewanella oneidensis* MR-1 biofilms. J Bacteriol 187:1014–1021. https://doi.org/10.1128/JB.187.3.1014-1021.2005
42. Schuetz B, Schicklberger M, Kuermann J, Spormann AM, Gescher J (2009) Periplasmic electron transfer via the *c*-type cytochromes MtrA and FccA of *Shewanella oneidensis* MR-1. Appl Environ Microbiol 75:7789–7796. https://doi.org/10.1128/AEM.01834-09
43. Pirbadian S, Barchinger SE, Leung KM, Byun HS, Jangir Y, Bouhenni RA, Reed SB, Romine MF, Saffarini DA, Shi L, Gorby YA, Golbeck JH, El-Naggar MY (2014) *Shewanella oneidensis* MR-1 nanowires are outer membrane and periplasmic extensions of the extracellular electron transport components. Proc Natl Acad Sci 111:12883–12888. https://doi.org/10.1073/pnas.1410551111
44. Schicklberger M, Sturm G, Gescher J (2013) Genomic plasticity enables a secondary electron transport pathway in *Shewanella oneidensis*. Appl Environ Microbiol 79:1150–1159. https://doi.org/10.1128/AEM.03556-12
45. Bücking C, Piepenbrock A, Kappler A, Gescher J (2012) Outer-membrane cytochrome-independent reduction of extracellular electron acceptors in *Shewanella oneidensis*. Microbiology 158:2144–2157. https://doi.org/10.1099/mic.0.058404-0
46. Schicklberger M, Bücking C, Schuetz B, Heide H, Gescher J (2011) Involvement of the *Shewanella oneidensis* Decaheme cytochrome MtrA in the periplasmic stability of the ß-barrel protein MtrB. Appl Environ Microbiol 77:1520–1523. https://doi.org/10.1128/AEM.01201-10
47. Schutz B, Seidel J, Sturm G, Einsle O, Gescher J (2011) Investigation of the electron transport chain to and the catalytic activity of the diheme cytochrome *c* peroxidase CcpA of *Shewanella oneidensis*. Appl Environ Microbiol 77:6172–6180. https://doi.org/10.1128/AEM.00606-11
48. Coursolle D, Gralnick JA (2012) Reconstruction of extracellular respiratory pathways for iron (III) reduction in *Shewanella oneidensis* strain MR-1. Front Microbiol 3. https://doi.org/10.3389/fmicb.2012.00056
49. Gralnick JA, Vali H, Lies DP, Newman DK (2006) Extracellular respiration of dimethyl sulfoxide by *Shewanella oneidensis* strain MR-1. Proc Natl Acad Sci 103:4669–4674. https://doi.org/10.1073/pnas.0505959103
50. Coursolle D, Baron DB, Bond DR, Gralnick JA (2010) The Mtr respiratory pathway is essential for reducing Flavins and electrodes in *Shewanella oneidensis*. J Bacteriol 192:467–474. https://doi.org/10.1128/JB.00925-09
51. Marsili E, Baron DB, Shikhare ID, Coursolle D, Gralnick JA, Bond DR (2008) *Shewanella* secretes flavins that mediate extracellular electron transfer. Proc Natl Acad Sci 105:3968–3973. https://doi.org/10.1073/pnas.0710525105
52. Gao H, Barua S, Liang Y, Wu L, Dong Y, Reed S, Chen J, Culley D, Kennedy D, Yang Y, He Z, Nealson KH, Fredrickson JK, Tiedje JM, Romine M, Zhou J (2010) Impacts of *Shewanella oneidensis* *c*-type cytochromes on aerobic and anaerobic respiration. J Microbial Biotechnol 3:455–466. https://doi.org/10.1111/j.1751-7915.2010.00181.x
53. Gorby YA, Yanina S, McLean JS, Rosso KM, Moyles D, Dohnalkova A, Beveridge TJ, Chang IS, Kim BH, Kim KS, Culley DE, Reed SB, Romine MF, Saffarini DA, Hill EA, Shi L, Elias DA, Kennedy DW, Pinchuk G, Watanabe K, Ishii S, Logan B, Nealson KH, Fredrickson JK (2006) Electrically conductive bacterial nanowires produced by *Shewanella oneidensis* strain MR-1 and other microorganisms. Proc Natl Acad Sci 103:11358–11363. https://doi.org/10.1073/pnas.0604517103
54. Thormann KM, Duttler S, Saville RM, Hyodo M, Shukla S, Hayakawa Y, Spormann AM (2006) Control of formation and cellular detachment from *Shewanella oneidensis* MR-1 biofilms by cyclic di-GMP. J Bacteriol 188:2681–2691. https://doi.org/10.1128/JB.188.7.2681-2691.2006
55. Kane AL, Brutinel ED, Joo H, Maysonet R, VanDrisse CM, Kotloski NJ, Gralnick JA (2016) Formate metabolism in *Shewanella oneidensis* generates proton motive force and prevents

growth without an electron acceptor. J Bacteriol 198:1337–1346. https://doi.org/10.1128/JB.00927-15
56. Delgado VP, Paquete CM, Sturm G, Gescher J (2019) Improvement of the electron transfer rate in *Shewanella oneidensis* MR-1 using a tailored periplasmic protein composition. Bioelectrochemistry 129:18–25. https://doi.org/10.1016/j.bioelechem.2019.04.022
57. Li F-H, Tang Q, Fan Y-Y, Li Y, Li J, Wu J-H, Luo C-F, Sun H, Li W-W, Yu H-Q (2020) Developing a population-state decision system for intelligently reprogramming extracellular electron transfer in *Shewanella oneidensis*. Proc Natl Acad Sci 117:23001–23010. https://doi.org/10.1073/pnas.2006534117
58. Li F-H, Min D, Cheng Z-H, Li J, Wu J-H, Tang Q, Yu H-Q (2021) Systematically assessing genetic strategies for engineering electroactive bacterium to promote bioelectrochemical performances and pollutant removal. Sustain Energy Technol Assess 47. https://doi.org/10.1016/j.seta.2021.101506
59. Li F, Tang R, Zhang B, Qiao C, Yu H, Liu Q, Zhang J, Shi L, Song H (2023) Systematic full-cycle engineering microbial biofilms to boost electricity production in *Shewanella oneidensis*. Research 6. https://doi.org/10.34133/research.0081
60. Enyeart PJ, Chirieleison SM, Dao MN, Perutka J, Quandt EM, Yao J, Whitt JT, Keatinge-Clay AT, Lambowitz AM, Ellington AD (2013) Generalized bacterial genome editing using mobile group II introns and Crelox. Mol Syst Biol 9. https://doi.org/10.1038/msb.2013.41
61. Fan Y, Tang Q, Li Y, Li F, Wu J, Li W, Yu H (2021) Rapid and highly efficient genomic engineering with a novel iEditing device for programming versatile extracellular electron transfer of electroactive bacteria. Environ Microbiol 23:1238–1255. https://doi.org/10.1111/1462-2920.15374
62. Corts AD, Thomason LC, Gill RT, Gralnick JA (2019) A new recombineering system for precise genome-editing in *Shewanella oneidensis* strain MR-1 using single-stranded oligonucleotides. Sci Rep 9. https://doi.org/10.1038/s41598-018-37025-4
63. Corts AD, Thomason LC, Gill RT, Gralnick JA (2019) Efficient and precise genome editing in *Shewanella* with recombineering and CRISPR/Cas9-mediated counter-selection. ACS Synth Biol 8:1877–1889. https://doi.org/10.1021/acssynbio.9b00188
64. Chen Y, Cheng M, Feng X, Niu X, Song H, Cao Y (2022) Genome editing by CRISPR/Cas12 recognizing AT-rich PAMs in *Shewanella oneidensis* MR-1. ACS Synth Biol 11:2947–2955. https://doi.org/10.1021/acssynbio.2c00208
65. Fan YY, Tang Q, Li FH, Sun H, Min D, Wu JH, Li Y, Li WW, Yu HQ (2021) Enhanced bioreduction of radionuclides by driving microbial extracellular electron pumping with an engineered CRISPR platform. Environ Sci Technol 55:11997–12008. https://doi.org/10.1021/acs.est.1c03713
66. Suzuki Y, Kouzuma A, Watanabe K (2020) CRISPR/Cas9-mediated genome editing of *Shewanella oneidensis* MR-1 using a broad host-range pBBR1-based plasmid. J Gen Appl Microbiol 66:41–45. https://doi.org/10.2323/jgam.2019.04.007
67. Strecker J, Ladha A, Gardner Z, Schmid-Burgk JL, Makarova KS, Koonin EV, Zhang F (2019) RNA-guided DNA insertion with CRISPR-associated transposases. Science 365:48–53. https://doi.org/10.1126/science.aax9181
68. Cheng ZH, Wu J, Liu JQ, Min D, Liu DF, Li WW, Yu HQ (2022) Repurposing CRISPR RNA-guided integrases system for one-step, efficient genomic integration of ultra-long DNA sequences. Nucleic Acids Res 50:7739–7750. https://doi.org/10.1093/NAR/GKAC554
69. Cao Y, Li X, Li F, Song H (2017) CRISPRi–sRNA: transcriptional–translational regulation of extracellular electron transfer in *Shewanella oneidensis*. ACS Synth Biol 6:1679–1690. https://doi.org/10.1021/acssynbio.6b00374
70. Yi YC, Ng IS (2021) Redirection of metabolic flux in *Shewanella oneidensis* MR-1 by CRISPRi and modular design for 5-aminolevulinic acid production. Bioresour Bioprocess 8:1–11. https://doi.org/10.1186/s40643-021-00366-6

71. Yu H, Lu Y, Lan F, Wang Y, Hu C, Mao L, Wu D, Li F, Song H (2023) Engineering outer membrane vesicles to increase extracellular electron transfer of *Shewanella oneidensis*. ACS Synth Biol 12:1645–1656. https://doi.org/10.1021/acssynbio.2c00636
72. Li J, Tang Q, Li Y, Fan YY, Li FH, Wu JH, Min D, Li WW, Lam PKS, Yu HQ (2020) Rediverting electron flux with an engineered CRISPR-ddAsCpf1 system to enhance the pollutant degradation capacity of *Shewanella oneidensis*. Environ Sci Technol 54:3599–3608. https://doi.org/10.1021/acs.est.9b06378
73. Chen Y, Cheng M, Song H, Cao Y (2022) Type I-F CRISPR-PAIR platform for multi-mode regulation to boost extracellular electron transfer in *Shewanella oneidensis*. iScience 25: 104491. https://doi.org/10.1016/j.isci.2022.104491
74. Chen Y, Niu X, Cheng M, Wang L, Sun P, Song H, Cao Y (2022) CRISPR/dCas9-RpoD-mediated simultaneous transcriptional activation and repression in *Shewanella oneidensis* MR-1. ACS Synth Biol 11:2184–2192. https://doi.org/10.1021/acssynbio.2c00149
75. Ford KC, Kaste JAM, Shachar-Hill Y, TerAvest MA (2022) Flux-balance analysis and Mobile CRISPRi-guided deletion of a conditionally essential gene in *Shewanella oneidensis* MR-1. ACS Synth Biol 11:3405–3413. https://doi.org/10.1021/acssynbio.2c00323
76. Cheng L, Min D, He R, Cheng Z, Liu D, Yu H (2020) Developing a base-editing system to expand the carbon source utilization spectra of *Shewanella oneidensis* MR-1 for enhanced pollutant degradation. Biotechnol Bioeng 117:2389–2400. https://doi.org/10.1002/bit.27368
77. Chen Y, Fang L, Ying X, Cheng M, Wang L, Sun P, Zhang Z, Shi L, Cao Y, Song H (2022) Development of whole genome-scale base editing toolbox to promote efficiency of extracellular electron transfer in *Shewanella oneidensis* MR-1. Adv Biol 6:e2101296. https://doi.org/10.1002/ADBI.202101296
78. Chen Y, Cheng M, Li Y, Wang L, Fang L, Cao Y, Song H (2023) Highly efficient multiplex base editing: one-shot deactivation of eight genes in *Shewanella oneidensis* MR-1. Synth Syst Biotechnol 8:1–10. https://doi.org/10.1016/j.synbio.2022.09.005
79. Wang T, Zhang J, Wei L, Zhao D, Bi C, Liu Q, Xu N, Liu J (2023) Developing a PAM-flexible CRISPR-mediated dual-deaminase base editor to regulate extracellular electron transport in *Shewanella oneidensis*. ACS Synth Biol 12:1727–1738. https://doi.org/10.1021/acssynbio.3c00045
80. Golitsch F, Bücking C, Gescher J (2013) Proof of principle for an engineered microbial biosensor based on *Shewanella oneidensis* outer membrane protein complexes. Biosens Bioelectron 47:285–291. https://doi.org/10.1016/j.bios.2013.03.010
81. Webster DP, TerAvest MA, Doud DFR, Chakravorty A, Holmes EC, Radens CM, Sureka S, Gralnick JA, Angenent LT (2014) An arsenic-specific biosensor with genetically engineered *Shewanella oneidensis* in a bioelectrochemical system. Biosens Bioelectron 62:320–324. https://doi.org/10.1016/j.bios.2014.07.003
82. Zhang J, Chen Z, Liu C, Li J, An X, Wu D, Sun X, Zhang B, Fu L, Li F, Song H (2021) Construction of an acetate metabolic pathway to enhance electron generation of engineered *Shewanella oneidensis*. Front Bioeng Biotechnol 9. https://doi.org/10.3389/fbioe.2021.757953
83. Min D, Cheng L, Zhang F, Huang X-N, Li D-B, Liu D-F, Lau T-C, Mu Y, Yu H-Q (2017) Enhancing extracellular electron transfer of *Shewanella oneidensis* MR-1 through coupling improved Flavin synthesis and metal-reducing conduit for pollutant degradation. Environ Sci Technol 51:5082–5089. https://doi.org/10.1021/acs.est.6b04640
84. Li F, Li Y, Sun L, Chen X, An X, Yin C, Cao Y, Wu H, Song H (2018) Modular engineering intracellular NADH regeneration boosts extracellular electron transfer of *Shewanella oneidensis* MR-1. ACS Synth Biol 7:885–895. https://doi.org/10.1021/acssynbio.7b00390
85. Li F, Li Y-X, Cao Y-X, Wang L, Liu C-G, Shi L, Song H (2018) Modular engineering to increase intracellular NAD(H/+) promotes rate of extracellular electron transfer of *Shewanella oneidensis*. Nat Commun 9. https://doi.org/10.1038/s41467-018-05995-8

86. Hajimorad M, Gralnick JA (2016) Towards enabling engineered microbial-electronic systems: RK2-based conjugal transfer system for *Shewanella* synthetic biology. Electron Lett 52:426–428. https://doi.org/10.1049/el.2015.3226
87. Cao Y, Song M, Li F, Li C, Lin X, Chen Y, Chen Y, Xu J, Ding Q, Song H (2019) A synthetic plasmid toolkit for *Shewanella oneidensis* MR-1. Front Microbiol 10. https://doi.org/10.3389/fmicb.2019.00410
88. Ng I-S, Guo Y, Zhou Y, Wu J-W, Tan S-I, Yi Y-C (2018) Turn on the Mtr pathway genes under pLacI promoter in *Shewanella oneidensis* MR-1. Bioresour Bioprocess 5:35. https://doi.org/10.1186/s40643-018-0221-9
89. Fan Y, Tang Q, Sun H, Yu H (2022) A designed plasmid-transition strategy enables rapid construction of robust and versatile synthetic exoelectrogens for environmental applications. Environ Microbiol 24:5292–5305. https://doi.org/10.1111/1462-2920.16181
90. Yi YC, Ng IS (2020) Establishment of toolkit and T7RNA polymerase/promoter system in *Shewanella oneidensis* MR-1. J Taiwan Inst Chem Eng 109:8–14. https://doi.org/10.1016/J.JTICE.2020.02.003
91. Cheng L, He R-L, Min D, Li W-W, Liu D-F, Yu H-Q (2021) Engineering a Rhamnose-inducible system to enhance the extracellular electron transfer ability of *Shewanella* genus for improved Cr(VI) reduction. ACS ES&T Eng 1:842–850. https://doi.org/10.1021/acsestengg.0c00125
92. Heidelberg JF, Paulsen IT, Nelson KE, Gaidos EJ, Nelson WC, Read TD, Eisen JA, Seshadri R, Ward N, Methe B, Clayton RA, Meyer T, Tsapin A, Scott J, Beanan M, Brinkac L, Daugherty S, DeBoy RT, Dodson RJ, Durkin AS, Haft DH, Kolonay JF, Madupu R, Peterson JD, Umayam LA, White O, Wolf AM, Vamathevan J, Weidman J, Impraim M, Lee K, Berry K, Lee C, Mueller J, Khouri H, Gill J, Utterback TR, McDonald LA, Feldblyum TV, Smith HO, Venter JC, Nealson KH, Fraser CM (2002) Genome sequence of the dissimilatory metal ion–reducing bacterium *Shewanella oneidensis*. Nat Biotechnol 20: 1118–1123. https://doi.org/10.1038/nbt749
93. Zeng Z, Liu X, Yao J, Guo Y, Li B, Li Y, Jiao N, Wang X (2016) Cold adaptation regulated by cryptic prophage excision in *Shewanella oneidensis*. ISME J 10:2787–2800. https://doi.org/10.1038/ismej.2016.85
94. Gödeke J, Paul K, Lassak J, Thormann KM (2010) Phage-induced lysis enhances biofilm formation in *Shewanella oneidensis* MR-1. ISME J 5:613–626. https://doi.org/10.1038/ismej.2010.153
95. Guo Q, Chen B, Tu Y, Du S, Chen X (2019) Prophage LambdaSo uses replication interference to suppress reproduction of coexisting temperate phage MuSo2 in *Shewanella oneidensis* MR-1. Environ Microbiol 21:2079–2094. https://doi.org/10.1111/1462-2920.14592
96. Yao J, Guo Y, Wang P, Zeng Z, Li B, Tang K, Liu X, Wang X (2018) Type II toxin/antitoxin system ParE SO /CopA SO stabilizes prophage CP4So in *Shewanella oneidensis*. Environ Microbiol 20:1224–1239. https://doi.org/10.1111/1462-2920.14068
97. Binnenkade L, Teichmann L, Thormann KM (2014) Iron triggers λSo prophage induction and release of extracellular DNA in *Shewanella oneidensis* MR-1 biofilms. Appl Environ Microbiol 80:5304–5316. https://doi.org/10.1128/AEM.01480-14
98. Kreienbaum M, Dörrich AK, Brandt D, Schmid NE, Leonhard T, Hager F, Brenzinger S, Hahn J, Glatter T, Ruwe M, Briegel A, Kalinowski J, Thormann KM (2020) Isolation and characterization of *Shewanella* phage Thanatos infecting and lysing *Shewanella oneidensis* and promoting nascent biofilm formation. Front Microbiol 11. https://doi.org/10.3389/fmicb.2020.573260
99. Schmid N, Brandt D, Walasek C, Rolland C, Wittmann J, Fischer D, Müsken M, Kalinowski J, Thormann K (2024) An autonomous plasmid as an inovirus phage satellite. Appl Environ Microbiol 90:e00246–e00224. https://doi.org/10.1128/aem.00246-24
100. Logan BE, Hamelers B, Rozendal R, Schröder U, Keller J, Freguia S, Aelterman P, Verstraete W, Rabaey K (2006) Microbial fuel cells: methodology and technology. Environ Sci Technol 40:5181–5192. https://doi.org/10.1021/ES0605016

101. Watson VJ, Logan BE (2010) Power production in MFCs inoculated with *Shewanella oneidensis* MR-1 or mixed cultures. Biotechnol Bioeng 105:489–498. https://doi.org/10.1002/BIT.22556
102. Lee SH, Ban JY, Oh CH, Park HK, Choi S (2016) A solvent-free microbial-activated air cathode battery paper platform made with pencil-traced graphite electrodes. Sci Rep 6:28588. https://doi.org/10.1038/srep28588
103. Hu H, Fan Y, Liu H (2008) Hydrogen production using single-chamber membrane-free microbial electrolysis cells. Water Res 42:4172–4178. https://doi.org/10.1016/J.WATRES.2008.06.015
104. Wang VB, Yam JKH, Chua SL, Zhang Q, Cao B, Chye JLS, Yang L (2014) Synergistic microbial consortium for bioenergy generation from complex natural energy sources. Sci World J:139653. https://doi.org/10.1155/2014/139653
105. Engel C, Schattenberg F, Dohnt K, Schröder U, Müller S, Krull R (2019) Long-term behavior of defined mixed cultures of *Geobacter sulfurreducens* and *Shewanella oneidensis* in bioelectrochemical systems. Front Bioeng Biotechnol 7:443540. https://doi.org/10.3389/FBIOE.2019.00060
106. Lin T, Bai X, Hu Y, Li B, Yuan YJ, Song H, Yang Y, Wang J (2017) Synthetic *Saccharomyces cerevisiae-Shewanella oneidensis* consortium enables glucose-fed high-performance microbial fuel cell. AICHE J 63:1830–1838. https://doi.org/10.1002/AIC.15611
107. Su L, Jia W, Hou C, Lei Y (2011) Microbial biosensors: a review. Biosens Bioelectron 26:1788–1799. https://doi.org/10.1016/J.BIOS.2010.09.005
108. Si RW, Zhai DD, Liao ZH, Gao L, Yong YC (2015) A whole-cell electrochemical biosensing system based on bacterial inward electron flow for fumarate quantification. Biosens Bioelectron 68:34–40. https://doi.org/10.1016/J.BIOS.2014.12.035
109. Reuillard B, Ly KH, Hildebrandt P, Jeuken LJC, Butt JN, Reisner E (2017) High performance reduction of H_2O_2 with an electron transport Decaheme cytochrome on a porous ITO electrode. J Am Chem Soc 139:3324–3327. https://doi.org/10.1021/JACS.6B12437
110. Hu Y, Rehnlund D, Klein E, Gescher J, Niemeyer CM (2020) Cultivation of exoelectrogenic bacteria in conductive DNA nanocomposite hydrogels yields a programmable biohybrid materials system. ACS Appl Mater Interfaces 12:14806–14813. https://doi.org/10.1021/ACSAMI.9B22116
111. Zajdel TJ, Baruch M, Méhes G, Stavrinidou E, Berggren M, Maharbiz MM, Simon DT, Ajo-Franklin CM (2018) PEDOT:PSS-based multilayer bacterial-composite films for bioelectronics. Sci Rep 8:15293. https://doi.org/10.1038/s41598-018-33521-9
112. Qi X, Wang S, Jiang Y, Liu P, Hao W, Han J, Zhou Y, Huang X, Liang P (2021) Additional polypyrrole as conductive medium in artificial electrochemically active biofilm (EAB) to increase the sensitivity of EAB based biosensor in water quality early-warning. Biosens Bioelectron 190:113453. https://doi.org/10.1016/J.BIOS.2021.113453
113. Abu-Ali H, Nabok A, Smith TJ (2019) Electrochemical inhibition bacterial sensor array for detection of water pollutants: artificial neural network (ANN) approach. Anal Bioanal Chem 411:7659–7668. https://doi.org/10.1007/S00216-019-01853-8
114. Schievano A, Pepé Sciarria T, Vanbroekhoven K, De Wever H, Puig S, Andersen SJ, Rabaey K, Pant D (2016) Electro-fermentation – merging electrochemistry with fermentation in industrial applications. Trends Biotechnol 34:866–878. https://doi.org/10.1016/j.tibtech.2016.04.007
115. Moscoviz R, Toledo-Alarcón J, Trably E, Bernet N (2016) Electro-fermentation: how to drive fermentation using electrochemical systems. Trends Biotechnol 34:856–865. https://doi.org/10.1016/j.tibtech.2016.04.009
116. Flynn JM, Ross DE, Hunt KA, Bond DR, Gralnick JA (2010) Enabling unbalanced fermentations by using engineered electrode-interfaced bacteria. MBio 1:190–200. https://doi.org/10.1128/MBIO.00190-10

117. Bursac T, Gralnick JA, Gescher J (2017) Acetoin production via unbalanced fermentation in *Shewanella oneidensis*. Biotechnol Bioeng 114:1283–1289. https://doi.org/10.1002/BIT.26243
118. Werpy T, Petersen G (2004) Top value added chemicals from biomass volume I-results of screening for potential candidates from sugars and synthesis gas
119. Wohlers H, Zentgraf L, van der Sande L, Holtmann D (2024) Metabolic engineering of *Shewanella oneidensis* to produce glutamate and itaconic acid. Appl Microbiol Biotechnol 108:1–10. https://doi.org/10.1007/S00253-023-12879-5
120. Uno M, Phansroy N, Aso Y, Ohara H (2017) Starch-fueled microbial fuel cells by two-step and parallel fermentation using *Shewanella oneidensis* MR-1 and *Streptococcus bovis* 148. J Biosci Bioeng 124:189–194. https://doi.org/10.1016/J.JBIOSC.2017.03.016
121. Rabaey K, Rozendal RA (2010) Microbial electrosynthesis — revisiting the electrical route for microbial production. Nat Rev Microbiol 8:706–716. https://doi.org/10.1038/nrmicro2422
122. Ross DE, Flynn JM, Baron DB, Gralnick JA, Bond DR (2011) Towards electrosynthesis in *Shewanella*: energetics of reversing the Mtr pathway for reductive metabolism. PLoS One 6: e16649. https://doi.org/10.1371/JOURNAL.PONE.0016649
123. Strycharz SM, Glaven RH, Coppi MV, Gannon SM, Perpetua LA, Liu A, Nevin KP, Lovley DR (2011) Gene expression and deletion analysis of mechanisms for electron transfer from electrodes to *Geobacter sulfurreducens*. Bioelectrochemistry 80:142–150. https://doi.org/10.1016/J.BIOELECHEM.2010.07.005
124. Hwang H, Yeon YJ, Lee S, Choe H, Jang MG, Cho DH, Park S, Kim YH (2015) Electro-biocatalytic production of formate from carbon dioxide using an oxygen-stable whole cell biocatalyst. Bioresour Technol 185:35–39. https://doi.org/10.1016/J.BIORTECH.2015.02.086
125. Jeon JM, Park H, Seo HM, Kim JH, Bhatia SK, Sathiyanarayanan G, Song HS, Park SH, Choi KY, Sang BI, Yang YH (2015) Isobutanol production from an engineered *Shewanella oneidensis* MR-1. Bioprocess Biosyst Eng 38:2147–2154. https://doi.org/10.1007/S00449-015-1454-Z
126. Le QAT, Kim HG, Kim YH (2018) Electrochemical synthesis of formic acid from CO2 catalyzed by *Shewanella oneidensis* MR-1 whole-cell biocatalyst. Enzyme Microb Technol 116:1–5. https://doi.org/10.1016/J.ENZMICTEC.2018.05.005
127. La Cava E, Guionet A, Saito J, Okamoto A (2020) Involvement of proton transfer for carbon dioxide reduction coupled with extracellular electron uptake in *Shewanella oneidensis* MR-1. Electroanalysis 32:1659–1663. https://doi.org/10.1002/ELAN.201900686
128. Tefft NM, Teravest MA (2019) Reversing an extracellular electron transfer pathway for electrode-driven acetoin reduction. ACS Synth Biol 8:1590–1600. https://doi.org/10.1021/ACSSYNBIO.8B00498
129. Li Y, Qiao S, Guo M, Hou C, Wang J, Yu C, Zhou J, Quan X (2023) Microbial electrosynthetic nitrate reduction to ammonia by reversing the typical electron transfer pathway in *Shewanella oneidensis*. Cell Rep Phys Sci 4:101433. https://doi.org/10.1016/J.XCRP.2023.101433
130. Okamoto A, Hashimoto K, Nealson KH (2014) Flavin redox bifurcation as a mechanism for controlling the direction of electron flow during extracellular electron transfer. Angew Chem Int Ed 53:10988–10991. https://doi.org/10.1002/ANIE.201407004
131. Hemmerich P, Knappe W-R, Kramer HEA, Traber R (1980) Distinction of $2e^-$ and $1e^-$ reduction modes of the flavin chromophore as studied by flash photolysis. Eur J Biochem 104:511–520. https://doi.org/10.1111/J.1432-1033.1980.TB04453.X
132. Abuyen K, El-Naggar MY (2023) Soluble iron enhances extracellular electron uptake by *Shewanella oneidensis* MR-1. ChemElectroChem 10:e202200965. https://doi.org/10.1002/CELC.202200965
133. Quek G, Vázquez RJ, McCuskey SR, Lopez-Garcia F, Bazan GC (2023) An n-type conjugated oligoelectrolyte mimics transmembrane electron transport proteins for enhanced microbial

electrosynthesis. Angew Chem Int Ed 62:e202305189. https://doi.org/10.1002/ANIE.202305189
134. Luo D, Ding H, Guo T, Li X, Song T, Xie J (2023) Self-assembly of graphene oxide and *Shewanella oneidensis* MR-1 formed a conductive bio-abiotic composite for enhancing microbial electrosynthesis performance. Renew Energy 215:119018. https://doi.org/10.1016/J.RENENE.2023.119018
135. Quek G, Vázquez RJ, Mccuskey SR, Kundukad B, Bazan GC, Quek G, Vázquez RJ, Mccuskey SR, Bazan GC, Kundukad B (2022) Enabling electron injection for microbial electrosynthesis with n-type conjugated polyelectrolytes. Adv Mater 34:2203480. https://doi.org/10.1002/ADMA.202203480
136. Hau HH, Gilbert A, Coursolle D, Gralnick JA (2008) Mechanism and consequences of anaerobic respiration of cobalt by *Shewanella oneidensis* strain MR-1. Appl Environ Microbiol 74:6880–6886. https://doi.org/10.1128/AEM.00840-08
137. Bretschger O, Obraztsova A, Sturm CA, In SC, Gorby YA, Reed SB, Culley DE, Reardon CL, Barua S, Romine MF, Zhou J, Beliaev AS, Bouhenni R, Saffarini D, Mansfeld F, Kim BH, Fredrickson JK, Nealson KH (2007) Current production and metal oxide reduction by *Shewanella oneidensis* MR-1 wild type and mutants. Appl Environ Microbiol 73:7003–7012. https://doi.org/10.1128/AEM.01087-07
138. De Corte S, Hennebel T, Verschuere S, Cuvelier C, Verstraete W, Boon N (2011) Gold nanoparticle formation using *Shewanella oneidensis*: a fast biosorption and slow reduction process. J Chem Technol Biotechnol 86:547–553. https://doi.org/10.1002/JCTB.2549
139. Rotaru AE, Jiang W, Finster K, Skrydstrup T, Meyer RL (2012) Non-enzymatic palladium recovery on microbial and synthetic surfaces. Biotechnol Bioeng 109:1889–1897. https://doi.org/10.1002/BIT.24500
140. Yin Y, Liu C, Zhao G, Chen Y (2022) Versatile mechanisms and enhanced strategies of pollutants removal mediated by *Shewanella oneidensis*: a review. J Hazard Mater 440:129703. https://doi.org/10.1016/J.JHAZMAT.2022.129703
141. Schkolnik G, Schmidt M, Mazza MG, Harnisch F, Musat N (2015) *In situ* analysis of a silver nanoparticle-precipitating *Shewanella* biofilm by surface enhanced confocal Raman microscopy. PLoS One 10:e0145871. https://doi.org/10.1371/JOURNAL.PONE.0145871
142. Maes S, Props R, Fitts JP, De Smet R, Vanhaecke F, Boon N, Hennebel T (2017) Biological recovery of platinum complexes from diluted aqueous streams by axenic cultures. PLoS One 12:e0169093. https://doi.org/10.1371/JOURNAL.PONE.0169093
143. Kimber RL, Lewis EA, Parmeggiani F, Smith K, Bagshaw H, Starborg T, Joshi N, Figueroa AI, van der Laan G, Cibin G, Gianolio D, Haigh SJ, Pattrick RAD, Turner NJ, Lloyd JR (2018) Biosynthesis and characterization of copper nanoparticles using *Shewanella oneidensis*: application for click chemistry. Small 14:1703145. https://doi.org/10.1002/SMLL.201703145
144. Miller RB, Giai C, Iannuzzi M, Monty CN, Senko JM (2016) Microbial reduction of Cr(VI) in the presence of chromate conversion coating constituents. Biorem J 20:174–182. https://doi.org/10.1080/10889868.2016.1149443
145. Barceloux DG, Barceloux D (1999) Chromium. J Toxicol Clin Toxicol 37:173–194. https://doi.org/10.1081/CLT-100102418
146. Ribera D, Labrot F, Tisnerat G, Narbonne JF (1996) Uranium in the environment: occurrence, transfer, and biological effects. Rev Environ Contam Toxicol 146:53–89. https://doi.org/10.1007/978-1-4613-8478-6_3
147. Bernier-Latmani R, Veeramani H, Vecchia ED, Junier P, Lezama-Pacheco JS, Suvorova EI, Sharp JO, Wigginton NS, Bargar JR (2010) Non-uraninite products of microbial U(VI) reduction. Environ Sci Technol 44:9456–9462. https://doi.org/10.1021/ES101675A
148. Lovley DR, Phillips EJP, Gorby YA, Landa ER (1991) Microbial reduction of uranium. Nature 350:413–416. https://doi.org/10.1038/350413a0
149. Wu LC, Chen CY, Lin TK, Su YY, Chung YC (2020) Highly efficient removal of Victoria blue R and bioelectricity generation from textile wastewater using a novel combined dual

microbial fuel cell system. Chemosphere 258:127326. https://doi.org/10.1016/J.CHEMOSPHERE.2020.127326
150. Mohamed A, Yu L, Fang Y, Ashry N, Riahi Y, Uddin I, Dai K, Huang Q (2020) Iron mineral-humic acid complex enhanced Cr(VI) reduction by *Shewanella oneidensis* MR-1. Chemosphere 247:125902. https://doi.org/10.1016/J.CHEMOSPHERE.2020.125902
151. Ma L, Du Y, Chen S, Zhang F, Zhan W, Du D, Zhang TC (2021) Nanoscale zero-valent iron coupling with *Shewanella oneidensis* MR-1 for enhanced reduction/removal of aqueous Cr (VI). Sep Purif Technol 277:119488. https://doi.org/10.1016/J.SEPPUR.2021.119488
152. Springthorpe SK, Dundas CM, Keitz BK (2019) Microbial reduction of metal-organic frameworks enables synergistic chromium removal. Nat Commun 10:5212. https://doi.org/10.1038/s41467-019-13219-w
153. Springthorpe SK, Keitz BK (2021) Extraction of Au(III) by microbially reduced metal-organic frameworks. Langmuir 37:9078–9088. https://doi.org/10.1021/ACS.LANGMUIR.1C01180
154. Zhang Y, Pan T, Li CX, Zheng D, Sun Y, Zhang XZ (2022) Double bacteria synergistic catalytic reduction system for heavy metal detoxification treatment. Nano Lett 22:5575–5583. https://doi.org/10.1021/ACS.NANOLETT.2C01907
155. Law CKY, Bonin L, De Gusseme B, Boon N, Kundu K (2022) Biogenic synthesis of palladium nanoparticles: new production methods and applications. Nanotechnol Rev 11: 3104–3124. https://doi.org/10.1515/NTREV-2022-0482
156. Wu R, Cui L, Chen L, Wang C, Cao C, Sheng G, Yu H, Zhao F (2013) Effects of bio-Au nanoparticles on electrochemical activity of *Shewanella oneidensis* wild type and Δ*omcA/mtrC* mutant. Sci Rep 3:3307. https://doi.org/10.1038/srep03307
157. De Corte S, Hennebel T, De Gusseme B, Verstraete W, Boon N (2012) Bio-palladium: from metal recovery to catalytic applications. J Microbial Biotechnol 5:5–17. https://doi.org/10.1111/J.1751-7915.2011.00265.X
158. Hennebel T, De Gusseme B, Boon N, Verstraete W (2009) Biogenic metals in advanced water treatment. Trends Biotechnol 27:90–98. https://doi.org/10.1016/J.TIBTECH.2008.11.002
159. Ali I, Peng C, Khan ZM, Naz I, Sultan M, Ali M, Abbasi IA, Islam T, Ye T (2019) Overview of microbes based fabricated biogenic nanoparticles for water and wastewater treatment. J Environ Manage 230:128–150. https://doi.org/10.1016/J.JENVMAN.2018.09.073
160. Hennebel T, Van Nevel S, Verschuere S, De Corte S, De Gusseme B, Cuvelier C, Fitts JP, Van Der Lelie D, Boon N, Verstraete W (2011) Palladium nanoparticles produced by fermentatively cultivated bacteria as catalyst for diatrizoate removal with biogenic hydrogen. Appl Microbiol Biotechnol 91:1435–1445. https://doi.org/10.1007/S00253-011-3329-9
161. Yang ZN, Hou YN, Zhang B, Cheng HY, Yong YC, Liu WZ, Han JL, Liu SJ, Wang AJ (2020) Insights into palladium nanoparticles produced by *Shewanella oneidensis* MR-1: roles of NADH dehydrogenases and hydrogenases. Environ Res 191:110196. https://doi.org/10.1016/J.ENVRES.2020.110196
162. Ng CK, Karahan HE, Loo SCJ, Chen Y, Cao B (2019) Biofilm-templated heteroatom-doped carbon-palladium nanocomposite catalyst for hexavalent chromium reduction. ACS Appl Mater Interfaces 11:24018–24026. https://doi.org/10.1021/ACSAMI.9B04095
163. Wang W, Zhang B, He Z (2019) Bioelectrochemical deposition of palladium nanoparticles as catalysts by *Shewanella oneidensis* MR-1 towards enhanced hydrogen production in microbial electrolysis cells. Electrochim Acta 318:794–800. https://doi.org/10.1016/J.ELECTACTA.2019.06.038
164. Lehner BAE, Janssen VAEC, Spiesz EM, Benz D, Brouns SJJ, Meyer AS, van der Zant HSJ (2019) Creation of conductive graphene materials by bacterial reduction using *Shewanella Oneidensis*. ChemistryOpen 8:888–895. https://doi.org/10.1002/OPEN.201900186
165. Pearce CI, Pattrick RAD, Law N, Charnock JM, Coker VS, Fellowes JW, Oremland RS, Lloyd JR (2009) Investigating different mechanisms for biogenic selenite transformations: *Geobacter sulfurreducens*, *Shewanella oneidensis* and *Veillonella atypica*. Environ Technol 30:1313–1326. https://doi.org/10.1080/09593330902984751

166. Fu XZ, Wu J, Cui S, Wang XM, Liu HQ, He RL, Yang C, Deng X, Tan ZL, Li WW (2021) Self-regenerable bio-hybrid with biogenic ferrous sulfide nanoparticles for treating high-concentration chromium-containing wastewater. Water Res 206:117731. https://doi.org/10.1016/J.WATRES.2021.117731
167. Dang Z, Guan Y, Wu Z, Tao XY, Xiong Y, Bai HB, Shao CS, Liu G, Huang Q, Tian LJ, Tian YC (2021) Regulating the synthesis rate and yield of bio-assembled FeS nanoparticles for efficient cancer therapy. Nanoscale 13:18977–18986. https://doi.org/10.1039/D1NR03591F
168. Bijay-Singh CE (2021) Fertilizers and nitrate pollution of surface and ground water: an increasingly pervasive global problem. SN Appl Sci 3:518. https://doi.org/10.1007/S42452-021-04521-8
169. Barrett JH, Parslow RC, McKinney PA, Law GR, Forman D (1998) Nitrate in drinking water and the incidence of gastric, esophageal, and brain cancer in Yorkshire, England. Cancer Causes Control 9:153–159. https://doi.org/10.1023/A:1008878126535
170. Myers CR, Nealson KH (1988) Microbial reduction of manganese oxides: interactions with iron and sulfur. Geochim Cosmochim Acta 52:2727–2732. https://doi.org/10.1016/0016-7037(88)90041-5
171. Yu Q, Sun W, Gao H (2021) Thiosulfate oxidation in sulfur-reducing *Shewanella oneidensis* and its unexpected influences on the cytochrome *c* content. Environ Microbiol 23:7056–7072. https://doi.org/10.1111/1462-2920.15807
172. Toporek YJ, Mok JK, Shin HD, Lee BD, Lee MH, DiChristina TJ (2019) Metal reduction and protein secretion genes required for iodate reduction by *Shewanella oneidensis*. Appl Environ Microbiol 85:e02115–e02118. https://doi.org/10.1128/AEM.02115-18
173. Jia X, Ma L, Liu J, Liu P, Yu L, Zhou J, Li W, Zhou W, Dong Z (2022) Reduction of antimony mobility from Sb-rich smelting slag by *Shewanella oneidensis*: integrated biosorption and precipitation. J Hazard Mater 426:127385. https://doi.org/10.1016/J.JHAZMAT.2021.127385
174. Bafana A, Devi SS, Chakrabarti T (2011) Azo dyes: past, present and the future. Environ Rev 19:350–371. https://doi.org/10.1139/a11-018
175. Xiao X, Li TT, Lu XR, Feng XL, Han X, Li WW, Li Q, Yu HQ (2018) A simple method for assaying anaerobic biodegradation of dyes. Bioresour Technol 251:204–209. https://doi.org/10.1016/J.BIORTECH.2017.12.052
176. Ju K-S, Parales RE (2010) Nitroaromatic compounds, from synthesis to biodegradation. Microbiol Mol Biol Rev 74:250. https://doi.org/10.1128/MMBR.00006-10
177. Wang H, Zhao HP, Zhu L (2020) Structures of nitroaromatic compounds induce *Shewanella oneidensis* MR-1 to adopt different electron transport pathways to reduce the contaminants. J Hazard Mater 384:121495. https://doi.org/10.1016/J.JHAZMAT.2019.121495
178. Xiao X, Xi BD, He XS, Zhang H, Li YH, Pu S, Liu SJ, Da YM, Yang C (2019) Redox properties and dechlorination capacities of landfill-derived humic-like acids. Environ Pollut 253:488–496. https://doi.org/10.1016/J.ENVPOL.2019.07.044
179. Wang G, Han N, Liu L, Ke Z, Li B, Chen G (2020) Molecular density regulating electron transfer efficiency of *S. oneidensis* MR-1 mediated roxarsone biotransformation. Environ Pollut 262:114370. https://doi.org/10.1016/J.ENVPOL.2020.114370
180. Adelaja O, Keshavarz T, Kyazze G (2014) Enhanced biodegradation of phenanthrene using different inoculum types in a microbial fuel cell. Eng Life Sci 14:218–228. https://doi.org/10.1002/ELSC.201300089
181. Yeferni M, Ben Saïd O, Mahmoudi E, Essid N, Hedfi A, Almalki M, Pacioglu O, Dervishi A, Boufahja F (2022) Effectiveness of *Shewanella oneidensis* bioaugmentation in the bioremediation of phenanthrene-contaminated sediments and possible consortia with omnivore-carnivore meiobenthic nematodes. Environ Pollut 292:118281. https://doi.org/10.1016/J.ENVPOL.2021.118281
182. You J, Deng Y, Chen H, Ye J, Zhang S, Zhao J (2020) Enhancement of gaseous o-xylene degradation in a microbial fuel cell by adding *Shewanella oneidensis* MR-1. Chemosphere 252:126571. https://doi.org/10.1016/J.CHEMOSPHERE.2020.126571

183. Neyens E, Baeyens J (2003) A review of classic Fenton's peroxidation as an advanced oxidation technique. J Hazard Mater 98:33–50. https://doi.org/10.1016/S0304-3894(02)00282-0
184. Sekar R, Dichristina TJ (2014) Microbially driven fenton reaction for degradation of the widespread environmental contaminant 1,4-dioxane. Environ Sci Technol 48:12858–12867. https://doi.org/10.1021/ES503454A
185. Sekar R, Dichristina TJ (2017) Degradation of the recalcitrant oil spill components anthracene and pyrene by a microbially driven Fenton reaction. FEMS Microbiol Lett 364:203. https://doi.org/10.1093/FEMSLE/FNX203
186. Sekar R, Taillefert M, DiChristina TJ (2016) Simultaneous transformation of commingled trichloroethylene, tetrachloroethylene, and 1,4-dioxane by a microbially driven Fenton reaction in batch liquid cultures. Appl Environ Microbiol 82:6335–6343. https://doi.org/10.1128/AEM.02325-16
187. Grossman JN, Kahan TF, Grossman JN, Kahan TF (2016) Hydroxyl radical formation from bacteria-assisted Fenton chemistry at neutral pH under environmentally relevant conditions. Environ Chem 13:757–766. https://doi.org/10.1071/EN15256
188. Sekar R, Shin HD, DiChristina TJ (2016) Direct conversion of cellulose and hemicellulose to fermentable sugars by a microbially-driven Fenton reaction. Bioresour Technol 218:1133–1139. https://doi.org/10.1016/J.BIORTECH.2016.07.087
189. Sukovich DJ, Seffernick JL, Richman JE, Hunt KA, Gralnick JA, Wackett LP (2010) Structure, function, and insights into the biosynthesis of a head-to-head hydrocarbon in *Shewanella oneidensis* strain MR-1. Appl Environ Microbiol 76:3842–3849. https://doi.org/10.1128/AEM.00433-10
190. Blatter M, Sugnaux M, Comninellis C, Nealson K, Fischer F (2016) Modeling of sustainable base production by microbial electrolysis cell. ChemSusChem 9:1570–1574. https://doi.org/10.1002/CSSC.201600459
191. Liu X, Ding J, Ren N, Tong Q, Zhang L (2016) The detoxification and degradation of benzothiazole from the wastewater in microbial electrolysis cells. Int J Environ Res Public Health 13:1259. https://doi.org/10.3390/IJERPH13121259
192. Hang MN, Gunsolus IL, Wayland H, Melby ES, Mensch AC, Hurley KR, Pedersen JA, Haynes CL, Hamers RJ (2016) Impact of nanoscale lithium nickel manganese cobalt oxide (NMC) on the bacterium *Shewanella oneidensis* MR-1. Chem Mater 28:1092–1100. https://doi.org/10.1021/ACS.CHEMMATER.5B04505
193. Philipp LA, Bühler K, Ulber R, Gescher J (2023) Beneficial applications of biofilms. Nat Rev Microbiol 22:276–290. https://doi.org/10.1038/s41579-023-00985-0
194. Edel M, Horn H, Gescher J (2019) Biofilm systems as tools in biotechnological production. Appl Microbiol Biotechnol 103:5095–5103. https://doi.org/10.1007/S00253-019-09869-X
195. Rowe AR, Rajeev P, Jain A, Pirbadian S, Okamoto A, Gralnick JA, El-Naggar MY, Nealson KH (2018) Tracking electron uptake from a cathode into *Shewanella* cells: implications for energy acquisition from solid-substrate electron donors. MBio 9. https://doi.org/10.1128/mbio.02203-17
196. Sriram S, Olivan LA, White RJ, Rowe AR (2024) Quantifying cathode biofilm resistance in a *cdrAB* modified *Shewanella oneidensis* MR-1 using electrochemical impedance spectroscopy. Electrochim Acta 481:143967. https://doi.org/10.1016/J.ELECTACTA.2024.143967

Adv Biochem Eng Biotechnol (2025) 192: 155–192
https://doi.org/10.1007/10_2025_282

Published online: 25 March 2025

Biotechnological Applications of Cyanobacteria: *Synechocystis* and *Synechococcus* Strains

Paul Bolay, Jörg Toepel, and Bruno Bühler

Contents

Abstract Cyanobacteria as phototrophic microorganisms bear great potential for biotechnological application and a truly sustainable bioeconomy. Besides production of biomass and natural compounds, CO_2-based production of diverse value-added compounds with engineered strains enjoys ever-growing interest. Representatives of the genera *Synechocystis* and *Synechococcus* are the most used cyanobacterial model organisms for this purpose, with studies ranging from basic research to their utilization as cell factories. For both genera, genetic tools become more and more established, being, however, still far less advanced compared to those

P. Bolay
WG Microbial Chemistry, Molecular Biomimetics, Department of Chemistry – Ångström Laboratory, Uppsala University, Uppsala, Sweden

J. Toepel and B. Bühler (✉)
WG Applied Biocatalysis, Department of Microbial Biotechnology, Helmholtz Center for Environmental Research, Leipzig, Germany
e-mail: bruno.buehler@ufz.de

available for heterotrophic workhorse strains. Production of CO_2-based compounds, typically established on a proof-of-concept basis, ranges from highly complex products such as pigments, proteins, and hormones to more simple bulk products such as biofuels and commodity chemicals. For some small molecules, e.g., isobutyraldehyde, 2,3-butanediol, L-lactic acid, sucrose, and ethanol, the gram per liter scale has been achieved. The general benefits of cyanobacterial photobiotechnology are the use of light as energy source and the capacity to use CO_2 via photosynthetic carbon fixation. Additionally, the photosynthetic apparatus offers the opportunity to directly utilize electrons derived from photosynthetic water oxidation for redox biotransformations. In this respect, several enzymes have successfully been implemented in cyanobacterial strains, and high specific rates comparable to those achieved with heterotrophs have been reached. Moreover, oxygenic photosynthesis provides an ideal framework to implement oxyfunctionalization reactions also benefitting from the intracellular in situ supply of O_2. This chapter summarizes the recent advances in cyanobacterial biotechnology with a focus on *Synechocystis* and *Synechococcus* strains, encompassing both biotransformation reactions and CO_2-based product formation. Additionally, we discuss advantages and limitations of cyanobacterial chassis strains and give perspectives for future research and required measures to establish this unique group of bacteria in industrial biotechnology.

Graphical Abstract

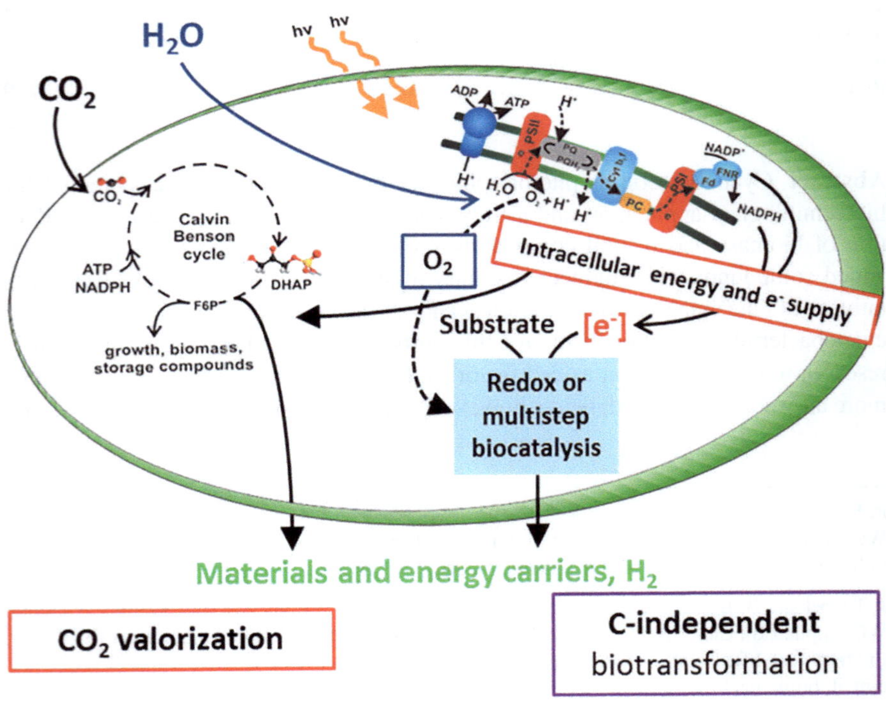

Keywords CO_2-based product formation, Cyanobacteria, Photobiotechnology, Photosynthetic efficiency, Redox biocatalysis

1 Introduction

Cyanobacteria are the only prokaryotes capable of oxygenic photosynthesis and significantly shaped life on earth since billions of years [1]. Cyanobacterial strains bear tremendous potential as chassis strains for truly sustainable microbial production profiting from their photoautotrophic lifestyle. Cyanobacteria are capable of producing biomass as well as target compounds utilizing sunlight as energy source, water as electron source, and CO_2 as carbon source with very low requirements for other nutrients, enabling carbon-negative production processes. Furthermore, the photosynthetic apparatus can be utilized to directly supply activated electrons derived from water splitting for carbon-independent production settings, e.g., redox biotransformations or hydrogen production.

In contrast to heterotrophic chassis strains, cyanobacteria do not require an organic carbon and/or energy source. Thus, they bear big potential to avoid competition for agriculturally derived feedstocks and for arable land in general. A suite of cyanobacterial strains, of which some will be discussed in more detail in this chapter, are naturally competent, i.e., amenable to genetic modifications which considerably broadens their application spectrum.

One of the key differences to well-established hosts for microbial biotechnology is the fact that the maximal CO_2 uptake rate, which sets boundaries for biomass accumulation as well as product formation from CO_2, is more than 20 times lower than uptake rates of heterotrophic chassis strains like *E. coli* for organic carbon such as glucose [2]. Consequently, the space-time yields achieved so far with cyanobacteria are low. However, in order to allow a fair and reasonable comparison, the land use and time required to generate plant-derived feedstock for heterotrophs should be considered [3].

Another paramount difference to processes based on heterotrophic strains, especially regarding up-scaling, is that cyanobacteria require light as energy source. With increasing cell density and culture depth, light penetration becomes a prominent factor limiting photo-biocatalyst performance. Thus, one of the major focal points for photobioreactor design is the maximization of the surface-to-volume ratio, in order to facilitate optimal illumination. By contrast, processes with heterotrophs mostly rely on organic feedstock as energy source, which can be administered at high rates more easily enabling the cultivation to and application of high cell densities.

A less obvious, yet important difference to well-established heterotrophs is the fact that the photoautotrophic lifestyle of cyanobacteria has evoked accordingly unique regulatory mechanisms absent in other clades. Intriguingly, even strongly conserved elements are regulated very differently. This for instance is true for the regulation of glutamine synthetase (GS), a very ancient enzyme crucial for nitrogen assimilation found in all domains of life (for a review, see [4]). Another example for the unique regulatory nature of cyanobacteria among bacteria is their distribution of signal transduction proteins. An evaluation of the signal transduction capacities of diverse bacterial clades has put forward the idea that cyanobacteria can be considered the only "introverts" among bacteria, meaning that they mainly sense the internal state of the cell rather than probing the external environment for changing conditions [5]. A possible explanation could be that key resources for cyanobacteria such as sunlight and CO_2 are mostly ubiquitous; thus, maintaining intracellular homeostasis appears to be more crucial than reacting to external changes. As a consequence, insights gained from heterotrophic systems can often not be directly transferred to the poorly understood regulatory networks of cyanobacteria which require separate evaluation.

While the phototrophic lifestyle appears to complicate the application of cyanobacteria, their redox metabolism can be highly attractive for reactions catalyzed by NADPH- or ferredoxin-dependent enzymes [6–9]. Moreover, O_2 removal is an important factor as opposed to processes based on heterotrophs [10], which often suffer from limiting O_2 supply [11, 12].

1.1 Cyanobacteria Overview

Cyanobacteria constitute a very ancient bacterial phylum and represent a highly diverse group of photoautotrophic Gram-negative bacteria [13]. They display great genetic, metabolic, as well as morphologic diversity ranging from simple, genome streamlined unicellular representatives to multicellular filamentous and colonial forms [14] including the ones featuring cellular differentiation and/or branching morphologies [15].

Cyanobacteria are of special interest as they can perform efficient oxygenic photosynthesis, of which the light reaction gives rise to the high-energy molecules ATP and NADPH as the main drivers of CO_2 assimilation in the Calvin–Benson–Bassham (CBB) cycle. In the CBB cycle, the enzyme ribulose-1,5-bisphosphate carboxylase/oxygenase (RuBisCO) catalyzes CO_2 fixation, i.e., its combination with ribulose-1,5-bisphosphate to form two molecules of the C3 compound 3-phosphoglycerate, which, upon cycling, finally is converted to the CBB product glyceraldehyde 3-phosphate (G3P). The central metabolite G3P then serves as a precursor for the synthesis of all CO_2-based biomass building blocks and storage compounds like glycogen. Some cyanobacterial representatives are capable of photoheterotrophic growth [16], while others adopted parasitic [17] or symbiotic lifestyles with a variety of different partners [18]. Moreover, cyanobacteria are the only prokaryotes that have evolved a circadian clock allowing them to establish and maintain a diurnal rhythm [19]. They occur in the entire light-exposed biosphere, occupying hostile environments like arctic and arid deserts as well as hypersaline lakes, hot springs, and rock surfaces [20–26]. Extensive sequencing efforts in recent years have rendered cyanobacteria the most thoroughly sampled group of photosynthetic organisms and revealed a staggering variability in genome size ranging from 2 to 11 Mbp [27–29]. This illustrates that cyanobacterial genomes have undergone both extension and reduction to facilitate the adaption to a variety of environments [30].

Historically, cyanobacteria have been mistaken as algae and also nowadays occasionally are referred to with the non-taxonomic term "microalgae." As a consequence, cyanobacterial cladistics was developed according to botanical nomenclature. In the late 1970s, it became apparent that cyanobacteria represent prokaryotes, and their inclusion in bacterial taxonomy was proposed [26]. Initially, they were classified based on morphological features that led to five sections, Sections I and II comprising unicellular species showing binary or multiple fission, respectively, and Sections III–V referring to multicellular species. With the advent of phylogenetic trees relying on whole-genome and 16S rRNA sequences, it was shown that several customary cyanobacterial genera are actually polyphyletic. For instance, the taxon *Synechococcus*, of which selected representatives will be discussed later in this chapter, was shown to be a polyphylum with freshwater and marine strains showing only very little kinship, while members of the genus *Prochlorococcus* appear to be more closely related to the marine *Synechococcus* representatives (see Fig. 1). As a consequence, cyanobacterial strains are nowadays not named strictly according to the binary nomenclature, but receive a number according to the culture collection they are disposed in. Figure 1 displays a phylogenetic tree for the *Synechocystis* and *Synechococcus* strains described in this study in comparison to the more ancient genus *Gloeobacter*. *Gloeobacter* is a sister group to all other cyanobacteria with specific and more ancient traits. Modern cyanobacteria are considered to have evolved after diverging from *Gloeobacter*. *Gloeobacter* do not contain thylakoid membranes, which, in other cyanobacteria, host the photosynthetic electron transport chain (PETC) including the photosystems (PSs) I and II and the light harvesting complexes. In *Gloeobacter* species instead, photosynthetic and respiratory electron transfer occurs in the cytoplasmic membrane. They contain structurally simple phycobilisomes with six rods forming a single bundle, whereas phycobilisomes in many other cyanobacteria are hemidiscoidal. The absence of the thylakoid membrane, differences in light harvesting, and missing photosynthesis proteins are used to define models for the evolution of oxygenic photosynthesis as well as the ecology and photochemistry of ancestral cyanobacteria [32].

Importantly, cyanobacteria possess ~5-fold higher photosynthetic activity than terrestrial plants, although the photosynthetic apparatus has the same evolutionary origin [33]. Cyanobacteria also show considerably shorter generation times than both plants and algae in combination with a lower degree of complexity and high metabolic plasticity.

1.2 Cyanobacterial Strains of Interest for Biotechnological Applications

In recent decades, the idea of using cyanobacteria as microbial cell factories gained increasing attention, with photosynthesis being the main attracting feature. Initially,

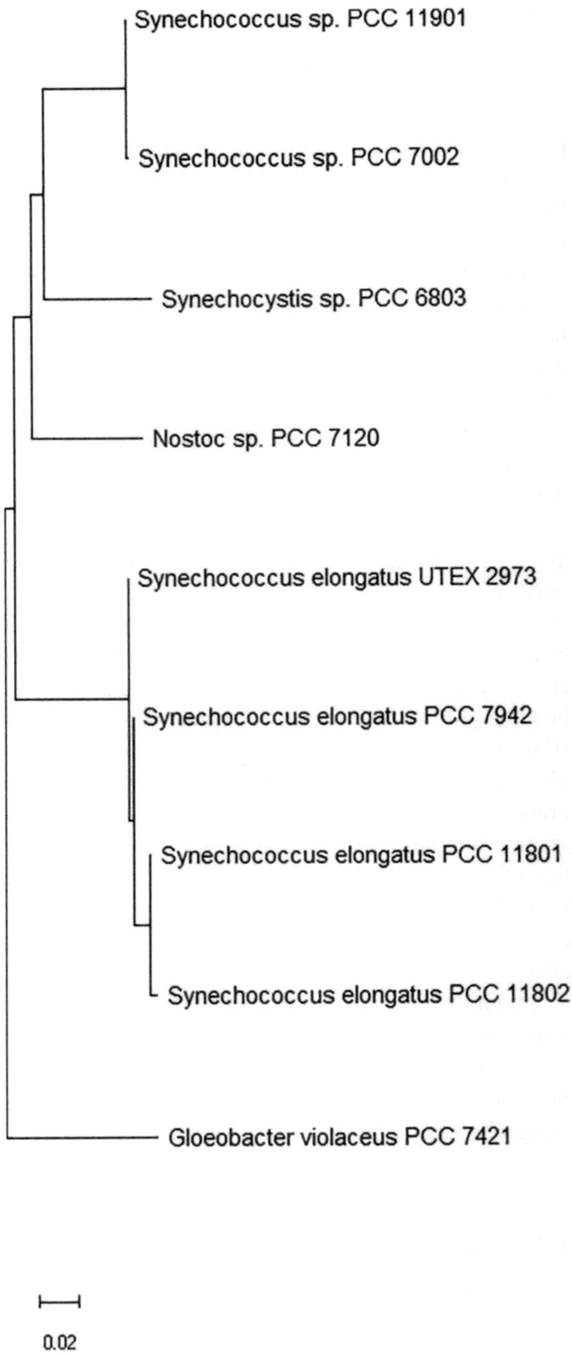

Fig. 1 Phylogenetic tree of selected cyanobacteria based on 16S rRNA gene sequences. The tree was generated with the MEGA11 [31] software package and the neighbor-joining method

cyanobacteria have mainly been considered as potential biofuel producers, bypassing the need for energy and arable land for the production of crops as feedstock for fermentative fuel production. However, in the last 15–20 years, the focus has shifted to include the sustainable production of different kinds of chemicals. Target products can be compounds natively produced by cyanobacteria, such as biomass itself, pigments, storage compounds such as cyanophycin or PHB, and molecular hydrogen. Bioactive natural products of cyanobacteria are also of biotechnological interest [34, 35]. However, in recent years, a large focus has been set on genetic engineering of cyanobacteria and the introduction of new metabolic capabilities enabling the production of compounds not natively synthesized by the host cells. These efforts brought about proofs of concept for the production of a wide range of different compounds in cyanobacteria [36–39]. Most native products, such as pigments, vitamins, or biomass, find their use as food supplements or coloring agents in the food industry [40]. While the use of cyanobacteria for food and feed dates back thousands of years, large-scale cultivation of cyanobacteria for such commercial purposes has been in operation since ~50 years [41]. Further, molecular hydrogen, considered as a main energy carrier of the future, already was targeted in the early stages of algal and cyanobacterial biotechnology [42], whereas lipids, hydrocarbons, and alcohols were also considered as fuel products. For the generation of energy carriers, large production facilities and a high solar-to-chemical energy conversion efficiency are required to reach economic and ecological viability [43, 44]. Cyanobacteria have also been engineered for the production of a range of other compounds of interest for chemical and pharmaceutical industries, including small alkanes, alkenes, alcohols, and organic acids serving as platform chemicals for the production of polymer building blocks and diverse commodity and fine chemicals [36, 45]. For such products, a smaller scale compared to fuel production is required.

One of the most abundant representatives of the cyanobacteria phylum, and in fact of all photosynthetic microorganisms, are the picocyanobacteria, encompassing only two genera, *Prochlorococcus* and *Synechococcus* [46]. These organisms are among the smallest microbes known, with a minimum cell diameter of around 0.6 μm [47] resulting in a high surface-area-to-volume ratio, facilitating nutrient uptake. *Prochlorococcus*, on the basis of its unique pigmentation, is able to absorb blue light being dominant in deep waters and thus prosper in such typical low-light zones [48]. Nevertheless, *Prochlorococcus* is also often found in high-light zones on the water surface, which demonstrates its adaptability and thus flexibility. *Synechococcus* strains share a common ancestor with *Prochlorococcus*, but they developed into genetically different groups with own survival strategies in marine ecosystems and grow in complementary but overlapping niches [46, 47, 49]. They can be found in the entire marine environment from tropical regions up to the polar circle. Several *Synechococcus* strains are by now established in both basic and applied research, namely *Synechococcus* PCC 7002 and PCC 7942. Moreover, for instance, *Synechococcus* PCC 7002 does not compete for freshwater resources, can grow in seawater, and survive high light intensities and temperatures up to 40°C [50].

Surpassing *Synechococcus* strains, *Synechocystis* sp. PCC 6803 is the most extensively studied cyanobacterial species, providing many desirable traits for chassis strains in biotechnology [51]. *Synechocystis* strains can grow photoauto-, mixo-, and heterotrophically [52] and serve as easy-to-handle model organisms for stress responses in higher plants as they are biochemically highly similar to plant chloroplasts [53]. Strain PCC 6803 was the first photoautotrophic organism, which was sequenced [54]. Despite being well investigated, only around 1,200 coding sequences (30%) are annotated a function up to date, less than half compared with *E. coli* [51], and only a small portion has been characterized. Although there may be huge differences between phototrophic and heterotrophic bacteria, most annotations are based on homologous sequences, with several *Synechocystis* genes having been re-annotated after experimental characterization [51].

Most studies on the investigation of molecular basics and on the engineering of cyanobacteria have been performed in a handful of model strains. These include the filamentous *Nostoc* sp. PCC 7120 and the unicellular freshwater strains *Synechocystis* PCC 6803, *Synechococcus elongatus* PCC 7942, as well as the marine strain *Synechococcus* PCC 7002, for which also genetic tools have been developed. This includes transformation via conjugation with *E. coli*, natural transformation as in the case of the unicellular model strains mentioned, as well as replicative and integrative vectors and control elements such as promoters for recombinant gene expression [55–59]. However, neither vectors nor promoters are necessarily compatible among cyanobacterial strains and with those used in heterotrophic workhorse strains, hampering transferability of tools and sequences [60]. Nevertheless, different tools for the different model strains have been developed for regulation of gene expression, such as anti-sense RNA or riboswitch-based methods as well as CRISPRi for multiplexed downregulation of several genes simultaneously [61–64]. Whereas, as mentioned, the PCC 6803 genome was the first of a photosynthetic organism to be sequenced, *Nostoc* sp. PCC 7120 is the primary model organism for studying nitrogen fixation and cell differentiation, PCC 7942 for studying the cyanobacterial circadian clock, and PCC 7002 for studying salt and high light tolerance [19, 50]. Being the focus of many fundamental studies, these strains are now the best-characterized strains and therefore have been the first options as host organisms for biotechnological applications.

In recent years, a number of other unicellular strains with biotechnologically desirable traits have been described, and some methods for engineering them have been established. Fast growth and thereby high metabolic/photosynthesis rates are considered a central trait for cyanobacterial production strains and heavily depend on the conditions applied in terms of light and CO_2 availability and medium composition [40]. While widely used models such as PCC 6803 (4.3 h doubling time) [65] and PCC 7942 (4.9 h doubling time) [66, 67] show moderate growth rates, several *Synechococcus* strains, namely PCC 11801, 11802, and 11901 as well as UTEX 2973, show faster growth, with the latter reaching a minimal doubling time of 1.5 h [66, 68]. Further, these strains show tolerance to and ability to benefit from higher temperatures and light intensities compared to the model *Synechocystis* and *Synechococcus* strains making them more promising for outdoor cultivation [69–

72]. In-depth studies on the molecular basis of fast autotrophic growth revealed a correlation with increased production of ATP and NADPH, high CO_2 uptake rates, and elevated transcription of genes for precursor biosynthesis and protein translation [40]. Model-guided pathway design and host strain engineering based on such findings combined with advanced synthetic biology tools and high-cell-density (HCD) cultivation/bioprocessing technologies will be central to improving performance parameters. Thus, an integrated, multidisciplinary approach is required.

More and more cyanobacterial genomes are becoming sequenced. Currently, 290 draft genomes and 84 full genomes are available online in the CyanoBase database (http://genome.microbedb.jp/cyanobase [73]). This promotes the development of genome-scale models (GSMs) and allows to adopt a systems biology approach to propose and predict the outcomes of engineering strategies [74].

2 *Synechocystis* as Model and Production Organism

While first representatives of the genus *Synechocystis* were established as early as 1892 [75], strain PCC 6803, nowadays the most thoroughly studied cyanobacterial strain, was initially isolated from a Californian pond and subsequently introduced to Pasteur's Culture Collection (PCC) in 1968 [26]. PCC 6803 is unicellular, non-diazotrophic (incapable of N_2 fixation), and euryhaline (tolerating a wide range of salt concentrations) [76, 77]. Moreover, PCC 6803 has become – in cyanobacteria terms – highly amenable to genetic modifications on account of its natural competence and the possibility to introduce untreated DNA into its chromosome via homologous recombination [78]. Attributable to its comparably long history of laboratory cultivation, several sub-strains of the initial "wild-type" isolate emerged with varying geno- and phenotypic properties such as motility and mixotrophic growth ability [79]. The pioneering role of PCC 6803 for cyanobacterial and also general photosynthesis research has led to the development of a multitude of molecular tools to investigate and engineer this strain [80, 81]. However, despite being fully sequenced since almost three decades, a major portion of its open reading frames remains to be elucidated. This means, on the one hand, that our understanding of cyanobacteria is limited and, on the other hand, that there is great potential for novel discoveries. One major drawback for the genetic engineering of PCC 6803 is its polyploidy, which varies depending on its physiological/growth stage. For instance, PCC 6803 was shown to contain up to 218 copies of its chromosome during exponential growth, whereas 58 copies were reported for later growth stages, i.e., linear or stationary phases [82]. Thus, the process to obtain full segregation, i.e., all cells bearing the introduced modification on each chromosome copy, often is time intensive and sometimes cannot be achieved at all, which limits the strain engineering throughput.

2.1 Product Formation Based on Carbon Derived from CO_2

Naturally, cyanobacteria are a large source of diverse bioactive compounds, which are produced to cope with various conditions and offer a suite of potential applications (for a review, see [83]). These include primary metabolites such as lipids, antioxidants (e.g., carotenoids) [84], as well as more complex secondary metabolites like hormones [85] and toxins [86]. Given the photoautotrophic lifestyle of cyanobacteria and the possibility to genetically engineer them, it is theoretically possible to tap their metabolism and produce virtually any enzyme-borne compound that can be found in the biosphere with CO_2 as the carbon source.

Up to date, PCC 6803 has been the most common target for such approaches, tapping a suite of different metabolites of central carbon metabolism (Fig. 2b). For instance, PCC 6803 has been modified to produce commodity chemicals such as isobutanol, isobutyraldehyde, 2,3-butanediol, lactate, or ethanol, all derived from pyruvate as the precursor in central carbon metabolism (Fig. 2b) [87–89]. Moreover, mannitol and 1,2-propanediol can be produced by directly withdrawing precursors from the CBB cycle. Several other compounds are withdrawn further away from primary metabolism, such as isobutene, derived via the branched chain amino acid (BCAA) synthesis metabolites α-ketoisovalerate and α-ketoisocaproate [90], as well as ethylene, which is derived from the TCA cycle metabolite 2-oxoglutarate [91] (2-OG, see Fig. 2b). While production of some compounds such as isoprene, which can be generated via a single enzymatic step from the MEP pathway product dimethylallylpyrophosphate, only requires the introduction of a single heterologous gene [92], other chemicals can only be produced via the introduction of multiple enzymes. For instance, 1-butanol production with PCC 6803 required the introduction of at least four non-native genes to establish a butanol biosynthetic pathway that generates 1-butanol from (*R*)-3-hydroxybutyryl-CoA as the native precursor metabolite [93]. In combination with optimized expression constructs, elimination of competing pathways, and rerouting of central carbon metabolism, a titer of 4.8 $g\ L^{-1}$ was achieved with PCC 6803 [93]. Lactate constitutes another product for which extensively engineered PCC 6803 strains have been developed [87, 94]. Here, a combinatorial approach of utilizing highly efficient lactate dehydrogenases and concomitant elimination of competing PHB and acetate synthesis has led to D-lactate titers in the $g\ L^{-1}$ scale (Table 1) [94]. For the stereoisomer L-lactate, a similar strategy was employed [87]. While the main bottleneck, the product formation through the final enzymatic step, was relieved by implementation of highly efficient ectopic expression constructs, competing pathways were attenuated by knockdown of PEP carboxylase. Moreover, additional product-forming capacity was introduced by overexpression of lactate dehydrogenase [87]. Similar to studies on heterotrophs, adapted laboratory evolution (ALE) has been applied for PCC 6803 to elicit desirable phenotypes. One remarkable result was the generation of phenylalanine excreting mutant strains via ALE for overproduction of the phenylpropanoids *p*-coumaric acid and trans-cinnamic acid. As both compounds are derived from aromatic amino acid biosynthesis (i.e., the shikimate pathway),

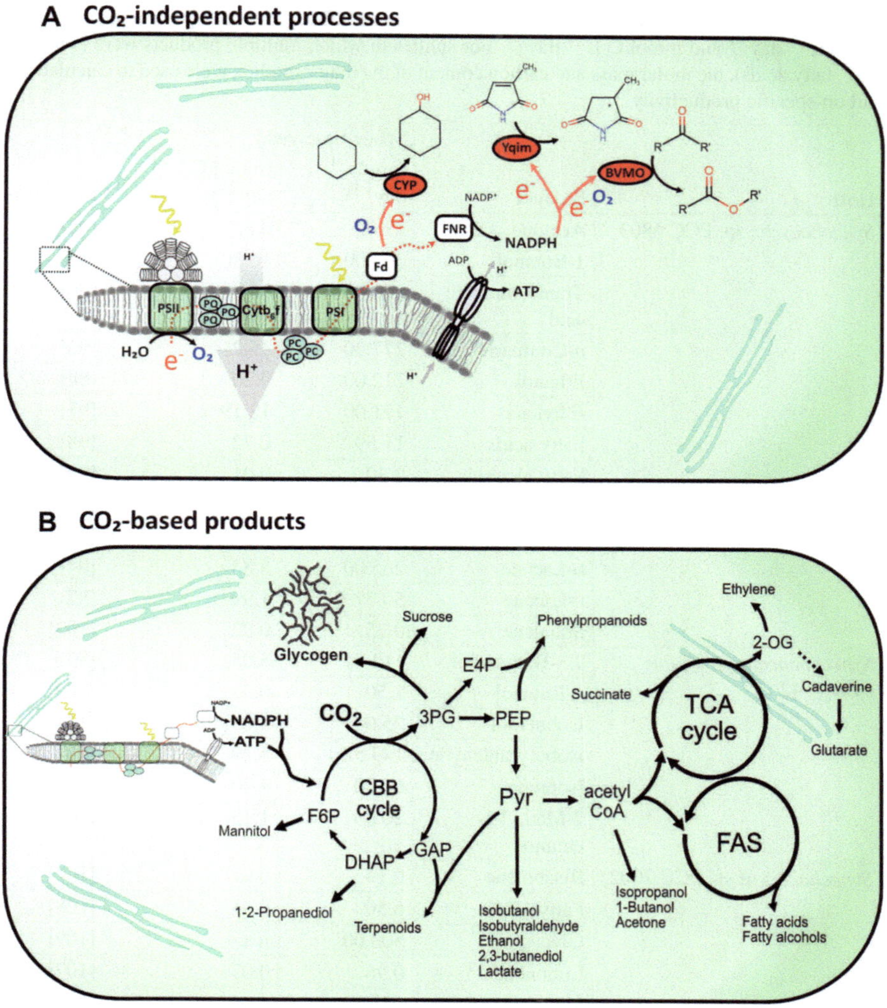

Fig. 2 Schematic overview of CO_2-independent and CO_2-based production approaches in cyanobacteria. (**a**) Redox biotransformations withdraw electrons from the photosynthetic electron transport chain. Heterologous redox enzymes are depicted in red. Oxygenases (e.g., BVMO, CYP) thereby also profit from photosynthetically produced O_2. (**b**) CO_2-based products that have been produced with cyanobacteria, connected to their source metabolite in host metabolism. Abbreviations: *PS II* photosystem II, *PQ* plastoquinone, *Cytb6f* cytochrome b_6f complex, *PC* phycocyanin, *PS I* photosystem, *Fd* ferredoxin, *FNR* ferredoxin-NADPH-reductase, *F6P* fructose-6-phosphate, *DHAP* dihydroxyacetone phosphate, *GAP* glyceraldehyde-3-phosphate, *CBB cycle* Calvin-Benson-Bassham cycle, *3PG* 3-phosphoglycerate, *PEP* phosphoenolpyruvate, *Pyr* pyruvate, *E4P* erythrose-4-phosphate, *TCA cycle* tricarboxylic acid cycle, *FAS* fatty acid synthesis, *2-OG* 2-oxoglutarate

mutant strains excreting aromatic amino acids may provide additional precursors. Indeed, by overexpressing the biosynthesis genes for *p*-coumaric acid and *trans*-cinnamic acid in phenylalanine-excreting mutants, i.e., phenylalanine ammonia

Table 1 CO_2-based product formation by *Synechocystis* sp. PCC 6803 and *Synechococcus* strains in mg L^{-1} day^{-1} and mmol C L^{-1} day^{-1}. For studies in which multiple products were generated (e.g., fatty acids), the molar mass and carbon content of the major product were used to calculate the carbon-specific productivity

Host	Product	Space-time yield		Reference
		(mg L^{-1} day^{-1})	(mmol C L^{-1} day^{-1})	
Synechocystis sp. PCC 6803	Acetone	12.00	0.62	[95]
	1-Butanol	302.00	16.30	[93]
	Trans-cinnamic acid	279.00	16.95	[96]
	p-Coumaric acid	277.20	15.21	[96]
	Ethanol	212.00	9.20	[88]
	Ethylene	171.00	12.19	[97]
	Fatty acids	11.59	0.72	[98]
	Fatty alcohols	0.30	0.01	[99]
	Isobutanol	43.60	2.35	[100]
	Isoprene	1.10	0.08	[101]
	D-Lactate	265.00	8.83	[94]
	L-Lactate	59.77	1.99	[87]
	Squalene	0.25	0.02	[102]
Synechococcus elongatus PCC 7942	2.3-Butanediol	113.81	5.05	[89]
	1-Butanol	1.50	0.08	[103]
	Isobutanol	75.00	4.05	[104]
	Isobutyraldehyde	149.52	8.29	[104]
	Isoprene	60.00	4.40	[105]
	2-Methyl-1-butanol	20.00	1.13	[106]
Synechococcus sp. PCC 7002	Bisabolene	0.14	0.01	[107]
	Fatty acids	6.50	0.41	[108]
	Glycogen	500.00	n.a.	[109]
	Limonene	0.96	0.07	[107]
	Mannitol	150.00	4.94	[110]
Synechococcus sp. PCC 11901	Free fatty acids	308.00	19.22	[71]
Synechococcus sp. PCC 11801	Ethylene	0.56	0.04	[111]
	Succinate	120.00	4.07	[112]
Synechococcus elongatus UTEX 2973	Cadaverine			
	Glutarate	16.88	0.64	[113]
	Inositol	37.51	1.25	[114]
	Limonene	8.20	0.60	[115]
	Sucrose	1900	66.61	[116]

n.a. not applicable

lyase and tyrosine ammonia lyase, respectively, the authors were able to reach g L^{-1} scale production in only 4 days [96]. However, it should be noted that these high titers were achieved with small-scale high-density cultivation systems [117].

These studies illustrate that once enzymatic steps towards product formation with high flux control have been optimized, the productivity of any compound derived from CO_2 is limited by the competition with host metabolism and, accordingly, the allocation of fixed carbon towards the compound of interest as a key factor [118]. Cyanobacteria, just like other microorganisms that thrive in ever-changing environments, have evolved stringent metabolic regulation on every conceivable level to maintain intracellular homeostasis. This can cause regulatory effects that counteract productivity and thus complicate streamlined overproduction of a given compound. For instance, it was shown that ethylene production from the central metabolite 2-OG in PCC 6803 results in considerable perturbation of C/N homeostasis [119, 120]. This is due to the fact that cyanobacteria sense the C/N ratio via 2-OG levels, as this compound acts as the carbon framework in the GS-GOGAT cycle, the major entry point for nitrogen into cyanobacterial metabolism (for a review, see [4]). Consequently, 2-OG levels inversely correlate with N availability, and 2-OG consumption by the ethylene-forming enzyme signals N-rich conditions. This leads to a downregulation of N assimilation and therefore causes N limitation on the long run. Another example for undesired regulatory effects is the production of aromatic amino acids in PCC 6803. Here, it was shown that, for unknown reasons, aromatic amino acid overproduction causes a strong increase of glycogen formation, an undesired carbon sink that competes with product formation [121].

A key feature of non-diazotrophic cyanobacteria such as PCC 6803 is the process of chlorosis, a developmental program leading to dormancy induced by limitation of key nutrients such as nitrogen or phosphate. A trademark of chlorosis is a swift proteolysis of the N-rich phycobillisome light-harvesting antenna which leads to the loss of the characteristic green color and bleaching of the cells [122]. This mechanism releases amino acids to transiently sustain protein synthesis and redox balancing, both required for acclimation to N limitation. Moreover, the C storage polymers glycogen and polyhydroxybutyrate (PHB) accumulate and act as electron and carbon sinks/storage [123].

One interesting approach that has gathered attention in recent years was the use of *glgC* knockout strains, i.e., strains incapable of glycogen synthesis, which feature the so-called overflow metabolism [124]. In contrast to WT strains, *glgC* knockout strains feature a non-bleaching phenotype when facing N limitation, which is accompanied by the secretion of organic acids such as pyruvate and 2-OG [125, 126]. The *glgC* gene encodes a glucose-1-phosphate adenylyltransferase, which is essential for the synthesis of glycogen, the major carbon and energy storage polymer of cyanobacteria [123]. While the molecular basis of the non-bleaching phenotype of *glgC* knockout strains remains enigmatic, it is believed that the lack of energy and carbon sinks strongly dysregulates the energy charge of cells, which eventually leads to the excretion of organic acids to siphon off energy and carbon. Several studies have tried to make use of the *glgC* knockout background and to utilize these strains as bona fide biocatalysts that do not accumulate biomass, but invest virtually all resources into product formation [125, 127, 128]. However, depending on the product of interest, the success of these strategies remained rather limited. In general, deletion of glycogen synthesis is critical, as these strains struggle under fluctuating light supply, given that, at night, cyanobacteria respire carbon storage compounds that have been accumulated during the day. Therefore, *glgC* knockout strains cannot be utilized under outdoor conditions.

PHB, a potential precursor for bioplastics production, is synthesized in an NADPH-consuming multistep process from acetyl-CoA by enzymes encoded by the *phaAB* and *phaEC* genes, which, in PCC 6803, are organized in two operons that are upregulated during N starvation. Several publications have shown that, under all conditions tested so far, PHB synthesis is dispensable for cellular viability, while its physiological role remains a subject of debate [129]. First attempts to overproduce PHB were successful and included redistribution of carbon flux by knockout of the recently discovered carbon flux regulator protein PirC [130] combined with overexpression of PHB synthesis genes derived from *Cupriavidus necator* [131].

Another highly interesting class of products that are naturally produced by cyanobacteria is sugars. Here, the inherent regulatory features are used to maximize sugar production and secretion. In order to cope with high salt concentrations, cyanobacteria synthesize sugars as compatible solutes to maintain osmotic pressure and to avoid desiccation.

2.2 Carbon-Independent Processes

Product formation based on carbon derived from CO_2 always has the disadvantage that the implemented pathway competes with biomass growth and thus internal metabolism for carbon. CO_2 fixation is typically the limiting factor for growth and production of CO_2-derived products, whereas the photosynthetic electron transport chain (PETC) in principle is able to provide more reduction equivalents than needed for carbon fixation. Under most conditions with illumination, light harvesting and photosynthetic water oxidation lead to an oversupply of reduction equivalents. Numerous mechanisms have evolved in phototrophic organisms to recycle the captured light energy with cyclic electron transfer as prime example. The unused potential can theoretically also be exploited to drive biocatalytic reactions via direct withdrawal of reduction equivalents from the photosynthetic apparatus and thus via reduction equivalents derived from water as the most sustainable electron donor for redox cofactor regeneration (Fig. 2a). Major challenges consist of competing light protection mechanisms and the efficient coupling of biocatalytic redox reactions via an appropriate redox cofactor supply. Thus, the efficiency of biocatalytic redox reactions implemented in cyanobacteria depends not only on suitable enzymes, but also on optimal electron transfer partners in terms of both redox potential and supply rate. Most commonly, NADH- or NADPH-dependent enzymes are applied, whereas NADPH, but not NADH, serves as the primary electron currency in phototrophs. However, cyanobacteria also allow the direct utilization of ferredoxin (Fd), the primary electron acceptor of photosystem I. The relatively low redox potential of Fd [6, 132, 133] can be an advantage provided by phototrophs for the application of Fd-dependent enzymes. Several Fd-dependent oxygenases have been coupled to photosynthesis in cyanobacteria (see below). Thereby, both enzyme-specific and host-intrinsic Fds have been found to shuttle electrons, differing in efficiency. PCC 6803 contains at least nine host-intrinsic Fds [134]. This complicates the prediction of electron transfer efficiencies towards Fd-dependent enzymes. In Table 2, an overview is given of enzymes applied for biotransformation reactions in PCC 6803 and other cyanobacterial strains.

NADPH [140, 144], NADH, flavodoxins, or plastoquinones also have the potential to be used as electron donors. Enzymes depending on NADPH as the main electron acceptor of the light reaction, thus constituting the main electron currency in cyanobacteria, may benefit from the relatively large NADPH pool and high NADPH regeneration rates, but may suffer from competition with electron-consuming metabolic routes, especially CO_2 fixation and nitrate assimilation [145]. Plastoquinone could also be used to exploit the reducing power of photosynthesis, with native transfer to the terminal oxidase as an example [146]. Its relatively positive redox potential however limits the scope of reactions to be potentially fueled with plastoquinone-derived electrons. The implementation of biocatalytic redox reactions into phototrophs was achieved in stages. Several oxidoreductases including dehydrogenases, hydrogenases, and oxygenases have been introduced into phototrophs. As with heterotrophic microbes, substrate and product inhibition, toxicity, and volatility have to be considered with substrate supply and in situ product removal as biochemical engineering options [147–149].

With dehydrogenases, the highest light-fueled specific activities of phototrophs have been reported. Testing different imine reductases, a specific activity of up to 22 U g_{CDW}^{-1} has been achieved, namely with an NADPH-dependent IRED from *Streptomyces* sp. GF 3587 in PCC 6803 for *R*-pyrrolidine formation from the corresponding imine [141]. Several studies focused on PCC 6803 containing the NAD(P)H-dependent ene-reductase YqjM from *Bacillus subtilis* catalyzing the reduction of 2-methylmaleimide to 2-methylsuccinimide as model reaction (Fig. 2a). Achieved activities depended on the substrate applied and, importantly, on the light intensity and/or cell density [137, 138, 150]. Activities of 150–180 U g_{CDW}^{-1} have been reported on the short term at a low cell density (~0.5 $g_{CDW}\,L^{-1}$), whereas on the long term, activities appeared to be restricted below 30 U g_{CDW}^{-1}. Especially at higher cell densities, the knockout of enzymes catalyzing competing (protective) reactions such as flavodiiron proteins and the soluble hydrogenase had a clearly beneficial effect on the achieved activities [137, 150]. Somewhat contradicting, 90 U g_{CDW}^{-1} has recently been proposed as reaching the maximal capacity of PCC 6803 for light-driven NAD(P)H supply in vivo, which may however also have been light-limited at the applied light intensities and cell densities [151]. These studies nicely show how sensitive these systems react to light availability and metabolic competition. Further, dehydrogenase-based systems have also been transferred to a continuous-flow microreactor system [152] and applied to drive

Table 2 Recombinant redox biotransformations in cyanobacteria (adapted from Theodosiou et al. [135] and [136] [136] and updated)

Host	Enzyme	Tapping point	Product	Max. activity, scale, time[a]	Reference
Synechocystis sp. PCC 6803	Yqjm ene reductase (*Bacillus subtilis*)	NADPH	2-Methylsuccinimide	150 U g_{CDW}^{-1}, 1 mL, 5–30 min 25.7 U g_{CDW}^{-1} [b], 200 mL, 22 h	[137, 138]
	CYP (*Acidovorax* sp. CHX100)	Fd	Cyclohexanol	39.2 U g_{CDW}^{-1}, 2 L, 52 h	[6]
	Alkane monooxygenase (*Pseudomonas putida*)	NADH	H-NAME	1.5 U g_{CDW}^{-1}, 1 mL, 30 min	[139]
	Baeyer-Villiger monooxygenase (*Acidovorax* sp. CHX100)	NADPH	ε-Caprolactone	60.9 U g_{CDW}^{-1}, 2 L, 24 h	[140]
	Imine reductase (*Streptomyces* sp. GF3587)	NADPH	Cyclic amines	21.8 U g_{CDW}^{-1} [b], 2 mL, 1 h	[141]
	CYP79A1 (*Sorghum bicolor*)	Fd	*p*-Hydroxyphenylacetaldoxime	1.0 U g_{CDW}^{-1}, ~mL-L, 24 h	[133]
	CYP79A1, CYP71E1, and glycosyltransferase (*Sorghum bicolor*)	Fd	Dhurrin	0.03 U g_{CDW}^{-1}, ~mL-L, 12 h	[133]
Synechococcus elongatus PCC 7942	Alcohol dehydrogenase (*Lactobacillus kefir*)	NADPH	(R)-1-phenylethanol	74–95 U g_{CDW}^{-1}, 50 mL, 4–6 h	[142]
Synechococcus sp. PCC 7002	CYP79A1 (*Sorghum bicolor*)	Fd	*p*-Hydroxyphenylacetaldoxime	26 μM titer with $OD_{730} = 2.5$[c]	[143]

CYP cytochrome P450 monooxygenase, *H-NAME* ω-hydroxynonanoic acid methyl ester, *Fd* ferredoxin

[a] Performance parameters given if available in literature

[b] Calculated from substrate consumption

[c] Biomass titer and time of biotransformation not available

extracellular redox biocatalysis with isopropanol as electron shuttle being regenerated from acetone by light-driven reduction in recombinant PCC 6803 [153].

The coupling of hydrogenases to photosynthesis constitutes the oldest but nevertheless highly attractive approach to utilize photosynthetically energized electrons, in this case to produce H_2 considered as one of the most important energy carriers of the future [154, 155]. The production of the so-called white H_2 can be achieved via two approaches: biophotolysis and biophotovoltaics (BPV). Biophotolysis implies direct coupling of oxygenic photosynthesis to H_2 formation via hydrogenases [156], whereas BPV is based on microbial electrolysis. Thereby, photosynthetically energized electrons from phototrophs are transferred via a carrier to an anode, finally enabling proton reduction at the cathode. Biophotolysis and BPV have the advantage to withdraw electrons directly from the light reaction, whereas other biological approaches such as fermentative H_2 production depend on carbon metabolism and respective energy loss. Whereas the efficiency of electron transfer to the anode is a critical factor for BPV, biophotolysis typically suffers from the O_2 sensitivity of most hydrogenases. Further, the electron transfer between PETC and hydrogenases constitutes a bottleneck [157, 158]. Besides NADPH, Fd constitutes an interesting electron carrier due to its low redox potential ($Fd_{red/ox}$: −430 mV; H_2/H^+: −413 mV) [159]. Recent advances in addressing these issues include the successful expression of an O_2-tolerant hydrogenase in PCC 6803 [160], with cofactor specificity as the remaining engineering target, and hydrogenase-PS I fusion, which was achieved in PCC 6803 for its native hydrogenase module enabling extended photo-H_2 production under anaerobic conditions [161]. As a strategy to avoid intracellular competition for redox equivalents, the affinity of the Fd-$NADP^+$-reductase (FNR) for Fd has been attenuated, thereby making more Fd available for target reactions, e.g., H_2 formation catalyzed by Fd-accepting hydrogenases [162]. Together with solving these obstacles, reaction engineering and downstream processing need to be implemented to enable efficient product recovery and overcome O_2 accumulation, which in combination with H_2 forms explosive oxyhydrogen gas [154].

Besides providing reduction power, the photosynthetic light reaction can also supply O_2 for O_2-dependent reactions [139, 163] (Fig. 2a). Thereby, a major limitation in the application of oxygenases in heterotrophs can be overcome. Reliance on PS II-derived O_2 was shown, e.g., for a *Pseudomonas*-derived alkane monooxygenase AlkBGT in PCC 6803 for terminal fatty acid methyl ester hydroxylation [139, 164]. Thereby, however, rather low rates (1–2 U g_{CDW}^{-1}) were reached, which can be attributed to the NADH dependency of AlkBGT. In contrast to NADPH, NADH is not the primary electron carrier in phototrophs. Application of such NADH-dependent enzymes in cyanobacteria can benefit from either a change in enzymatic cofactor preference [87, 165, 166] or redirection of electrons towards NADH via transhydrogenases [167].

With an *Acidovorax* cytochrome P450 monooxygenase (CYP) in PCC 6803, external O_2 supply could even be completely omitted with the reaction fueled by photosynthesis-derived O_2 [6]. This avoided substrate evaporation and allowed a specific activity of 40 U g_{CDW}^{-1} in a two-liquid phase system operated in a lab-scale stirred-tank photobioreactor, the highest CYP activity reported for cyanobacteria so far. O_2 supply was also demonstrated in mixed-species biofilms, in which PCC 6803 provided inherent heterologous CYP as well as heterotrophic whole-cell biocatalysts with O_2 [168, 169]. Testosterone 15β-hydroxylation has also been shown in PCC 6803 containing CYP110D1 from *Nostoc* sp. PCC 7120 [170]. Without recombinantly expressing electron transfer partners, a specific activity of 1 U g_{CDW}^{-1} was achieved. A study investigating the most suitable electron carrier for photosynthesis-driven CYP catalysis showed that (1) the type of available electron carrier determines the biocatalytic reaction rate and (2) this rate profits from electron carrier to CYP450 fusion [171]. Methods to improve intracellular CYP catalysis, including enzyme, redox partner, and substrate engineering, have recently been reviewed [172]. Direct linkage of enzymes to the PETC can significantly increase production rates as shown for a thylakoid-linked CYP [143]. Interestingly, PCC 6803-intrinsic enzymes were found capable of completely substituting the function of the FNR of the *Acidovorax* three-component CYP mentioned above, with transfer of electrons from PS I to the CYP-inherent ferredoxin as the main electron transfer pathway [9]. To a certain extent (20%), CYP-inherent ferredoxin was also substituted by PCC 6803-intrinsic electron shuttles.

With NADPH-dependent Baeyer-Villiger monooxygenases (BVMOs) in PCC 6803, specific activities of 2 up to 61 U g_{CDW}^{-1} have been achieved for the conversion of cyclic ketones [173]. For the highest activity achieved for cyclohexanone conversion, kinetic constraints have been found to be more relevant in the longer term than photosynthesis-related effects such as light limitation [140]. The identified product inhibition could be overcome by implementing a first cascade into PCC 6803, which involves the further conversion of caprolactone to the corresponding hydroxy acid catalyzed by a lactonase [8, 140]. Interestingly, electron withdrawal via BVMO catalysis appeared to cause an ATP/NADPH imbalance, affecting the host metabolism and limiting light-driven redox biocatalysis, especially in the long run. Engineering approaches to combine pathways relying on ATP supply with pathways relying on reducing power showed a way to avoid such imbalance [174, 175]. Thus, the balance of sources and sinks should be considered.

Photosynthetic in situ O_2 supply can also be considered attractive for O_2-dependent enzymes not relying on reduction power, e.g., alcohol oxidases, amine oxidases, laccases, or dioxygenases. The oxidation of alcohols to aldehydes or ketones by alcohol oxidases is of particular interest for the production of flavors and fragrances. Examples include benzaldehyde (bitter almond), cinnamaldehyde (cinnamon), octanal (citrus), 2-heptanone (banana), ionones (rose), and vanillin (vanilla) [176, 177]. Laccases, which gained interest in recent years [178], are present in phototrophic species and have the potential to be used for degradation processes [179, 180].

3 *Synechococcus* Strains Interesting for Productive Applications

While one of the major limitations of cyanobacterial applications is the relatively slow growth and concomitantly low CO_2 uptake rates, several highly interesting strains with very promising features now have been isolated and characterized that might turn the tide for commercial photobiotechnology. These include the marine strains *Synechococcus* PCC 11901 [71] and PCC 7002 [181] as well as the freshwater strains *S. elongatus* PCC 11802 [69], UTEX 2973 [72], and PCC 11801 [70]. A great benefit of marine strains is that they can potentially be cultivated in seawater not competing for freshwater resources. The most notorious is PCC 11901, a marine cyanobacterium isolated in Singapore that made headlines for its ability to swiftly accumulate large amounts of biomass while being able to thrive under high temperatures, salinities, and light intensities. Intriguingly, PCC 11901 can accumulate up to 33 g_{CDW} L^{-1} [71], thus being in the same range as what can be achieved with *E. coli* upon stirred-tank bioreactor-based fed-batch cultivation [182]. While having a seven-times elevated O_2 evolution rate per chlorophyll molecule when compared to PCC 6803 (a measure for photosynthetic efficiency) [183], PCC 11901 was also shown to be genetically engineerable with tools that have already been developed for other, closely related strains [71]. In a proof-of-concept study, Włodarczyk et al. were able to engineer 11901 to produce significant amounts of free fatty acids in the g L^{-1} range, similar to what can be achieved with heterotrophic hosts [71]. This has sparked a hype in the cyanobacterial biotech community with several laboratories currently trying to make use of these new potential "superhosts" while providing a toolbox for convenient and straightforward engineering. Recent studies report the development of CRISPR-based engineering approaches, allowing marker-less modification of the PCC 11901 genome, as well as the development of a suite of both constitutive and inducible promoters for PCC 11901, both crucial for metabolic engineering [184]. Taken together, these high-performance strains hold high promises for the commercialization of cyanobacterial photobiotechnology being able to compete with processes based on heterotrophic chassis strains.

Up to now, just a few biotransformation reactions have been implemented in *Synechococcus* strains. A plant CYP79A1 from *Sorghum bicolor* was transferred into PCC 7002 and was anchored to the thylakoid membrane, which improved the rate for tyrosine conversion to *p*-hydroxyphenylacetaldoxime [143]. CYP79A1 was shown to receive electrons directly from native ferredoxins. Further studies in algae gave interesting insights into redox partner preferences [145, 171, 185]. Light-driven catalysis has also been shown for mammalian CYP1A1 in PCC 7002 [132, 186]. Thereby, CYP1A1 activity towards ethoxyresorufin giving rise to fluorescent resorufin could be enhanced by deletion of the strain-inherent cytochrome c. Overall, these activities achieved with CYPs of eukaryotic origin remained rather low. For the asymmetric conversion of acetophenone to chiral 1-phenylethanol by recombinant PCC 7942 that expresses an NADPH-dependent *L. kefir* alcohol dehydrogenase, significantly higher rates have been achieved (Table 2) [142]. Besides light, carbon dioxide levels were found to play a critical role possibly correlating with enhanced growth involving improved cofactor availability at elevated CO_2 levels. The faster photosynthetic apparatus of fast-growing *Synechocystis* strains promises even higher specific activities of the implemented enzymes. However, a possible ATP/NADPH imbalance at high rates of electron withdrawal remains to be considered [175].

The energy/redox balance also has to be contemplated in case of CO_2-based product formation [187]. Further, restrictions in genetic toolboxes available to produce enzymes in sufficient amounts currently hinder the engineering and application of cyanobacteria. However, toolboxes have recently been amended. This is especially true for CRISPR/CAS-based generation of knockout strains [68, 183, 184, 188]. Further, synthetic biology approaches to delete large parts of the genome nonessential for survival and thus reduce genome size deserve mentioning [189, 190], with the potential to enable faster growth and create optimized cyanobacterial chassis strains for biotechnological applications. Further basic research towards this goal includes the identification of neutral sites for genomic integrations and the screening, identification, and testing of expression elements like promoters and RBSs [184, 188]. Victoria et al. developed tools based on the CyanoGate MoClo system for PCC 11901. They characterized neutral sites and tested a suite of constitutive promoters, terminators, and inducible promoters including a 2,4-diacetylphloroglucinol (DAPG)-inducible PhlF repressor system, which has not previously been demonstrated in cyanobacteria, and showed tight regulation and a 228-fold dynamic range of induction [184]. Sun et al. adapted the genome of PCC 7942 to a hypermutation system [191], increasing mutation rates to produce high light- and high temperature-tolerant strains, needed for applications in outdoor systems. DNA replication fidelity mechanisms (*mutS*, *mutL*, *dam*, *xthA*, *uvrB*, *uvrC*, *mutY*, and *mutM*) were knocked out, and four genes contributing to error-prone DNA replication (*umuC*, *umuD*, *recN*, and *recA*) were overexpressed individually or in combination.

Up to date, PCC 6803 and PCC 7942 are the most used chassis strains to create “light-driven autotrophic cell factories,” which enabled the production of several dozens of biofuels and/or fine chemicals directly from CO_2. Both are model organisms for basic as well as applied research [51, 68]. Particularly, strains engineered for sucrose production have been investigated regarding production constraints, such as the influence of environmental factors like light and CO_2 availability on production and regulation processes involved, as well as regarding mitigation approaches such as the use of microbial consortia. Further, knockout of a glucokinase in PCC 7942 was shown to enable intracellular glucose accumulation. In combination with spontaneous mutations, the secretion of 1.5–5 g L^{-1} glucose was achieved [192]. However, to overcome issues related to slow growth and low biomass accumulation, UTEX 2973, PCC 11801, PCC 11802, and PCC 11901 are moving into focus [68].

As shown for PCC 11801 and PCC 11802 via ^{13}C metabolic flux analysis, fast-growing *Synechococcus* strains feature a higher CBB cycle flux, negligible fluxes through the oxidative pentose phosphate and the photorespiratory pathways, and lower anaplerotic fluxes under photoautotrophic conditions [193]. The fast-growing strain PCC 11801 seems to be ideal for the large-scale production of TCA cycle-derived chemicals due to a negligible flux through the malate dehydrogenase in the TCA cycle and its possible diversion to respective products. Fast-growing UTEX 2973 has been applied for sucrose production. Making use of the capability of cyanobacteria to produce sugars as compatible solutes to maintain osmotic pressure and avoid desiccation, the introduction of a sucrose exporter into this strain has led to a productivity of 1.9 g L^{-1} day^{-1}, the highest reported so far for CO_2-based production via photobiotechnology [116]. Further research on fast-growing *Synechococcus* strains can be expected to advance and expand their huge potential for photobiotechnological application.

4 Applications, Limitations, Critical Factors, and Challenges

Industrial scale applications of cyanobacteria are scarce. Up to date, no commercial production processes are established with PCC 6803 and the *Synechococcus* strains described in this chapter. Commercialization mainly comprises low-volume high-cost products produced by wild-type strains such as antiaging products, for which low product yields and high costs are tolerable. Large-scale production processes are established for biomass or pigment production, mainly with *Spirulina* and similar strains. The high protein content (up to 50% [194]) and the optimal protein composition make them ideal for application as food supplements [195]. Pigments and other intracellular metabolites, including those with pharmaceutical relevance, are preferentially produced in large-scale outdoor facilities [41, 196]. A prerequisite for application thereby is cheap biomass production.

To produce biofuels or chemicals in a large scale, several hurdles have to be taken. Whereas model and emerging strains discussed in this chapter are most suitable for such production processes, their successful implementation relies on the relief of bottlenecks in CO_2 fixation, improved carbon yields, enhanced photosynthetic efficiency, genetic toolbox enlargements, and development and implementation of new light-efficient reactor systems [197]. Another factor that has to be considered especially for large-scale outdoor cultivation is biosafety. Here, it is paramount to ensure that neither genetically modified cyanobacteria nor potentially harmful genetic constructs such as broad-range plasmids or antibiotic resistance genes persist in the environment. A potential approach is the implementation of intrinsic suicide switches that are activated upon spillage, i.e., when the controlled environment of the bioreactor is left [198].

For the envisioned photosynthesis-driven redox biocatalysis, the light-to-product efficiency is of special interest. Major questions to be answered are the following:

1. How efficiently does the whole-cell biocatalyst use the incident light? What electron transfer rates (ETRs) can be reached?
2. At which point can photosynthesis be tapped at which efficiency?
3. Which strains or strain traits are most feasible?

The light-to-biomass efficiency is maximal at the light saturation point E_k, where the incident light is just sufficient to supply the metabolism with the reductive power needed for biomass production. In natural environments, phototrophic organisms face a "light usage dilemma" [199], consisting of the incident light being either below or above E_k and thus impairing overall photosynthetic efficiency. In fact, 82% of the incident photosynthetically active radiation (PAR) reaches the earth surface with light intensities above E_k [199]. Consequently, the photosynthetic apparatus is evolutionarily optimized to handle over-excitation rather than to maximize light usage efficiency.

Within the photosynthetic apparatus, (excess) energy is lost at different levels, often by mechanisms important to protect the apparatus from over-excitation:

1. Depending on their pigmentation, phototrophic organisms absorb only a share of PAR, expressed as the share of photosynthetically absorbed quanta Q_{phar}.
2. At PS II, part of the energy is dissipated via heat or fluorescence emission and non-photochemical quenching, which decreases the operative quantum yield at PS II (Φ_{PSII}). Φ_{PSII} can be assessed via pulse-amplitude-modulation (PAM) fluorometry to calculate the ETR in PS II.
3. Alternative electron fluxes via terminal oxidases and Mehler- or Mehler-type reactions further compromise the light-to-biomass efficiency.

As a result, only a fraction of the incident light energy can be used for biomass formation or productive (biocatalytic) reactions. Due to the energy losses along the PETC, it is obvious that the closer a reaction is situated to the water splitting, the higher are electron supply rate and light-use efficiency. About 50–60% of the light energy can in principle be tapped at PS II or directly downstream to it, at the plastoquinone pool, where losses only result from heat and fluorescence dissipation and non-photochemical quenching [199]. Tapping reductive power at PS I (doubling the amount of photons required) or at its acceptor side (via Fd or NADPH as described above) involves electron losses to respiratory elements like oxidases or Mehler- and Mehler-type reactions. Taken together, these losses account for 20–30%, reducing the light-use efficiency to 15–20%. Besides these theoretical efficiencies, the redox potentials of tapped PETC cofactors or electron carriers are of practical importance as reviewed by Mellor et al. [145]: PS II (~−110 mV in PCC 6803 [200]) and plastoquinone (+80 to +110 mV [201]) have a by far lower reductive power than already used electron carriers like Fd (−430 to −300 mV [202]) and NADPH (−320 mV), which makes the latter ones highly suitable electron donors for redox biocatalysis.

CO_2-based production further depends on the efficiencies of the CBB cycle and involved metabolic pathways as well as on the carbon partitioning at the relevant branching points after the CBB cycle. Angermayr et al. nicely correlated products and corresponding branching points to achieved fluxes and partitioning [36]. It becomes clear that achieved rates and partitionings highly differ and that production efficiency not only depends on the branching point. Engineering of driving forces, utilization of cofactors generated, and understanding of the regulation of the carbon flow are essential to achieve efficient product formation [187]. Furthermore, studies showed that the so-called sink and source hypothesis needs to be investigated and considered in detail. This means that investigating and determining the potential to increase photosynthetic efficiency just by providing a suitable sink can help to identify optimal strategies and starting points to improve photosynthetic efficiency and streamline carbon flow.

In agreement with these theoretical considerations regarding light-to-product efficiencies, potential hosts for light-driven biotechnology require special traits to enable powerful and robust processes. High-light and high-temperature tolerances are especially important for outdoor applications. ALE in combination with systems biology approaches was successfully applied for *Synechocystis* and *Synechococcus* strains [203–207]. PCC 6803, for instance, was adapted to cope with additional (light) energy on account of a higher light harvesting capacity [203], auguring well for an effective use of high light intensities. As detailed above, fast-growing *Synechococcus* such as UTEX 2973 [66] and PCC 11901, 11801, and 11802 are prone to enable significant progress in terms of production rates [71]. For the production of fuels or chemicals with (phototrophic) whole-cell factories, solvent tolerance constitutes another desirable trait, for which considerable ALE-based progress has recently been reported for *Synechococcus* [206].

As mentioned above, phototrophs experience light excess under normal/natural conditions and have evolved multiple mechanisms for the dissipation of excessive energy [208, 209]. Approaches are needed to avoid such energy loss by overflow valves and to control electron flow. Several studies tackled this aspect in cyanobacteria and green algae. The major protection mechanisms in cyanobacteria are cyclic electron flow around PS I consuming up to 35% of the photosynthetically captured electrons [210], flavodiiron proteins, orange carotenoid protein as the main energy dissipater in cyanobacterial antenna systems, and adaptations of the photosynthetic apparatus, e.g., state transitions and antenna sizes. Flavodiiron proteins were targeted in several studies [211]. Knockout strategies increased electron flow and desired product formation [137, 212]. As recently reviewed [213, 214], numerous mechanisms protect green algae from photoinhibition, which should be tackled in a holistic approach [215].

5 Perspectives

Overall, typical limitations of photobiotechnological processes include light limitation, inefficient light utilization, low biomass concentrations, and low production rates. As light supply is a key aspect and a major limiting factor for high-density cultivation of phototrophs, measures are needed to improve light supply and utilization [216]. Major factors are the limited absorption capacities and self-shading effects in outdoor applications [217]. Systems and synthetic biology approaches aim at the engineering of strains with improved photosynthetic performance based on, e.g., improved photosystems, optimal protein allocation, and/or efficient carbon concentration and fixation [218–223]. Strategies include the reduction of antenna systems for more efficient light absorption in high-density cultures [224], attenuation of protection mechanisms [211, 213, 225], and streamlining of carbon, energy, and redox metabolism [37, 175, 223, 226, 227]. Conventional photobioreactor systems are often limited by low cell densities [228]. Therefore, innovative approaches to overcome light limitation have been developed recently [228–230], including bioreactors with internal illumination [138, 231], biofilm-based concepts [232], and cell immobilization approaches [233, 234]. Biofilm bioreactors enable high surface-to-volume ratios and thus high biomass concentrations and high light utilization efficiency. Bioreactors with internal light supply enabled a reduced light limitation of suspended cells and a more than twofold increase in volumetric productivity for photosynthesis-driven 2-methylmaleimide reduction [138].

Furthermore, operating costs and costs for product removal have to be considered. Substrate and product toxicity constitute a general critical factor for industrial application, which can be mitigated by in situ product removal (ISPR) concepts. As an example, a two-liquid-phase system has been demonstrated to stabilize a cyanobacterial CYP-based process by reducing substrate toxicity and volatility [6].

Further, natural or synthetic consortia recently attracted attention [235, 236], offering benefits regarding process stability, multistep enzyme cascades, cofactor supply, and light utilization [237]. It was demonstrated that a two-species biofilm approach enables a process duration of several weeks for cyclohexane hydroxylation at a rate of up to 3.76 $g_{cyclohexanol}$ m^{-2} day^{-1} with reduction equivalents and O_2 provided by the photosynthetic light reaction in cyanobacteria [169].

Other critical factors in photobiotechnology, not further addressed here, include temperature and pH control, gas (CO_2, O_2) mass transfer, strain-specific nutrient supply, water evaporation, harvesting costs, and sterility. Interested readers are referred to relevant reviews [238–241].

6 Conclusion: Future Research Demands

To apply cyanobacteria in production processes and to overcome the abovementioned limitations, the following aspects have to be considered and partially are under investigation. Production processes need to be competitive with existing structures. Therefore, production processes need to be large scale and cost saving. Outdoor cultivation thereby constitutes the most promising approach. For this purpose, however, production strains need to be stress tolerant, and, ideally, waste or seawater should be used. This requires high salt and light tolerance as well as a high efficiency in light to chemical energy conversion, for which manipulation of the photosynthetic machinery is needed. Further, systems biology approaches promise a streamlining of carbon and electron fluxes in cyanobacteria, which, in combination with genome reduction, is also prone to enhance growth. Thereby, it is important to note that regulation of most pathways in cyanobacteria differs from that of heterotrophic bacteria, necessitating research on regulatory network structure and operation. The implementation of artificial CO_2-fixing pathways can be envisaged to circumvent the limited CO_2 fixation capacity of cyanobacteria. Finally, to overcome growth limitation and limited production capacity, co-cultivation with heterotrophic production strains can be considered. Furthermore, photomixotrophy, with growth relying not only on photosynthesis but also on residues, may be exploited to enhance growth, productivities, and product titers.

References

1. Sanchez-Baracaldo P, Cardona T (2020) On the origin of oxygenic photosynthesis and cyanobacteria. New Phytol 225:1440–1446
2. Zavrel T, Faizi M, Loureiro C, Poschmann G, Stuhler K, Sinetova M, Zorina A, Steuer R, Cerveny J (2019) Quantitative insights into the cyanobacterial cell economy. elife 8:e42508
3. Brandenburg F, Theodosiou E, Bertelmann C, Grund M, Klähn S, Schmid A, Krömer JO (2021) Trans-4-hydroxy-L-proline production by the cyanobacterium *Synechocystis* sp. PCC 6803. Metab Eng Commun 12:e00155
4. Bolay P, Muro-Pastor MI, Florencio FJ, Klähn S (2018) The distinctive regulation of cyanobacterial glutamine synthetase. Life 8:52
5. Galperin MY (2005) A census of membrane-bound and intracellular signal transduction proteins in bacteria: bacterial IQ, extroverts and introverts. BMC Microbiol 5:35
6. Hoschek A, Toepel J, Hochkeppel A, Karande R, Bühler B, Schmid A (2019) Light-dependent and aeration-independent gram-scale hydroxylation of cyclohexane to cyclohexanol by CYP450 harboring *Synechocystis* sp. PCC 6803. Biotechnol J 14:e1800724
7. Toepel J, Karande R, Klähn S, Bühler B (2023) Cyanobacteria as whole-cell factories: current status and future prospectives. Curr Opin Biotechnol 80:102892
8. Tüllinghoff A, Djaya-Mbissam H, Toepel J, Bühler B (2023) Light-driven redox biocatalysis on gram-scale in *Synechocystis* sp. PCC 6803 via an *in vivo* cascade. Plant Biotechnol J 21: 2074–2083
9. Tüllinghoff A, Toepel J, Bühler B (2024) Enlighting electron routes in oxyfunctionalizing *Synechocystis* sp. PCC 6803. ChemBioChem 25:e202300475
10. Kazbar A, Cogne G, Urbain B, Marec H, Le-Gouic B, Tallec J, Takache H, Ismail A, Pruvost J (2019) Effect of dissolved oxygen concentration on microalgal culture in photobioreactors. Algal Res 39:101432
11. Duetz WA, van Beilen JB, Witholt B (2001) Using proteins in their natural environment: potential and limitations of microbial whole-cell hydroxylations in applied biocatalysis. Curr Opin Biotechnol 12:419–425
12. Marques MPC, Cabral JMS, Fernandes P (2010) Bioprocess scale-up: quest for the parameters to be used as criterion to move from microreactors to lab-scale. J Chem Technol Biotechnol 85:1184–1198
13. Demoulin CF, Lara YJ, Cornet L, Francois C, Baurain D, Wilmotte A, Javaux EJ (2019) Cyanobacteria evolution: insight from the fossil record. Free Radic Biol Med 140:206–223
14. Rippka R, Deruelles J, Waterbury JB, Herdman M, Stanier RY (1979) Generic assignments, strain histories and properties of pure cultures of cyanobacteria. J Gen Microbiol 111:1–61
15. Herrero A, Stavans J, Flores E (2016) The multicellular nature of filamentous heterocyst-forming cyanobacteria. FEMS Microbiol Rev 40:831–854

16. Joset-Espardellier F, Astier C, Evans EH, Carr NG (1978) Cyanobacteria grown under photoautotrophic, photoheterotrophic, and heterotrophic regimes: sugar metabolism and carbon dioxide fixation. FEMS Microbiol Lett 4:261–264
17. Kristiansen A (2019) *Sarcinastrum urosporae*, a colourless parasitic blue-green alga. Phycologia 4:19–22
18. Gehringer MM, Pengelly JJ, Cuddy WS, Fieker C, Forster PI, Neilan BA (2010) Host selection of symbiotic cyanobacteria in 31 species of the Australian cycad genus: *Macrozamia* (Zamiaceae). Mol Plant Microbe Interact 23:811–822
19. Cohen SE, Golden SS (2015) Circadian rhythms in cyanobacteria. Microbiol Mol Biol Rev 79: 373–385
20. Cires S, Casero MC, Quesada A (2017) Toxicity at the edge of life: a review on cyanobacterial toxins from extreme environments. Mar Drugs 15:233
21. Gaysina LA, Saraf A, Singh P (2019) Cyanobacteria in diverse habitats. In: Mishra AK, Tiwari DN, Rai AN (eds) Cyanobacteria. Academic Press, pp 1–28
22. Hauer T, Mühlsteinová R, Bohunická M, Kastovsky J, Mares J (2015) Diversity of cyanobacteria on rock surfaces. Biodivers Conserv 24:759–779
23. Jungblut AD, Vincent WF (2017) Cyanobacteria in polar and alpine ecosystems. In: Margesin R (ed) Psychrophiles: from biodiversity to biotechnology. Springer International Publishing, Cham, pp 181–206
24. Makhalanyane TP, Valverde A, Velázquez D, Gunnigle E, Van Goethem MW, Quesada A, Cowan DA (2015) Ecology and biogeochemistry of cyanobacteria in soils, permafrost, aquatic and cryptic polar habitats. Biodivers Conserv 24:819–840
25. Rasul F, You D, Jiang Y, Liu X, Daroch M (2024) Thermophilic cyanobacteria-exciting, yet challenging biotechnological chassis. Appl Microbiol Biotechnol 108:270
26. Stanier RY, Kunisawa R, Mandel M, Cohen-Bazire G (1971) Purification and properties of unicellular blue-green algae (order Chroococcales). Bacteriol Rev 35:171–205
27. Berta-Thompson JW, Thomas E, Cubillos-Ruiz A, Hackl T, Becker JW, Coe A, Biller SJ, Berube PM, Chisholm SW (2023) Draft genomes of three closely related low light-adapted *Prochlorococcus*. BMC Genomic Data 24:11
28. Bombar D, Heller P, Sanchez-Baracaldo P, Carter BJ, Zehr JP (2014) Comparative genomics reveals surprising divergence of two closely related strains of uncultivated UCYN-A cyanobacteria. ISME J 8:2530–2542
29. Dagan T, Roettger M, Stucken K, Landan G, Koch R, Major P, Gould SB, Goremykin VV, Rippka R, Tandeau de Marsac N, Gugger M, Lockhart PJ, Allen JF, Brune I, Maus I, Puhler A, Martin WF (2013) Genomes of stigonematalean cyanobacteria (subsection V) and the evolution of oxygenic photosynthesis from prokaryotes to plastids. Genome Biol Evol 5:31–44
30. Shih PM, Wu D, Latifi A, Axen SD, Fewer DP, Talla E, Calteau A, Cai F, Tandeau de Marsac N, Rippka R, Herdman M, Sivonen K, Coursin T, Laurent T, Goodwin L, Nolan M, Davenport KW, Han CS, Rubin EM, Eisen JA, Woyke T, Gugger M, Kerfeld CA (2013) Improving the coverage of the cyanobacterial phylum using diversity-driven genome sequencing. Proc Natl Acad Sci USA 110:1053–1058
31. Tamura K, Stecher G, Kumar S (2021) MEGA11: molecular evolutionary genetics analysis version 11. Mol Biol Evol 38:3022–3027
32. Grettenberger CL, Sumner DY, Wall K, Brown CT, Eisen JA, Mackey TJ, Hawes I, Jospin G, Jungblut AD (2020) A phylogenetically novel cyanobacterium most closely related to *Gloeobacter*. ISME J 14:2142–2152
33. Melis A (2009) Solar energy conversion efficiencies in photosynthesis: minimizing the chlorophyll antennae to maximize efficiency. Plant Sci 177:272–280
34. Demay J, Bernard C, Reinhardt A, Marie B (2019) Natural products from cyanobacteria: focus on beneficial activities. Mar Drugs 17:320
35. Tan LT (2007) Bioactive natural products from marine cyanobacteria for drug discovery. Phytochemistry 68:954–979

36. Angermayr SA, Gorchs Rovira A, Hellingwerf KJ (2015) Metabolic engineering of cyanobacteria for the synthesis of commodity products. Trends Biotechnol 33:352–361
37. Jaiswal D, Sahasrabuddhe D, Wangikar PP (2022) Cyanobacteria as cell factories: the roles of host and pathway engineering and translational research. Curr Opin Biotechnol 73:314–322
38. Savakis P, Hellingwerf KJ (2015) Engineering cyanobacteria for direct biofuel production from CO_2. Curr Opin Biotechnol 33:8–14
39. Treece TR, Gonzales JN, Pressley JR, Atsumi S (2022) Synthetic biology approaches for improving chemical production in cyanobacteria. Front Bioeng Biotechnol 10:869195
40. Liu D, Liberton M, Hendry JI, Aminian-Dehkordi J, Maranas CD, Pakrasi HB (2021) Engineering biology approaches for food and nutrient production by cyanobacteria. Curr Opin Biotechnol 67:1–6
41. Borowitzka MA (1999) Commercial production of microalgae: ponds, tanks, tubes and fermenters. J Biotechnol 70:313–321
42. Weaver PF, Lien S, Seibert M (1980) Photobiological production of hydrogen. Sol Energy 24: 3–45
43. Nilsson A, Shabestary K, Brandao M, Hudson EP (2020) Environmental impacts and limitations of third-generation biobutanol: life cycle assessment of-butanol produced by genetically engineered cyanobacteria. J Ind Ecol 24:205–216
44. Wijffels RH, Barbosa MJ (2010) An outlook on microalgal biofuels. Science 329:796–799
45. Knoot CJ, Ungerer J, Wangikar PP, Pakrasi HB (2018) Cyanobacteria: promising biocatalysts for sustainable chemical production. J Biol Chem 293:5044–5052
46. Flombaum P, Gallegos JL, Gordillo RA, Rincon J, Zabala LL, Jiao N, Karl DM, Li WK, Lomas MW, Veneziano D, Vera CS, Vrugt JA, Martiny AC (2013) Present and future global distributions of the marine cyanobacteria *Prochlorococcus* and *Synechococcus*. Proc Natl Acad Sci USA 110:9824–9829
47. Veldhuis MJW, Timmermans KR, Croot P, van der Wagt B (2005) Picophytoplankton; a comparative study of their biochemical composition and photosynthetic properties. J Sea Res 53:7–24
48. Ralf G, Repeta DJ (2003) The pigments of *Prochlorococcus marinus*: the presence of divinylchlorophyll a and b in a marine procaryote. Limnol Oceanogr 37:425–433
49. Scanlan DJ, Ostrowski M, Mazard S, Dufresne A, Garczarek L, Hess WR, Post AF, Hagemann M, Paulsen I, Partensky F (2009) Ecological genomics of marine picocyanobacteria. Microbiol Mol Biol Rev 73:249–299
50. Ludwig M, Bryant DA (2012) *Synechococcus* sp. strain PCC 7002 transcriptome: acclimation to temperature, salinity, oxidative stress, and mixotrophic growth conditions. Front Microbiol 3:354
51. Mills LA, McCormick AJ, Lea-Smith DJ (2020) Current knowledge and recent advances in understanding metabolism of the model cyanobacterium *Synechocystis* sp. PCC 6803. Biosci Rep 40:BSR20193325
52. Vermaas W (1996) Molecular genetics of the cyanobacterium *Synechocystis* sp. PCC 6803: principles and possible biotechnology applications. J Appl Phycol 8:263–273
53. Los DA, Zorina A, Sinetova M, Kryazhov S, Mironov K, Zinchenko VV (2010) Stress sensors and signal transducers in cyanobacteria. Sensors 10:2386–2415
54. Kaneko T, Sato S, Kotani H, Tanaka A, Asamizu E, Nakamura Y, Miyajima N, Hirosawa M, Sugiura M, Sasamoto S, Kimura T, Hosouchi T, Matsuno A, Muraki A, Nakazaki N, Naruo K, Okumura S, Shimpo S, Takeuchi C, Wada T, Watanabe A, Yamada M, Yasuda M, Tabata S (1996) Sequence analysis of the genome of the unicellular cyanobacterium *Synechocystis* sp. strain PCC6803. II. Sequence determination of the entire genome and assignment of potential protein-coding regions. DNA Res 3:109–136
55. Heidorn T, Camsund D, Huang HH, Lindberg P, Oliveira P, Stensjö K, Lindblad P (2011) Synthetic biology in cyanobacteria engineering and analyzing novel functions. Methods Enzymol 497:539–579

56. Klähn S, Opel F, Hess WR (2024) Customized molecular tools to strengthen metabolic engineering of cyanobacteria. Green Carbon 2:149–163
57. Markley AL, Begemann MB, Clarke RE, Gordon GC, Pfleger BF (2015) Synthetic biology toolbox for controlling gene expression in the cyanobacterium *Synechococcus* sp. strain PCC 7002. ACS Synth Biol 4:595–603
58. Opel F, Axmann IM, Klähn S (2023) The molecular toolset and techniques required to build cyanobacterial cell factories. Adv Biochem Eng Biotechnol 183:65–103
59. Till P, Toepel J, Bühler B, Mach RL, Mach-Aigner AR (2020) Regulatory systems for gene expression control in cyanobacteria. Appl Microbiol Biotechnol 104:1977–1991
60. Vavitsas K, Kugler A, Satta A, Hatzinikolaou DG, Lindblad P, Fewer DP, Lindberg P, Toivari M, Stensjö K (2021) Doing synthetic biology with photosynthetic microorganisms. Physiol Plant 173:624–638
61. Sun T, Li S, Song X, Pei G, Diao J, Cui J, Shi M, Chen L, Zhang W (2018) Re-direction of carbon flux to key precursor malonyl-CoA via artificial small RNAs in photosynthetic *Synechocystis* sp. PCC 6803. Biotechnol Biofuels 11:26
62. Taton A, Ma AT, Ota M, Golden SS, Golden JW (2017) NOT gate genetic circuits to control gene expression in cyanobacteria. ACS Synth Biol 6:2175–2182
63. Yao L, Cengic I, Anfelt J, Hudson EP (2016) Multiple gene repression in cyanobacteria using CRISPRi. ACS Synth Biol 5:207–212
64. Yao L, Shabestary K, Björk SM, Asplund-Samuelsson J, Joensson HN, Jahn M, Hudson EP (2020) Pooled CRISPRi screening of the cyanobacterium *Synechocystis* sp PCC 6803 for enhanced industrial phenotypes. Nat Commun 11:1666
65. van Alphen P, Abedini Najafabadi H, Branco Dos Santos F, Hellingwerf KJ (2018) Increasing the photoautotrophic growth rate of *Synechocystis* sp. PCC 6803 by identifying the limitations of its cultivation. Biotechnol J 13:e1700764
66. Ungerer J, Lin PC, Chen HY, Pakrasi HB (2018) Adjustments to photosystem stoichiometry and electron transfer proteins are key to the remarkably fast growth of the cyanobacterium *Synechococcus elongatus* UTEX 2973. mBio 9:e02327-17
67. Ungerer J, Wendt KE, Hendry JI, Maranas CD, Pakrasi HB (2018) Comparative genomics reveals the molecular determinants of rapid growth of the cyanobacterium *Synechococcus elongatus* UTEX 2973. Proc Natl Acad Sci USA 115:E11761–E11770
68. Li Z, Li S, Chen L, Sun T, Zhang W (2024) Fast-growing cyanobacterial chassis for synthetic biology application. Crit Rev Biotechnol 44:414–428
69. Jaiswal D, Sengupta A, Sengupta S, Madhu S, Pakrasi HB, Wangikar PP (2020) A novel cyanobacterium *Synechococcus elongatus* PCC 11802 has distinct genomic and metabolomic characteristics compared to its neighbor PCC 11801. Sci Rep 10:191
70. Jaiswal D, Sengupta A, Sohoni S, Sengupta S, Phadnavis AG, Pakrasi HB, Wangikar PP (2018) Genome features and biochemical characteristics of a robust, fast growing and naturally transformable cyanobacterium *Synechococcus elongatus* PCC 11801 isolated from India. Sci Rep 8:16632
71. Wlodarczyk A, Selao TT, Norling B, Nixon PJ (2020) Newly discovered *Synechococcus* sp. PCC 11901 is a robust cyanobacterial strain for high biomass production. Commun Biol 3: 215
72. Yu J, Liberton M, Cliften PF, Head RD, Jacobs JM, Smith RD, Koppenaal DW, Brand JJ, Pakrasi HB (2015) *Synechococcus elongatus* UTEX 2973, a fast growing cyanobacterial chassis for biosynthesis using light and CO_2. Sci Rep 5:8132
73. Fujisawa T, Narikawa R, Maeda SI, Watanabe S, Kanesaki Y, Kobayashi K, Nomata J, Hanaoka M, Watanabe M, Ehira S, Suzuki E, Awai K, Nakamura Y (2017) CyanoBase: a large-scale update on its 20th anniversary. Nucleic Acids Res 45:D551–D554
74. Gale GAR, Schiavon Osorio AA, Mills LA, Wang B, Lea-Smith DJ, McCormick AJ (2019) Emerging species and genome editing tools: future prospects in cyanobacterial synthetic biology. Microorganisms 7:409

75. Sauvageau MC (2014) Sur Les Algues D'Eau Douce Récoltées En Algérie Pendant La Session De La Société Botanique En 1892. Bull Soc Bot France 39:CIV–CXXVIII
76. Iijima H, Nakaya Y, Kuwahara A, Hirai MY, Osanai T (2015) Seawater cultivation of freshwater cyanobacterium *Synechocystis* sp. PCC 6803 drastically alters amino acid composition and glycogen metabolism. Front Microbiol 6:326
77. Klähn S, Mikkat S, Riediger M, Georg J, Hess WR, Hagemann M (2021) Integrative analysis of the salt stress response in cyanobacteria. Biol Direct 16:26
78. Barten R, Lill H (1995) DNA-uptake in the naturally competent cyanobacterium, *Synechocystis* sp PCC 6803. FEMS Microbiol Lett 129:83–88
79. Zavrel T, Ocenasova P, Cerveny J (2017) Phenotypic characterization of *Synechocystis* sp. PCC 6803 substrains reveals differences in sensitivity to abiotic stress. PLoS One 12: e0189130
80. Ferreira EA, Pacheco CC, Pinto F, Pereira J, Lamosa P, Oliveira P, Kirov B, Jaramillo A, Tamagnini P (2018) Expanding the toolbox for *Synechocystis* sp. PCC 6803: validation of replicative vectors and characterization of a novel set of promoters. Synth Biol 3:ysy014
81. Yu Y, You L, Liu D, Hollinshead W, Tang YJ, Zhang F (2013) Development of *Synechocystis* sp. PCC 6803 as a phototrophic cell factory. Mar Drugs 11:2894–2916
82. Griese M, Lange C, Soppa J (2011) Ploidy in cyanobacteria. FEMS Microbiol Lett 323:124–131
83. Nandagopal P, Steven AN, Chan LW, Rahmat Z, Jamaluddin H, Mohd Noh NI (2021) Bioactive metabolites produced by cyanobacteria for growth adaptation and their pharmacological properties. Biology 10:1061
84. Pagels F, Vasconcelos V, Guedes AC (2021) Carotenoids from cyanobacteria: biotechnological potential and optimization strategies. Biomolecules 11:735
85. Santini G, Biondi N, Rodolfi L, Tredici MR (2021) Plant biostimulants from cyanobacteria: an emerging strategy to improve yields and sustainability in agriculture. Plants 10:643
86. Ricciardelli A, Pollio A, Costantini M, Zupo V (2023) Harmful and beneficial properties of cyanotoxins: two sides of the same coin. Biotechnol Adv 68:108235
87. Angermayr SA, van der Woude AD, Correddu D, Vreugdenhil A, Verrone V, Hellingwerf KJ (2014) Exploring metabolic engineering design principles for the photosynthetic production of lactic acid by *Synechocystis* sp. PCC 6803. Biotechnol Biofuels 7:99
88. Gao ZX, Zhao H, Li ZM, Tan XM, Lu XF (2012) Photosynthetic production of ethanol from carbon dioxide in genetically engineered cyanobacteria. Energy Environ Sci 5:9857–9865
89. Oliver JW, Machado IM, Yoneda H, Atsumi S (2013) Cyanobacterial conversion of carbon dioxide to 2,3-butanediol. Proc Natl Acad Sci USA 110:1249–1254
90. Mustila H, Kugler A, Stensjö K (2021) Isobutene production in *Synechocystis* sp. PCC 6803 by introducing alpha-ketoisocaproate dioxygenase from *Rattus norvegicus*. Metab Eng Commun 12:e00163
91. Vajravel S, Sirin S, Kosourov S, Allahverdiyeva Y (2020) Towards sustainable ethylene production with cyanobacterial artificial biofilms. Green Chem 22:6404–6414
92. Lindberg P, Park S, Melis A (2010) Engineering a platform for photosynthetic isoprene production in cyanobacteria, using *Synechocystis* as the model organism. Metab Eng 12:70–79
93. Liu XF, Miao R, Lindberg P, Lindblad P (2019) Modular engineering for efficient photosynthetic biosynthesis of 1-butanol from CO_2 in cyanobacteria. Energy Environ Sci 12:2765–2777
94. Zhou J, Zhang HF, Meng HK, Zhang YP, Li Y (2014) Production of optically pure D-lactate from CO_2 by blocking the PHB and acetate pathways and expressing D-lactate dehydrogenase in the cyanobacterium *Synechocystis* sp. PCC 6803. Process Biochem 49:2071–2077
95. Zhou J, Zhang H, Zhang Y, Li Y, Ma Y (2012) Designing and creating a modularized synthetic pathway in cyanobacterium *Synechocystis* enables production of acetone from carbon dioxide. Metab Eng 14:394–400
96. Kukil K, Englund E, Crang N, Hudson EP, Lindberg P (2023) Laboratory evolution of *Synechocystis* sp. PCC 6803 for phenylpropanoid production. Metab Eng 79:27–37

97. Ungerer J, Tao L, Davis M, Ghirardi M, Maness P-C, Yu J (2012) Sustained photosynthetic conversion of CO_2 to ethylene in recombinant cyanobacterium *Synechocystis* sp. PCC 6803. Energy Environ Sci 5:8998–9006
98. Liu X, Sheng J, Curtiss 3rd R (2011) Fatty acid production in genetically modified cyanobacteria. Proc Natl Acad Sci USA 108:6899–6904
99. Yao L, Qi F, Tan X, Lu X (2014) Improved production of fatty alcohols in cyanobacteria by metabolic engineering. Biotechnol Biofuels 7:94
100. Miao R, Xie H, Lindblad P (2018) Enhancement of photosynthetic isobutanol production in engineered cells of *Synechocystis* sp. PCC 6803. Biotechnol Biofuels 11:267
101. Chaves JE, Melis A (2018) Engineering isoprene synthesis in cyanobacteria. FEBS Lett 592: 2059–2069
102. Englund E, Pattanaik B, Ubhayasekera SJ, Stensjö K, Bergquist J, Lindberg P (2014) Production of squalene in *Synechocystis* sp. PCC 6803. PLoS One 9:e90270
103. Lan EI, Liao JC (2012) ATP drives direct photosynthetic production of 1-butanol in cyanobacteria. Proc Natl Adad Sci U S A 109:6018–6023
104. Atsumi S, Higashide W, Liao JC (2009) Direct photosynthetic recycling of carbon dioxide to isobutyraldehyde. Nat Biotechnol 27:1177–1180
105. Gao X, Gao F, Liu D, Zhang H, Nie XQ, Yang C (2016) Engineering the methylerythritol phosphate pathway in cyanobacteria for photosynthetic isoprene production from CO_2. Energy Environ Sci 9:1400–1411
106. Shen CR, Liao JC (2012) Photosynthetic production of 2-methyl-1-butanol from CO in cyanobacterium *Synechococcus* PCC 7942 and characterization of the native acetohydroxyacid synthase. Energy Environ Sci 5:9574–9583
107. Davies FK, Work VH, Beliaev AS, Posewitz MC (2014) Engineering limonene and bisabolene production in wild type and a glycogen-deficient mutant of *Synechococcus* sp. PCC 7002. Front Bioeng Biotechnol 2:21
108. Ruffing AM (2014) Improved free fatty acid production in cyanobacteria with *Synechococcus* sp. PCC 7002 as host. Front Bioeng Biotechnol 2:17
109. Aikawa S, Nishida A, Ho SH, Chang JS, Hasunuma T, Kondo A (2014) Glycogen production for biofuels by the euryhaline cyanobacteria *Synechococcus* sp. strain PCC 7002 from an oceanic environment. Biotechnol Biofuels 7:88
110. Jacobsen JH, Frigaard NU (2014) Engineering of photosynthetic mannitol biosynthesis from CO_2 in a cyanobacterium. Metab Eng 21:60–70
111. Sengupta A, Pritam P, Jaiswal D, Bandyopadhyay A, Pakrasi HB, Wangikar PP (2020) Photosynthetic co-production of succinate and ethylene in a fast-growing cyanobacterium, *Synechococcus elongatus* PCC 11801. Metabolites 10:250
112. Sengupta S, Jaiswal D, Sengupta A, Shah S, Gadagkar S, Wangikar PP (2020) Metabolic engineering of a fast-growing cyanobacterium *Synechococcus elongatus* PCC 11801 for photoautotrophic production of succinic acid. Biotechnol Biofuels 13:89
113. Dookeran ZA, Nielsen DR (2021) Systematic engineering of *Synechococcus elongatus* UTEX 2973 for photosynthetic production of L-lysine, cadaverine, and glutarate. ACS Synth Biol 10: 3561–3575
114. Sun T, Li Z, Li S, Chen L, Zhang W (2023) Exploring and validating key factors limiting cyanobacteria-based CO_2 bioconversion: case study to maximize myo-inositol biosynthesis. Chem Eng J 452:139158
115. Lin PC, Zhang F, Pakrasi HB (2021) Enhanced limonene production in a fast-growing cyanobacterium through combinatorial metabolic engineering. Metab Eng Commun 12: e00164
116. Lin PC, Zhang F, Pakrasi HB (2020) Enhanced production of sucrose in the fast-growing cyanobacterium *Synechococcus elongatus* UTEX 2973. Sci Rep 10:390
117. Bähr L, Wüstenberg A, Ehwald R (2015) Two-tier vessel for photoautotrophic high-density cultures. J Appl Phycol 28:783–793

118. Nielsen AZ, Mellor SB, Vavitsas K, Wlodarczyk AJ, Gnanasekaran T, de Jesus MPRH, King BC, Bakowski K, Jensen PE (2016) Extending the biosynthetic repertoires of cyanobacteria and chloroplasts. Plant J 87:87–102
119. Durall C, Lindberg P, Yu J, Lindblad P (2020) Increased ethylene production by overexpressing phosphoenolpyruvate carboxylase in the cyanobacterium *Synechocystis* sp. PCC 6803. Biotechnol Biofuels 13:16
120. Kallio P, Kugler A, Pyytovaara S, Stensjö K, Allahverdiyeva Y, Gao X, Lindblad P, Lindberg P (2021) Photoautotrophic production of renewable ethylene by engineered cyanobacteria: steering the cell metabolism towards biotechnological use. Physiol Plant 173:579–590
121. Brey LF, Wlodarczyk AJ, Bang Thofner JF, Burow M, Crocoll C, Nielsen I, Zygadlo Nielsen AJ, Jensen PE (2020) Metabolic engineering of *Synechocystis* sp. PCC 6803 for the production of aromatic amino acids and derived phenylpropanoids. Metab Eng 57:129–139
122. Karradt A, Sobanski J, Mattow J, Lockau W, Baier K (2008) NblA, a key protein of phycobilisome degradation, interacts with ClpC, a HSP100 chaperone partner of a cyanobacterial Clp protease. J Biol Chem 283:32394–32403
123. Diaz-Troya S, Lopez-Maury L, Sanchez-Riego AM, Roldan M, Florencio FJ (2014) Redox regulation of glycogen biosynthesis in the cyanobacterium *Synechocystis* sp. PCC 6803: analysis of the AGP and glycogen synthases. Mol Plant 7:87–100
124. Cano M, Holland SC, Artier J, Burnap RL, Ghirardi M, Morgan JA, Yu J (2018) Glycogen synthesis and metabolite overflow contribute to energy balancing in cyanobacteria. Cell Rep 23:667–672
125. Carrieri D, Broadbent C, Carruth D, Paddock T, Ungerer J, Maness PC, Ghirardi M, Yu J (2015) Enhancing photo-catalytic production of organic acids in the cyanobacterium *Synechocystis* sp. PCC 6803 Δ*glgC*, a strain incapable of glycogen storage. Microb Biotechnol 8:275–280
126. Gründel M, Scheunemann R, Lockau W, Zilliges Y (2012) Impaired glycogen synthesis causes metabolic overflow reactions and affects stress responses in the cyanobacterium *Synechocystis* sp. PCC 6803. Microbiology (Reading, UK) 158:3032–3043
127. Carrieri D, Lombardi T, Paddock T, Cano M, Goodney GA, Nag A, Old W, Maness PC, Seibert M, Ghirardi M, Yu JP (2017) Transcriptome and proteome analysis of nitrogen starvation responses in 6803 Δ*glgC*, a mutant incapable of glycogen storage. Algal Res 21: 64–75
128. Carrieri D, Paddock T, Maness PC, Seibert M, Yu JP (2012) Photo-catalytic conversion of carbon dioxide to organic acids by a recombinant cyanobacterium incapable of glycogen storage. Energy Environ Sci 5:9457–9461
129. Koch M, Berendzen KW, Forchhammer AK (2020) On the role and production of polyhydroxybutyrate (PHB) in the cyanobacterium *Synechocystis* sp. PCC 6803. Life 10:47
130. Orthwein T, Scholl J, Spat P, Lucius S, Koch M, Macek B, Hagemann M, Forchhammer K (2021) The novel P(II)-interactor PirC identifies phosphoglycerate mutase as key control point of carbon storage metabolism in cyanobacteria. Proc Natl Acad Sci USA 118:e2019988118
131. Koch M, Bruckmoser J, Scholl J, Hauf W, Rieger B, Forchhammer K (2020) Maximizing PHB content in *Synechocystis* sp. PCC 6803: a new metabolic engineering strategy based on the regulator PirC. Microb Cell Factories 19:231
132. Berepiki A, Hitchcock A, Moore CM, Bibby TS (2016) Tapping the unused potential of photosynthesis with a heterologous electron sink. ACS Synth Biol 5:1369–1375
133. Wlodarczyk A, Gnanasekaran T, Nielsen AZ, Zulu NN, Mellor SB, Luckner M, Thofner JFB, Olsen CE, Mottawie MS, Burow M, Pribil M, Feussner I, Møller BL, Jensen PE (2016) Metabolic engineering of light-driven cytochrome P450 dependent pathways into *Synechocystis* sp. PCC 6803. Metab Eng 33:1–11
134. Cassier-Chauvat C, Chauvat F (2014) Function and regulation of ferredoxins in the cyanobacterium, *Synechocystis* sp. PCC 6803: recent advances. Life 4:666–680

135. Theodosiou E, Tüllinghoff A, Toepel J, Bühler B (2022) Exploitation of hetero- and phototrophic metabolic modules for redox-intensive whole-cell biocatalysis. Front Bioeng Biotechnol 10:855715
136. Jodlbauer J, Rohr T, Spadiut O, Mihovilovic MD, Rudroff F (2021) Biocatalysis in green and blue: cyanobacteria. Trends Biotechnol 39:875–889
137. Assil-Companioni L, Büchsenschütz HC, Solymosi D, Dyczmons-Nowaczyk NG, Bauer KKF, Wallner S, Macheroux P, Allahverdiyeva Y, Nowaczyk MM, Kourist R (2020) Engineering of NADPH supply boosts photosynthesis-driven biotransformations. ACS Catal 10: 11864–11877
138. Hobisch M, Spasic J, Malihan-Yap L, Barone GD, Castiglione K, Tamagnini P, Kara S, Kourist R (2021) Internal illumination to overcome the cell density limitation in the scale-up of whole-cell photobiocatalysis. ChemSusChem 14:3219–3225
139. Hoschek A, Bühler B, Schmid A (2017) Overcoming the gas-liquid mass transfer of oxygen by coupling photosynthetic water oxidation with biocatalytic oxyfunctionalization. Angew Chem Int Ed 56:15146–15149
140. Tüllinghoff A, Uhl MB, Nintzel FE, Schmid A, Bühler B, Toepel J (2022) Maximizing photosynthesis-driven Baeyer-Villiger oxidation efficiency in recombinant *Synechocystis* sp. PCC6803. Front Catal 1:780474
141. Büchsenschütz HC, Vidimce-Risteski V, Eggbauer B, Schmidt S, Winkler CK, Schrittwieser JH, Kroutil W, Kourist R (2019) Stereoselective biotransformations of cyclic imines in recombinant cells of *Synechocystis* sp. PCC 6803. ChemCatChem 12:726–730
142. Sengupta A, Sunder AV, Sohoni SV, Wangikar PP (2019) The effect of CO_2 in enhancing photosynthetic cofactor recycling for alcohol dehydrogenase mediated chiral synthesis in cyanobacteria. J Biotechnol 289:1–6
143. Lassen LM, Nielsen AZ, Olsen CE, Bialek W, Jensen K, Møller BL, Jensen PE (2014) Anchoring a plant cytochrome P450 via PsaM to the thylakoids in *Synechococcus* sp. PCC 7002: evidence for light-driven biosynthesis. PLoS One 9:e102184
144. Böhmer S, Köninger K, Gómez-Baraibar Á, Bojarra S, Mügge C, Schmidt S, Nowaczyk M, Kourist R (2017) Enzymatic oxyfunctionalization driven by photosynthetic water-splitting in the cyanobacterium *Synechocystis* sp. PCC 6803. Catalysts 7:240
145. Mellor SB, Vavitsas K, Nielsen AZ, Jensen PE (2017) Photosynthetic fuel for heterologous enzymes: the role of electron carrier proteins. Photosynth Res 134:329–342
146. Feilke K, Ajlani G, Krieger-Liszkay A (2017) Overexpression of plastid terminal oxidase in *Synechocystis* sp. PCC 6803 alters cellular redox state. Philos Trans R Soc B 372:20160379
147. Lin B, Tao Y (2017) Whole-cell biocatalysts by design. Microb Cell Factories 16:106
148. Schrewe M, Julsing MK, Bühler B, Schmid A (2013) Whole-cell biocatalysis for selective and productive C-O functional group introduction and modification. Chem Soc Rev 42:6346–6377
149. Willrodt C, Karande R, Schmid A, Julsing MK (2015) Guiding efficient microbial synthesis of non-natural chemicals by physicochemical properties of reactants. Curr Opin Biotechnol 35: 52–62
150. Spasic J, Oliveira P, Pacheco C, Kourist R, Tamagnini P (2022) Engineering cyanobacterial chassis for improved electron supply toward a heterologous ene-reductase. J Biotechnol 360: 152–159
151. Barone GD, Hubacek M, Malihan-Yap L, Grimm HC, Nikkanen L, Pacheco CC, Tamagnini P, Allahverdiyeva Y, Kourist R (2023) Towards the rate limit of heterologous biotechnological reactions in recombinant cyanobacteria. Biotechnol Biofuels Bioprod 16:4
152. Valotta A, Malihan-Yap L, Hinteregger K, Kourist R, Gruber-Woelfler H (2022) Design and investigation of a photocatalytic setup for efficient biotransformations within recombinant cyanobacteria in continuous flow. ChemSusChem 15:e202201468
153. Jurkas V, Weissensteiner F, De Santis P, Vrabl S, Sorgenfrei FA, Bierbaumer S, Kara S, Kourist R, Wangikar PP, Winkler CK, Kroutil W (2022) Transmembrane shuttling of photosynthetically produced electrons to propel extracellular biocatalytic redox reactions in a modular fashion. Angew Chem Int Ed 61:e202207971

154. Bühler K, Bühler B, Klähn S, Krömer JO, Dusny C, Schmid A (2021) Biocatalytic production of white hydrogen from water using cyanobacteria. In: Matthias R (ed) Photosynthesis. De Gruyter, Berlin, pp 279–306
155. Redding KE, Appel J, Boehm M, Schuhmann W, Nowaczyk MM, Yacoby I, Gutekunst K (2022) Advances and challenges in photosynthetic hydrogen production. Trends Biotechnol 40:1313–1325
156. Benemann JR, Weissman JC (1977) Biophotolysis: problems and prospects. In: Schlegel HG, Barnea J (eds) Microbial energy conversion. Pergamon, pp 413–426
157. Ding C, Yang KL, He J (2016) Biological and fermentative production of hydrogen. In: Luque R, Lin CSK, Wilson K, Clark J (eds) Handbook of biofuels production. Woodhead Publishing, pp 303–333
158. Rexroth S, Wiegand K, Rögner M (2015) Cyanobacterial design cell for the production of hydrogen from water. In: Matthias R (ed) Biohydrogen. De Gruyter, Berlin, pp 1–18
159. Gutekunst K, Chen X, Schreiber K, Kaspar U, Makam S, Appel J (2014) The bidirectional NiFe-hydrogenase in *Synechocystis* sp. PCC 6803 is reduced by flavodoxin and ferredoxin and is essential under mixotrophic, nitrate-limiting conditions. J Biol Chem 289:1930–1937
160. Lupacchini S, Appel J, Stauder R, Bolay P, Klähn S, Lettau E, Adrian L, Lauterbach L, Bühler B, Schmid A, Toepel J (2021) Rewiring cyanobacterial photosynthesis by the implementation of an oxygen-tolerant hydrogenase. Metab Eng 68:199–209
161. Appel J, Hueren V, Boehm M, Gutekunst K (2020) Cyanobacterial *in vivo* solar hydrogen production using a photosystem I-hydrogenase (PsaD-HoxYH) fusion complex. Nat Energy 5: 458–467
162. Kannchen D, Zabret J, Oworah-Nkruma R, Dyczmons-Nowaczyk N, Wiegand K, Löbbert P, Frank A, Nowaczyk MM, Rexroth S, Rögner M (2020) Remodeling of photosynthetic electron transport in *Synechocystis* sp. PCC 6803 for future hydrogen production from water. Biochim Biophys Acta Bioenerg 1861:148208
163. Hoschek A, Schmid A, Bühler B (2018) *In situ* O_2 generation for biocatalytic oxyfunctionalization reactions. ChemCatChem 10:5366–5371
164. Hoschek A, Bühler B, Schmid A (2019) Stabilization and scale-up of photosynthesis-driven omega-hydroxylation of nonanoic acid methyl ester by two-liquid phase whole-cell biocatalysis. Biotechnol Bioeng 116:1887–1900
165. Park J, Choi Y (2017) Cofactor engineering in cyanobacteria to overcome imbalance between NADPH and NADH: a mini review. Front Chem Sci Eng 11:66–71
166. Wang M, Chen B, Fang Y, Tan T (2017) Cofactor engineering for more efficient production of chemicals and biofuels. Biotechnol Adv 35:1032–1039
167. Angermayr SA, Paszota M, Hellingwerf KJ (2012) Engineering a cyanobacterial cell factory for production of lactic acid. Appl Environ Microbiol 78:7098–7106
168. Heuschkel I, Hoschek A, Schmid A, Bühler B, Karande R, Bühler K (2019) Mixed-trophies biofilm cultivation in capillary reactors. MethodsX 6:1822–1831
169. Hoschek A, Heuschkel I, Schmid A, Bühler B, Karande R, Bühler K (2019) Mixed-species biofilms for high-cell-density application of *Synechocystis* sp. PCC 6803 in capillary reactors for continuous cyclohexane oxidation to cyclohexanol. Bioresour Technol 282:171–178
170. Mascia F, Pereira SB, Pacheco CC, Oliveira P, Solarczek J, Schallmey A, Kourist R, Alphand V, Tamagnini P (2022) Light-driven hydroxylation of testosterone by *Synechocystis* sp. PCC 6803 expressing the heterologous CYP450 monooxygenase CYP110D1. Green Chem 24:6156–6167
171. Mellor SB, Vinde MH, Nielsen AZ, Hanke GT, Abdiaziz K, Roessler MM, Burow M, Motawia MS, Møller BL, Jensen PE (2019) Defining optimal electron transfer partners for light-driven cytochrome P450 reactions. Metab Eng 55:33–43
172. Li Z, Jiang Y, Guengerich FP, Ma L, Li S, Zhang W (2020) Engineering cytochrome P450 enzyme systems for biomedical and biotechnological applications. J Biol Chem 295:833–849
173. Erdem E, Malihan-Yap L, Assil-Companioni L, Grimm H, Barone GD, Serveau-Avesque C, Amouric A, Duquesne K, de Berardinis V, Allahverdiyeva Y, Alphand V, Kourist R (2022)

Photobiocatalytic oxyfunctionalization with high reaction rate using a Baeyer-Villiger monooxygenase from *Burkholderia xenovorans* in metabolically engineered cyanobacteria. ACS Catal 12:66–72
174. Berepiki A, Gittins JR, Moore CM, Bibby TS (2018) Rational engineering of photosynthetic electron flux enhances light-powered cytochrome P450 activity. Synth Biol 3:ysy009
175. Santos-Merino M, Torrado A, Davis GA, Rottig A, Bibby TS, Kramer DM, Ducat DC (2021) Improved photosynthetic capacity and photosystem I oxidation via heterologous metabolism engineering in cyanobacteria. Proc Natl Acad Sci USA 118:171–209
176. Berger RG (2009) Biotechnology of flavours – the next generation. Biotechnol Lett 31:1651–1659
177. Berger RG (2015) Biotechnology as a source of natural volatile flavours. Curr Opin Food Sci 1:38–43
178. Bassanini I, Ferrandi EE, Riva S, Monti D (2021) Biocatalysis with laccases: an updated overview. Catalysts 11:26
179. Afreen S, Shamsi TN, Baig MA, Ahmad N, Fatima S, Qureshi MI, Hassan MI, Fatma T (2017) A novel multicopper oxidase (laccase) from cyanobacteria: purification, characterization with potential in the decolorization of anthraquinonic dye. PLoS One 12:e0175144
180. Liang Y, Hou J, Liu Y, Luo Y, Tang J, Cheng JJ, Daroch M (2018) Textile dye decolorizing *Synechococcus* sp. PCC 7942 engineered with CotA laccase. Front Bioeng Biotechnol 6:95
181. Batterton Jr JC, Van Baalen C (1971) Growth responses of blue-green algae to sodium chloride concentration. Arch Microbiol 76:151–165
182. Ganjave SD, Dodia H, Sunder AV, Madhu S, Wangikar PP (2022) High cell density cultivation of *E. coli* in shake flasks for the production of recombinant proteins. Biotechnol Rep 33: e00694
183. Mills LA, Moreno-Cabezuelo JA, Wlodarczyk A, Victoria AJ, Mejias R, Nenninger A, Moxon S, Bombelli P, Selao TT, McCormick AJ, Lea-Smith DJ (2022) Development of a biotechnology platform for the fast-growing cyanobacterium *Synechococcus* sp. PCC 11901. Biomolecules 12:872
184. Victoria AJ, Selao TT, Moreno-Cabezuelo JA, Mills LA, Gale GAR, Lea-Smith DJ, McCormick AJ (2024) A toolbox to engineer the highly productive cyanobacterium *Synechococcus* sp. PCC 11901. Plant Physiol 196:1674–1690
185. Mellor SB, Nielsen AZ, Burow M, Motawia MS, Jakubauskas D, Møller BL, Jensen PE (2016) Fusion of ferredoxin and cytochrome P450 enables direct light-driven biosynthesis. ACS Chem Biol 11:1862–1869
186. Torrado A, Connabeer HM, Rottig A, Pratt N, Baylay AJ, Terry MJ, Moore CM, Bibby TS (2022) Directing cyanobacterial photosynthesis in a cytochrome c oxidase mutant using a heterologous electron sink. Plant Physiol 189:2554–2566
187. Santos-Merino M, Sakkos JK, Singh AK, Ducat DC (2024) Coordination of carbon partitioning and photosynthesis by a two-component signaling network in *Synechococcus elongatus* PCC 7942. Metab Eng 81:38–52
188. Madhu S, Sengupta A, Sarnaik AP, Wangikar PP (2024) Expanding the synthetic biology repertoire of a fast-growing cyanobacterium *Synechococcus elongatus* PCC 11801. Biotechnol Bioeng 121:2974–2980
189. Sengupta A, Bandyopadhyay A, Sarkar D, Hendry JI, Schubert MG, Liu D, Church GM, Maranas CD, Pakrasi HB (2024) Genome streamlining to improve performance of a fast-growing cyanobacterium *Synechococcus elongatus* UTEX 2973. mBio 15:e0353023
190. Sengupta A, Bandyopadhyay A, Schubert MG, Church GM, Pakrasi HB (2023) Antenna modification in a fast-growing cyanobacterium *Synechococcus elongatus* UTEX 2973 leads to improved efficiency and carbon-neutral productivity. Microbiol Spectr 11:e0050023
191. Sun H, Luan G, Ma Y, Lou W, Chen R, Feng D, Zhang S, Sun J, Lu X (2023) Engineered hypermutation adapts cyanobacterial photosynthesis to combined high light and high temperature stress. Nat Commun 14:1238

192. Zhang S, Sun J, Feng D, Sun H, Cui J, Zeng X, Wu Y, Luan G, Lu X (2023) Unlocking the potentials of cyanobacterial photosynthesis for directly converting carbon dioxide into glucose. Nat Commun 14:3425
193. Jaiswal D, Nenwani M, Wangikar PP (2023) Isotopically non-stationary ^{13}C metabolic flux analysis of two closely related fast-growing cyanobacteria, *Synechococcus elongatus* PCC 11801 and 11802. Plant J 116:558–573
194. Velasquez SF, Chan MA, Abisado RG, Traifalgar RFM, Tayamen MM, Maliwat GCF, Ragaza JA (2016) Dietary *Spirulina* replacement enhances performance of juvenile *Nile tilapia*. J Appl Phycol 28:1023–1030
195. Alagawany M, Taha AE, Noreldin A, El-Tarabily KA, El-Hack MEA (2021) Nutritional applications of species of *Spirulina* and *Chlorella* in farmed fish: a review. Aquaculture 542: 736841
196. Legrand J, Artu A, Pruvost J (2021) A review on photobioreactor design and modelling for microalgae production. React Chem Eng 6:1134–1151
197. Nozzi NE, Oliver JW, Atsumi S (2013) Cyanobacteria as a platform for biofuel production. Front Bioeng Biotechnol 1:7
198. Celesnik H, Tansek A, Tahirovic A, Vizintin A, Mustar J, Vidmar V, Dolinar M (2016) Biosafety of biotechnologically important microalgae: intrinsic suicide switch implementation in cyanobacterium *Synechocystis* sp. PCC 6803. Biol Open 5:519–528
199. Wilhelm C, Wagner H (2021) Chapter 6: Rate-limiting steps in algal energy conversion from sunlight to products – the role of photosynthesis. In: Matthias R (ed) Photosynthesis: biotechnological applications with microalgae. De Gruyter, pp 149–174
200. Allakhverdiev SI, Tsuchiya T, Watabe K, Kojima A, Los DA, Tomo T, Klimov VV, Mimuro M (2011) Redox potentials of primary electron acceptor quinone molecule (QA)- and conserved energetics of photosystem II in cyanobacteria with chlorophyll a and chlorophyll d. Proc Natl Acad Sci USA 108:8054–8058
201. Antal TK, Kovalenko IB, Rubin AB, Tyystjarvi E (2013) Photosynthesis-related quantities for education and modeling. Photosynth Res 117:1–30
202. Cammack R, Rao KK, Bargeron CP, Hutson KG, Andrew PW, Rogers LJ (1977) Midpoint redox potentials of plant and algal ferredoxins. Biochem J 168:205–209
203. Dann M, Ortiz EM, Thomas M, Guljamow A, Lehmann M, Schaefer H, Leister D (2021) Enhancing photosynthesis at high light levels by adaptive laboratory evolution. Nat Plants 7: 681–695
204. LaPanse AJ, Krishnan A, Posewitz MC (2021) Adaptive laboratory evolution for algal strain improvement: methodologies and applications. Algal Res 53:102122
205. Mukherjee B, Madhu S, Wangikar PP (2020) The role of systems biology in developing non-model cyanobacteria as hosts for chemical production. Curr Opin Biotechnol 64:62–69
206. Srivastava V, Amanna R, Rowden SJL, Sengupta S, Madhu S, Howe CJ, Wangikar PP (2021) Adaptive laboratory evolution of the fast-growing cyanobacterium *Synechococcus elongatus* PCC 11801 for improved solvent tolerance. J Biosci Bioeng 131:491–500
207. Zhang B, Wu J, Meng F (2021) Adaptive laboratory evolution of microalgae: a review of the regulation of growth, stress resistance, metabolic processes, and biodegradation of pollutants. Front Microbiol 12:737248
208. Canonico M, Konert G, Crepin A, Šedivá B, Kaňa R (2021) Gradual response of cyanobacterial thylakoids to acute high-light stress-importance of carotenoid accumulation. Cells 10:1916
209. Muramatsu M, Hihara Y (2012) Acclimation to high-light conditions in cyanobacteria: from gene expression to physiological responses. J Plant Res 125:11–39
210. Theune ML, Hildebrandt S, Steffen-Heins A, Bilger W, Gutekunst K, Appel J (2021) *In-vivo* quantification of electron flow through photosystem I – cyclic electron transport makes up about 35% in a cyanobacterium. Biochim Biophys Acta Bioenerg 1862:148353
211. Allahverdiyeva Y, Isojarvi J, Zhang P, Aro EM (2015) Cyanobacterial oxygenic photosynthesis is protected by flavodiiron proteins. Life (Basel) 5:716–743

212. Thiel K, Patrikainen P, Nagy C, Fitzpatrick D, Pope N, Aro EM, Kallio P (2019) Redirecting photosynthetic electron flux in the cyanobacterium *Synechocystis* sp. PCC 6803 by the deletion of flavodiiron protein Flv3. Microb Cell Factories 18:189
213. Bassi R, Dall'Osto L (2021) Dissipation of light energy absorbed in excess: the molecular mechanisms. Annu Rev Plant Biol 72:47–76
214. Magdaong NCM, Blankenship RE (2018) Photoprotective, excited-state quenching mechanisms in diverse photosynthetic organisms. J Biol Chem 293:5018–5025
215. Bao H, Melnicki MR, Kerfeld CA (2017) Structure and functions of orange carotenoid protein homologs in cyanobacteria. Curr Opin Plant Biol 37:1–9
216. Stephens S, Mahadevan R, Allen DG (2020) Engineering photosynthetic bioprocesses for sustainable chemical production: a review. Front Bioeng Biotechnol 8:610723
217. Socher ML, Löser C, Schott C, Bley T, Steingroewer J (2016) The challenge of scaling up photobioreactors: modeling and approaches in small scale. Eng Life Sci 16:598–609
218. Burnap RL (2015) Systems and photosystems: cellular limits of autotrophic productivity in cyanobacteria. Front Bioeng Biotechnol 3:1
219. Jahn M, Vialas V, Karlsen J, Maddalo G, Edfors F, Forsstrom B, Uhlen M, Kall L, Hudson EP (2018) Growth of cyanobacteria is constrained by the abundance of light and carbon assimilation proteins. Cell Rep 25:478–486
220. Luan G, Zhang S, Lu X (2020) Engineering cyanobacteria chassis cells toward more efficient photosynthesis. Curr Opin Biotechnol 62:1–6
221. Miao R, Xie H, Liu X, Lindberg P, Lindblad P (2020) Current processes and future challenges of photoautotrophic production of acetyl-CoA-derived solar fuels and chemicals in cyanobacteria. Curr Opin Chem Biol 59:69–76
222. Ort DR, Merchant SS, Alric J, Barkan A, Blankenship RE, Bock R, Croce R, Hanson MR, Hibberd JM, Long SP, Moore TA, Moroney J, Niyogi KK, Parry MA, Peralta-Yahya PP, Prince RC, Redding KE, Spalding MH, van Wijk KJ, Vermaas WF, von Caemmerer S, Weber AP, Yeates TO, Yuan JS, Zhu XG (2015) Redesigning photosynthesis to sustainably meet global food and bioenergy demand. Proc Natl Acad Sci USA 112:8529–8536
223. Roussou S, Albergati A, Liang F, Lindblad P (2021) Engineered cyanobacteria with additional overexpression of selected Calvin-Benson-Bassham enzymes show further increased ethanol production. Metab Eng Commun 12:e00161
224. Kirst H, Formighieri C, Melis A (2014) Maximizing photosynthetic efficiency and culture productivity in cyanobacteria upon minimizing the phycobilisome light-harvesting antenna size. Biochim Biophys Acta 1837:1653–1664
225. Kirilovsky D, Kaňa R, Prášil O (2014) Mechanisms modulating energy arriving at reaction centers in cyanobacteria. In: Demmig-Adams B, Garab G, Adams III W, Govindjee (eds) Non-photochemical quenching and energy dissipation in plants, algae and cyanobacteria. Springer, Dordrecht, pp 471–501
226. Veaudor T, Blanc-Garin V, Chenebault C, Diaz-Santos E, Sassi JF, Cassier-Chauvat C, Chauvat F (2020) Recent advances in the photoautotrophic metabolism of cyanobacteria: biotechnological implications. Life 10:71
227. Zhou J, Yang F, Zhang F, Meng H, Zhang Y, Li Y (2021) Impairing photorespiration increases photosynthetic conversion of CO_2 to isoprene in engineered cyanobacteria. Bioresour Bioprocess 8:42
228. Chanquia SN, Vernet G, Kara S (2021) Photobioreactors for cultivation and synthesis: specifications, challenges, and perspectives. Eng Life Sci 22:712–724
229. Johnson TJ, Katuwal S, Anderson GA, Gu L, Zhou R, Gibbons WR (2018) Photobioreactor cultivation strategies for microalgae and cyanobacteria. Biotechnol Prog 34:811–827
230. Kirnev PCS, Carvalho JC, Vandenberghe LPS, Karp SG, Soccol CR (2020) Technological mapping and trends in photobioreactors for the production of microalgae. World J Microbiol Biotechnol 36:42
231. Heining M, Sutor A, Stute SC, Lindenberger CP, Buchholz R (2015) Internal illumination of photobioreactors via wireless light emitters: a proof of concept. J Appl Phycol 27:59–66

232. Li T, Strous M, Melkonian M (2017) Biofilm-based photobioreactors: their design and improving productivity through efficient supply of dissolved inorganic carbon. FEMS Microbiol Lett 364:fnx218
233. Leonard A, Rooke JC, Meunier CF, Sarmento H, Descy JP, Su BL (2010) Cyanobacteria immobilised in porous silica gels: exploring biocompatible synthesis routes for the development of photobioreactors. Energy Environ Sci 3:370–377
234. Polakovic M, Svitel J, Bucko M, Filip J, Nedela V, Ansorge-Schumacher MB, Gemeiner P (2017) Progress in biocatalysis with immobilized viable whole cells: systems development, reaction engineering and applications. Biotechnol Lett 39:667–683
235. Hays SG, Yan LLW, Silver PA, Ducat DC (2017) Synthetic photosynthetic consortia define interactions leading to robustness and photoproduction. J Biol Eng 11:4
236. Weiss TL, Young EJ, Ducat DC (2017) A synthetic, light-driven consortium of cyanobacteria and heterotrophic bacteria enables stable polyhydroxybutyrate production. Metab Eng 44: 236–245
237. Fedeson DT, Ducat DC (2021) Symbiotic interactions of phototrophic microbes: engineering synthetic consortia for biotechnology. In: Seneviratne G, Zavahir JS (eds) Role of microbial communities for sustainability. Springer, Singapore, pp 37–62
238. Fabris M, Abbriano RM, Pernice M, Sutherland DL, Commault AS, Hall CC, Labeeuw L, McCauley JI, Kuzhiuparambil U, Ray P, Kahlke T, Ralph PJ (2020) Emerging technologies in algal biotechnology: toward the establishment of a sustainable, algae-based bioeconomy. Front Plant Sci 11:279
239. Fernandes BD, Mota A, Teixeira JA, Vicente AA (2015) Continuous cultivation of photosynthetic microorganisms: approaches, applications and future trends. Biotechnol Adv 33:1228–1245
240. Ullmann J, Grimm D (2021) Algae and their potential for a future bioeconomy, landless food production, and the socio-economic impact of an algae industry. Org Agric 11:261–267
241. van den Berg C, Eppink MHM, Wijffels RH (2019) Integrated product recovery will boost industrial cyanobacterial processes. Trends Biotechnol 37:454–463

Adv Biochem Eng Biotechnol (2025) 192: 193–214
https://doi.org/10.1007/10_2025_284

Published online: 18 March 2025

Cable Bacteria and Their Biotechnological Application

Judith Stiefelmaier

Contents

Abstract Cable bacteria grow as multicellular filaments several centimetres deep into the sediment of freshwaters and oceans. Hereby, cable bacteria show unique characteristics such as electrogenic sulphur oxidation, extremely high conductivity and ability for CO_2 fixation. This offers several possibilities of future applications in biotechnology with an outlook to sustainable processes. So far, research on cable bacteria is mostly concerning metabolism, electron transfer and effect on the surrounding sediment. Cultures are always performed on sediment from the natural habitat and in simple, small-scale reaction tubes, requiring further development for reproducible cultivation with scale-up capabilities. However, based on the known properties of cable bacteria, possible areas of application can already be derived. The use of cable bacteria in bioremediation is a promising approach, as the degradation of hydrocarbons has already been proven. Co-cultivation with plants could open up a

J. Stiefelmaier (✉)
Bioprocess Engineering, RPTU Kaiserslautern-Landau, Kaiserslautern, Germany
e-mail: judith.stiefelmaier@mv.rptu.de

further field of application, such as the described reduction of methane emissions from rice fields. Due to the extremely high conductivity of the filaments, cable bacteria are also very promising for incorporation into biodegradable microelectronics. By integrating electrodes into a suitable reactor system, bioelectrochemical processes could be implemented, either with the goal of electron uptake and product formation or for electricity generation.

Graphical Abstract

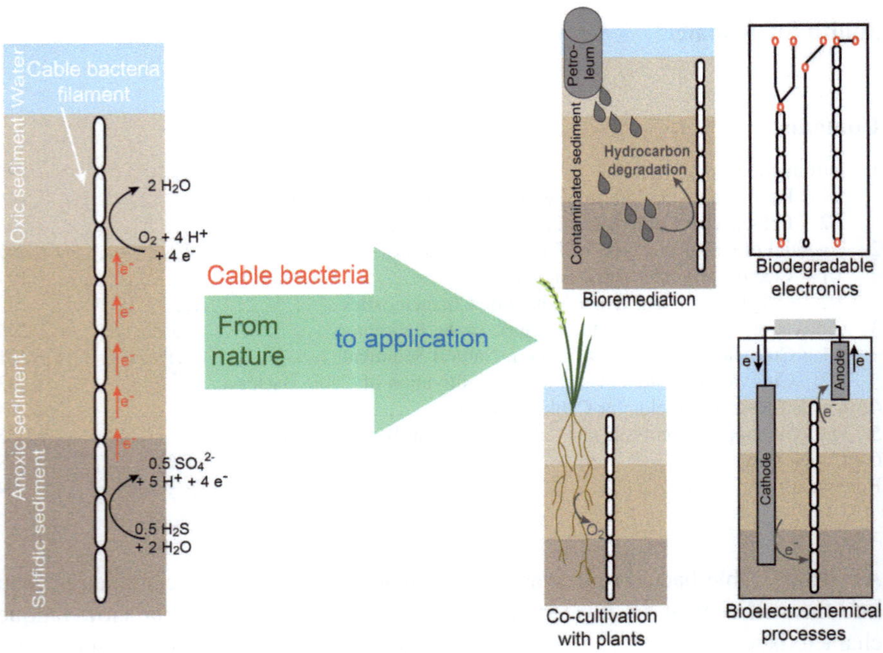

Keywords Biotechnological application, Cable bacteria, Conductive filaments, Electron transfer, Sediment

1 Introduction

Cable bacteria are gram-negative bacteria belonging to the family of Desulfobulbaceae [1] and grow as multicellular filaments (see Fig. 1a) in numerous fresh and salt water sediments [2, 3]. Despite their characteristic filamentous growth form of several centimetres in length and their known widespread occurrence in various sediments, cable bacteria were only discovered a few years ago, as explained in detail by Nielsen and Risgaard-Petersen [4]. Hereby, one peculiarity of cable bacteria and the reason for their filamentous growth is their use of inorganic sulphur compounds in the deeper anoxic sediment as electron donor. By oxidizing those sulphur compounds, electrons are released and conducted alongside the cable bacteria filament. The cells in the oxic sediment finally reduce oxygen to release electrons. An extensive overview of long-distance electron transfer (LDET) by cable bacteria is given by Meysman [5]. Thus, in this article, this topic is only described shortly, and the focus lies more on metabolism, used cultivation methods and the potential for future applications.

1.1 *Filamentous Growth in Marine and Freshwater Sediments*

One characteristic of cable bacteria is their growth from the oxic sediment surface into the deeper, anoxic sediment layers, which is consistent for marine and freshwater sediments. Thereby, depths of up to several centimetres have been described for the occurrence of cable bacteria filaments [6, 7]. The cells in the anoxic sediment oxidize sulphur compounds (see Fig. 1b). Released electrons are conducted along

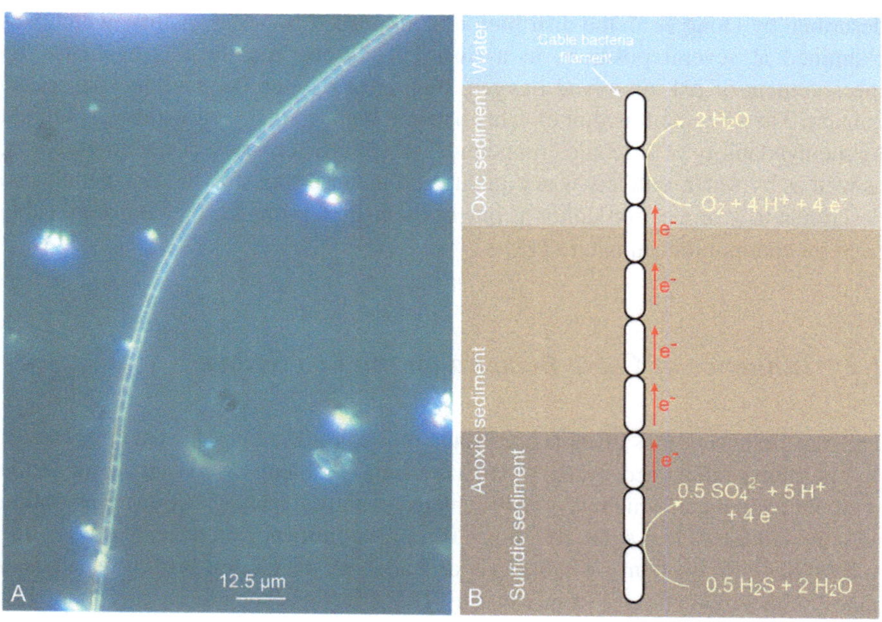

Fig. 1 (**a**) Microscopic picture of a cable bacteria filament of *Candidatus* Electronema aureum GS. (**b**) Overview of a cable bacteria filament with oxidation of sulphidic compounds in the anoxic, sulphidic sediment, transfer of electrons along the filament into the oxic sediment layer, where the cells reduce oxygen

the filament into the oxic sediment, where the cells reduce oxygen [8]. By transporting electrons from cell to cell, cable bacteria can use electron donors and acceptors that are spatially separated from each other. This offers cable bacteria an advantage for survival in aquatic sediments [5]. In order to grow under these conditions, the cable bacteria in the filament must cooperate well with each other and their metabolism must be adapted, depending on their position in the sediment. For example, less than 10% of cells in a filament have contact to oxygen and thus need to have an extremely high oxygen uptake rate, while the remaining over 90% of the cells can grow without oxidative stress [9]. Thereby, biosynthesis and growth mainly take place in the anoxic sediment [10]. However, the cells in the oxic and anoxic sediment are not permanently fixed to their function in the respective sediment layer but can adapt flexibly to changing environmental conditions. By using their ability for gliding motility, the different parts of a cable bacteria filament could move between the different sediment zones over time, allowing the cells to change between electron transfer to oxygen and growth in the absence of oxygen [10, 11]. In addition, the electron transfer and the connection of the cable bacteria from the anoxic to the oxic sediment have a positive effect on other microorganisms in the sediment [12, 13]. A possible electron transfer between cable bacteria and the so-called flocking bacteria is suspected here. For example, it has been described that aerobic microorganisms can also occur in anoxic sediment in the presence of cable bacteria, possibly by transferring electrons to the cable bacteria for respiration [12].

The occurrence of cable bacteria in sediments depends on different factors. As described by Dong et al. the distribution is influenced by water quality, which was examined at several positions in a river [14]. Water quality was evaluated by measurement of pH, dissolved oxygen, biological oxygen demand and phosphate content. Thereby, the function of cable bacteria but also their cell width was limited by the availability of sulphide compounds, which are necessary as electron donors, as well as by water with low water quality and the resulting stresses. Setaj et al. also describe a lack of electron donor as one of the main reasons for the demise of cable bacteria communities in nature [15].

1.2 *Influence of Cable Bacteria on the Eco System*

The metabolic activities of cable bacteria in the sediment hereby have a large impact on the surrounding eco system. An increasing problem in coastal areas is the seasonal hypoxia, in which oxygen is depleted in bottom waters [16]. In combination with the release of free sulphides, this has a strong influence on the eco system, due to its toxicity for the fauna, which can even lead to the development of so-called dead zones, with a total absence of marine life. This effect has been increasing in the last years due to a higher distribution of nutrients into the ocean as well as increasing water temperatures caused by climate change [16]. Seitay et al. investigated the influence of cable bacteria on euxinia in coastal waters, meaning anoxic and sulfidic water [15]. Hereby, they discovered that cable bacteria generate iron oxides in the sediment and remove free sulphides, preventing the development of euxinic zones. Since this process occurs in many hypoxic systems, the presence of cable bacteria is probably one reason for the reduced occurrence of euxinia in oceans. Similar findings have been described by van de Velde et al. [17]. They evaluated the effect of electrogenic sulphur oxidation on pore water geochemistry, with the highest impact on Ca^{2+} and SO_4^{2-} ions as well as on carbon and sulphur cycling. By analysing marine sediments in different seasons, Hermanns et al. could show that the conversion of iron sulphides to iron oxides by cable bacteria in spring probably leads to low sulphide contents in summer, which thus prevents euxinia and the loss of marine fauna [18]. Furthermore, the occurrence of cable bacteria has an impact on the enrichment of iron, manganese and phosphorus in the upper sediment [19].

This proves the importance of the occurrence of cable bacteria in sediments and their large impact on the eco system. However, the characteristics of cable bacteria not only make them relevant for their natural habitat, but could also be used in the future in biotechnological applications.

2 Potential of Cable Bacteria for Future Applications

Increasing research on cable bacteria in the last years concerning their growth, metabolism and filamentous, multicellular structure revealed some unique features of cable bacteria. Especially, first insights into their genome [20] and the unveiling of the conductivity mechanisms [21, 22] gave a glimpse of the possible potential of cable bacteria, which could be of interest for future applications.

2.1 Electron Transfer and Conductivity

In the microscopic examination of cable bacteria, the special nature of the surface of the common filament sheath, connecting single cells in cable bacteria filaments, was already very quickly established through the use of scanning electron microscopy [6, 23]. However, further investigations were necessary to resolve the role of these structures. Cornelissen et al. describe the extensive analysis of the cell envelope architecture and developed a structural cable bacteria model based on their results [22]. Thereby, the cell envelope is built up from parallel ridge compartments of a distinct size (see Fig. 2). An increased filament thickness is thus reached by an increased number of ridges (Fig. 2d). Each of these ridges contains a central fibre (Fig. 2b, c), which runs alongside the complete filament length (Fig. 2a), even across the cartwheel-like junctions between single cells in the filament (Fig. 2b). The ridges thereby as well give stability to the filament and strengthen cell-to-cell junctions [24]. As a result of the continuous fibres, the common filament sheath connecting the individual cells of a filament to each other enables the transfer of released electrons along the filament over a distance of several centimetres [25].

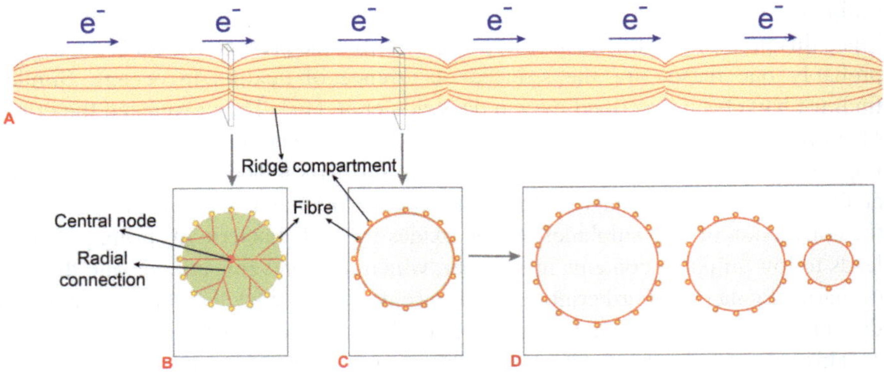

Fig. 2 Schematic representation of a cable bacteria filament: (**a**) Cable bacteria filament with continuous ridges. Fibres inside the ridges allow electron transfer along the filament. (**b**) Cross section of the junction between two single cells in the filament. (**c**) Cross section of a single cell in the filament. (**d**) Variations in filament thickness due to variations in the number of ridges. (Based upon Cornelissen et al. [22] and Thiruvallur Eachambadi et al. [25])

Table 1 Overview of measured conductivities described for cable bacteria strains from different extraction places

Strain	Extraction place	Conductivity	Comments	Literature
Not determined	Marine sediments	79 S cm^{-1}		[21]
Not determined	Salt marsh	26 S cm^{-1}	Measurement of extracted filament sheaths; given is an average, maximum was 73 S cm^{-1}	[26]
Ca. Electronema aureum GS	Freshwater sediment	48 ± 30 S cm^{-1}	Single ridge: 1.4 ± 0.9 S cm^{-1} [a]	[27]
Ca. Electrothrix communis RB	Marine sediments	0.45 ± 0.45 S cm^{-1}	Single ridge: 0.03 ± 0.03 S cm^{-1} [a]	[27]
Not determined	Marine sediment	Not determined	Single ridge: 1.3 ± 1.4 S cm^{-1}, number of ridges not given	[27]
Not determined	Salt marsh	Not determined	Single ridge: 4.2 ± 6.0 S cm^{-1}, number of ridges not given	[27]

[a] Measurements in the study were given for single ridges and were calculated for the whole filament by multiplication with the number of ridges for better comparability

Using AFM measurements on intact filaments and extracted fibre sheaths, Thiruvallur Eachambadi et al. were able to show that the fibre sheath is responsible for the conductivity of cable bacteria [25]. Thereby, the filament sheath exhibits the highest conductivity for biological materials described to date [21]. For cable bacteria filaments, conductivities of up to 79 S cm^{-1} were measured on explants from marine sediments by Meysman et al. (see overview in Table 1). A similar maximum of 73 S cm^{-1} was measured by Boschker et al. for a single extracted filament sheath from a salt marsh, with an average of 26 S cm^{-1} for all measurements [26]. Digel et al. analysed conductivity of cable bacteria from different isolation sources and brought it in connection with filament thickness by determining the conductivity of the single parallel longitudinal ridges [27]. For better comparability, as a part of this study, the values were as well calculated for the whole cable bacteria filament, assuming a number of ridges based on microscopic pictures given in the respective publication. Thereby, for the freshwater sediment strain *Candidatus* Electronema aureum GS and for *Candidatus* Electrothrix communis RB from marine sediments, conductivities of 48 ± 30 S cm^{-1} (1.4 ± 0.9 S cm^{-1} per single ridge) and 0.45 ± 0.45 S cm^{-1} (0.03 ± 0.03 S cm^{-1} per single ridge) were determined, showing large differences between investigated cable bacteria strains. For a not further determined strain from a salt marsh environment, even 4.2 ± 6.0 S cm^{-1} per single ridge was measured (number of ridges per filament not given), which would probably lead to an even higher total conductivity [27]. The physiological state of the examined filaments is discussed as an explanation for the high variations measured for the same strain, which could be influenced by the activity of a culture at the time of extraction, state of cell division or possible infections with bacteriophages. Differences between strains and the connection to their habitat need to be further analysed.

Concludingly, for different cable bacteria strains, extremely high conductivities were determined. For comparison with other electroactive microorganisms: *Shewanella* nanowires reach values of 0.03 S cm^{-1} [28] and wild-type *Geobacter* nanowires values of 0.09 S cm^{-1} [29]. With conductivities over 10 S cm^{-1}, cable bacteria are comparable to organic semiconductors [30], which has not been described for any other microorganisms so far. This, amongst other characteristics, makes cable bacteria promising microorganisms for potential application in microelectronics and bioelectrochemical systems, which will be discussed in more detail in the course of this chapter.

2.2 CO_2 Fixation and Other Metabolic Characteristics

Another point that is promising for the future use of cable bacteria is their ability to grow largely on inorganic compounds and use, for example, CO_2 or hydrogen carbonate as a carbon source [20]. However, hereby differences between cable bacteria strains from different origins were observed: Vasquez-Cardenas et al. saw only growth on propionate and just minor growth on hydrogen carbonate for a marine cable bacteria strain, which was examined by supplementing natural sediment with ^{13}C-propionate and ^{13}C-hydrogen carbonate [31]. However, Kjeldsen et al. could show incorporation of hydrogen carbonate into 43% of 51 experimentally investigated cable bacteria strains [20]. Furthermore, they could prove CO_2 fixation based on genetic as well as experimental data: It was shown that cable bacteria encode for the Wood-Ljungdahl pathway and the respective proteins were verified by proteome analysis of Ca. Electronema aureum GS, which could allow CO_2 fixation to acetyl-CoA [20, 32]. CO_2 uptake was experimentally confirmed by the addition of $^{14}CO_2$ to a cable bacterial culture and subsequent analysis by microautoradiography and fluorescence in situ hybridization [20]. Especially with regard to increasing CO_2 emissions and climate change, this makes cable bacteria interesting candidates for biotechnological applications with parallel CO_2 fixation. In general, based on their genome, the heterotrophic potential of cable bacteria is limited, with possible use of formate and acetate [20, 33]. Vasquez-Cardenas et al. investigated dark carbon fixation in coastal sediments in detail [34]. Hereby, dark carbon fixation describes the ability of chemoautotrophic microorganism to use dissolved inorganic carbon as carbon source. Vasquez-Cardenas et al. could show a connection between dark carbon fixation in the deep sediment with the electrogenic sulphur oxidation of cable bacteria. Furthermore, cable bacteria have the genetic ability for N_2 fixation, which however still needs to be proven experimentally [20]. By this, nitrogen availability for other organisms such as plants could be increased [35].

Cable bacteria cells growing in the oxic sediment layer of the filament are not generating energy; thus, those cells probably rely on storage compounds [36]. Since cable bacteria contain polyphosphates, these are possibly used as energy stores [20, 37, 38]. It is hypothesized that in the oxic zone, polyphosphates additionally protect against oxic stress, whereas in the anoxic sediment, polyphosphates could have a regulatory function [36]. Polyphosphates therefore enable the cable bacteria to better withstand stress situations and phases of energy scarcity, which is as well an advantage for possible biotechnological applications, since fluctuations in process conditions could be tolerated by cable bacteria up to a certain extent.

Although first studies on cable bacteria metabolism revealed some interesting details, such as CO_2 fixation, which could be further used in the development of sustainable processes, more research in this field is necessary. By this, a more complete understanding of cable bacteria metabolism could be achieved, which could be exploited for an effective application in biotechnology. With regard to the product formation of cable bacteria in particular, virtually nothing has been described in the literature to date. For a possible use in biotechnological production processes, this would therefore be a field of research that still needs to be investigated in more detail.

3 Cultivation Methods for Cable Bacteria in the Laboratory

Extraction from their natural habitat and cultivation of cable bacteria have been successfully described for several extraction places worldwide. Hereby, strains from different habitats such as sediment from marine areas, rivers, lakes or small ponds are being cultivated. Used cultivation conditions and cultivation set-ups will be summarized in the next sections.

3.1 Cultivation of Cable Bacteria in Natural Sediment

Cultivation of cable bacteria in lab scale has been described for several strains of marine as well as freshwater sediments. So far, all cultivations of cable bacteria in the lab are conducted in their natural sediment. This can be mostly drawn back to their occurrence in solid sediment, growth in centimetre-long filaments and unusual metabolism with dependence on gradients in the sediment, which makes transfer to standard cultivation conditions in liquid media not applicable. Although all strains are cultivated in their natural sediment, differences in the used cultivation methods have been described (exemplary overview in Table 2). In most cases, the sediment collected from the respective extraction spot is sieved and homogenized [26, 39–41]. In some cases, the sediment is additionally heat sterilized, and cable bacteria from an existing culture are used to inoculate the sterilized sediment [12, 42, 43]. Thorup et al. could establish a single-strain culture of the strain *Candidatus* Electronema aureum GS by repeated transfer of cable bacteria filaments to fresh homogenized and autoclaved natural sediment [42]. Dam et al. could show the different requirements of freshwater and marine strains to salinity by adjusting the salinity of the used sediments [44]. By embedding pillars made of agar into the sediment and measuring substrate concentrations, Sachs et al. could show the establishment of an equilibrium of the composition between agar and sediment and the subsequent colonization of the agar by cable bacteria [45]. Different authors describe the addition of FeS into the natural sediment as an additional electron donor, which can be used by the cable bacteria [45, 46]. Another approach is the incubation of natural sediment in artificial seawater, which is described in several publications [26, 40, 47]. Summarizing, there are already different procedures for cultivation of cable bacteria in the lab, which allow for example the examination of cable bacteria growth; however, all those methods are based on natural sediment.

Although cultivation on natural sediment offers advantages, such as cultivation under natural conditions and thus the possibility of researching cable bacteria growth and metabolism in their natural environment as well as their possible influence on the sediment, for a future biotechnological application of cable bacteria, the cultivation on natural sediment shows several disadvantages. On the one hand, natural sediment, even if taken from the same water body, is subject to fluctuations: Consequently, variations in parameters such as even small differences in the sampling site, season or weather lead to variations in sediment composition [17]. This limits the reproducibility of the results. In addition, the cultivation on natural sediment makes the exchange of locally obtained cable bacteria strains between research groups difficult, because the respective sediment is not permanently available for spatially distant research laboratories. Furthermore, preparation of the sediment for use in the laboratory is time consuming, which depends on the proportion of leaves and larger

Table 2 Overview of cable bacteria cultivated in the lab, their origin and the used cultivation methods. All sediments were collected from the natural extraction spot of the respective cable bacteria strains

Cable bacteria	Origin	Cultivation medium	Additional information	Literature
Ca. Electronema aureum GS	Freshwater pond	Homogenized, autoclaved sediment	Single-strain culture	[42]
Ca. Electronema Ca. Electrothrix	Freshwater lake Marine brackish water	Untreated sediment with varied salinity	Influence of salinity was investigated	[44]
Mixed culture	Different freshwaters (river, lake, aquifer)	Homogenized sediment with embedded agar pillars, additional FeS	Equilibrium between agar and sediment, colonization by cable bacteria	[45]
Mixed culture	Freshwater (aquifer)	Homogenized sediment with additional FeS	Sediment from an oil-contaminated site	[46]
Mixed culture	Salt marsh	Sieved, homogenized sediment in artificial seawater	Investigation of growth and cell division	[47]
Enrichment culture	Salt marsh	Sieved, homogenized sediment in artificial seawater	Investigation of filament conductivity	[26]
Ca. Electronema aureum GS	Freshwater pond	Heat-treated sediment	Interaction of cable bacteria with surrounding bacteria	[12]
Mixed culture	Marine	Homogenized sediment	Investigation of interaction of cable bacteria with microbial community	[48]
Ca. Electronema aureum GS	Freshwater pond	Sieved, homogenized, and autoclaved sediment	Investigation of microbial community	[13]
Mixed culture	Marine	Sieved sediment in filtered seawater	Co-cultivation with seagrass to investigate the interaction	[41]
Mixed culture	Freshwater, alkaline lakes	Sieved and homogenized sediment with in situ water	Investigation of sulphate reduction in freshwaters	[39]
Mixed culture	Salt marsh	Sieved and homogenized sediment in artificial seawater	Later isolation of cable bacteria filaments for extraction of fibre sheaths	[40]
Mixed culture	Freshwater, river delta	Sieved and homogenized sediment	Investigation of the influence of different oxygenation levels in sediments	[49]

(continued)

Table 2 (continued)

Cable bacteria	Origin	Cultivation medium	Additional information	Literature
Mixed culture	Marine mudbank deposits	Sieved sediment	Investigation of 2D development of pH and H_2S	[50]
Mixed enrichment culture	Groundwater	Sediment with fresh-water mineral medium	Investigation of sulphur disproportionation	[51]

stones contained in the extracted sediment. Another important point is a possible scale-up in the future: For larger scale experiments or applications, the removal of large volumes of natural sediment could have negative impacts on the ecosystem and is therefore not a suitable solution. Thus, natural sediment is only usable for small laboratory studies. In summary, for microbial and molecular biology studies of cable bacteria, the study in natural sediment is a good option. For possible use in larger scale biotechnological applications, other solutions must be found that still need to be researched.

3.2 Cultivation Systems Used for the Cultivation of Cable Bacteria

Depending on subsequent investigations, different cultivation set-ups are used for incubation of cable bacteria in natural sediment in the lab (exemplary overview in Table 3). Hereby, most systems have in common that cultivations are conducted in simple reaction tubes with diameters between 2.5 and 4.5 cm [2, 48], which are often integrated into liquid-filled aquaria with active aeration [2, 7, 26, 52]. By this, the natural habitat of the cable bacteria is mimicked, and sufficient oxygen to flare off electrons is supplied. Sediment heights between 6.8 and 10 cm are used [2, 11], which can be drawn back to the maximum described length of about 7 cm for cable bacteria filaments. Thus, higher filament columns would not bring an increased biomass formation. The used set-ups hereby offer the advantage of an easy-to-implement cultivation system. By using reaction tubes with small diameter which are integrated into larger aquaria, additionally several cultivations can be conducted in parallel, allowing replication of experiments.

Concludingly, currently used cultivation systems allow cultivation of cable bacteria in the lab in their natural sediment with additional aeration for sufficient oxygen supply. However, when thinking of future applications, the development of larger-scale sediment bioreactors seems necessary. These would allow control of process parameters, such as temperature or pH, and easy sampling of biomass and pore water.

Table 3 Overview of so far used cultivation systems for cable bacteria including diameter of the system and the used sediment height

Cultivation system	Sediment height	Diameter	Literature
Plexiglas liners in aquarium with tap water	Not stated	Not stated	[42]
PVC core liner tubes	10 cm	4 cm	[21]
Glass tube in aquarium with tap water	6.8 cm	4.5 cm	[2]
Glass liner cores in seawater aquarium	7 cm	4.5 cm	[7]
Plexiglas liners in aquarium with tap water	10 cm	4.3 cm	[53]
PVC core liner tubes with aerated artificial seawater	Not stated	4.0 cm	[26]
Plastic containers with humid air above	10 cm	Not stated	[11]
Plexiglas core liners in container with artificial seawater	Full core liner	4.0 cm	[54]
Plastic cores submerged in artificial seawater	Full core	3.6 cm	[38]
Polycarbonate core liners in aquarium with aerated artificial seawater	8 cm	2.5 cm	[48]
Plexiglas core liner in aquarium with tap water	Full core	Not stated	[13]
Acrylic box in aquarium with aerated seawater	Not stated	20 × 20 × 2 cm	[41]
10 L aquarium	Sediment + water: 10 cm	–	[39]
PVC core liner tubes in aerated, artificial seawater	Not stated	4.0 cm	[40]
Beaker (100 mL)	Not stated (80 mL sediment)	Not stated	[49]

4 Current Research Fields for Cable Bacteria

The increasing research field of cable bacteria in general is currently focusing on different important aspects concerning cable bacteria growth, metabolism and worldwide distribution. A major interest hereby lies on the occurrence of cable bacteria in different habitats such as marine and freshwater environments [2, 3] and on the impact cable bacteria have on the sediments and ecosystems there [17, 18]. Furthermore, the ability of cable bacteria for electrogenic sulphur oxidation and electron transfer along the filament [21, 26] as well as other metabolic characteristics such as possible electron sinks or carbon sources are investigated [20, 55]. Since the growth of cable bacteria in centimetre-long filaments is rather unusual for bacteria, the cell cycle and cell division are of interest [10, 47]. Different methods for possible quantification of growth and metabolic activity have been developed, since standard methods such as measurement of optical density are not applicable. Microsensors, specially developed for application in sediments, allow measurement of pH, H_2S and oxygen content in different sediment layers [56–58]. Methods for fluorescence staining of cable bacteria [2, 7, 23] and quantitative polymerase chain reaction (qPCR) [54] allow better visualization and determination of cable bacteria as well as quantification of the amount of cable bacteria in a mixed culture. By using so-called trench slides, the motility and thus vitality of cable bacteria can be investigated [8, 11, 42]. Hereby, small glass panes are glued onto a microscopic slide, forming a trench in their centre. This trench is then filled with sediment and covered with deoxygenated water and a cover slip. As the oxygen can diffuse in from the sides of the cover glass, the cable bacteria move with one end of their filaments towards the oxic zone that forms. Additionally, many other aspects such as exemplarily motility of cable bacteria [11] or influence of cable bacteria on other microorganisms in the sediment [12, 13] are currently under investigation.

Concludingly, cable bacteria are extensively investigated concerning diverse aspects, which helps with better understanding cable bacteria and identifying further interesting features with a future application potential. However, so far, most research is not aiming directly towards a possible biotechnological application. Which possibilities based on current knowledge can already be seen for cable bacteria in the future and what is recently investigated towards this goal will be presented and discussed in the following section.

5 Outlook: Application Fields of Cable Bacteria in Biotechnology

When thinking of a future application of cable bacteria, their unique characteristics and metabolic processes make their use in various fields of application conceivable. Thus, in the following section, the possibilities for the use of cable bacteria in the fields of bioremediation, co-cultivation with plants, biodegradable electronics and bioelectrochemical processes will be discussed (see overview in Fig. 3).

Hereby, one possible application, where cable bacteria could find use, lies in bioremediation of contaminated sediments (see Fig. 3a). Bioremediation hereby describes the use of (micro)organisms for the biological detoxification of contaminated ecosystems. The possible use of cable bacteria for this environmentally friendly procedure has already been investigated by different working groups. Yuan et al. summarize the possibilities of using long-distance electron transfer in bioremediation, highlighting the advantages of cable bacteria in that context [59]. Hereby, they focus on the advantages of LDET, which, by using remote electron acceptors, could be a sustainable, cost-effective solution for the treatment of contaminated sediments and soils. Marzocci et al. describe the use of cable bacteria for the decontamination of petroleum-containing sediments [52]. For that, they exemplarily tested the degradation of alkanes in marine sediments containing cable bacteria, with an increased degradation of 24% caused by the cable bacteria. They attributed the improvement to the LDET of the cable bacteria and the resulting improved removal of sulphides. Concludingly, cable bacteria could be further investigated concerning the development of new treatments for petroleum-contaminated sediments. Matturro et al. showed as well the influence of cable bacteria on hydrocarbon degradation by establishing a bioelectrochemical

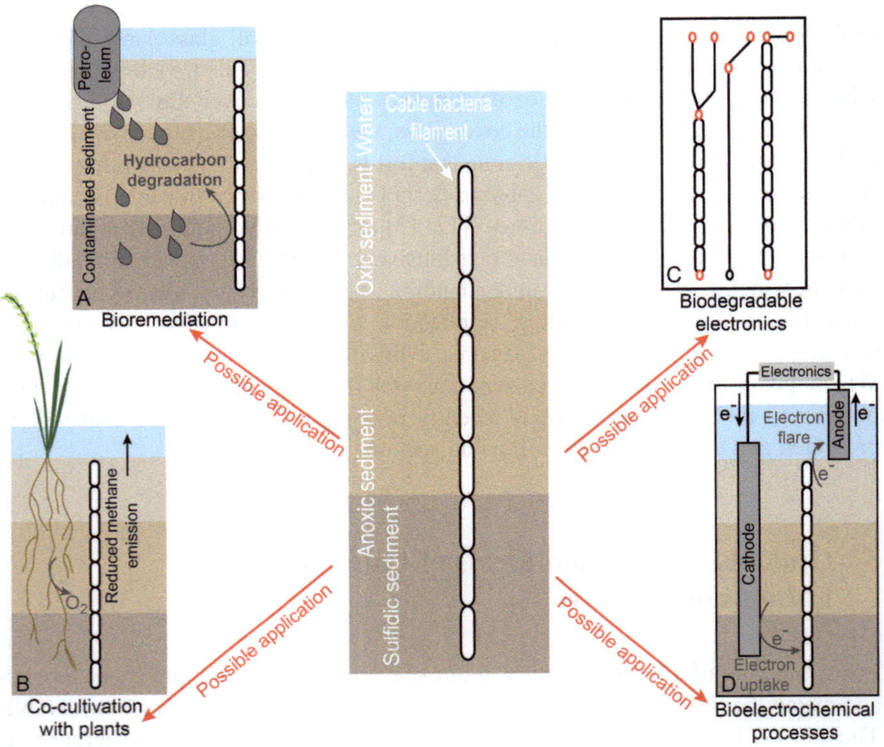

Fig. 3 Overview of the possible application fields for cable bacteria in the future. (**a**) Use of cable bacteria for bioremediation processes in contaminated soils, e.g., for the degradation of hydrocarbons. (**b**) Co-cultivation of cable bacteria with plants, e.g., for a reduced methane emission in rice. (**c**) Integration of cable bacteria into biodegradable bioelectronics, due to their high conductivity. (**d**) Application of cable bacteria in bioelectrochemical processes

connection between the oxic and anoxic sediment layers while maintaining an efficient sulphur cycle [60]. Furthermore, the effect of cable bacteria on organic pollutants in aquifer sediments was investigated by Müller et al. [46]. Hereby, aquifer sediments contaminated with hydrocarbons were enriched with FeS and incubated in the laboratory for several months. Microsensor measurements for the evaluation of geobiochemistry and microscopic analysis as well as DNA analysis revealed the presence of cable bacteria. By using sulphate as electron acceptor, the improvement of the biodegradation rate is discussed, which was proven by the detected LDET and degradation of toluene. Cable bacteria could as well give an advantage in degrading polycyclic aromatic hydrocarbons (PAHs) in sediments [61]. Hereby, indirect PAH degradation by cable bacteria is described; however, by influencing physiochemical properties of the sediment, cable bacteria can have a positive effect on PAH degraders and thus indirectly the reduction of PAHs in the sediment. Summarizing, cable bacteria offer a new possibility for a sustainable form of bioremediation, which however needs to be tested on its efficiency in larger-scale on-site experiments. Furthermore, methods for pre-cultivation of cable bacteria and subsequent application on contaminated sediments need to be developed.

Another possible environmentally friendly application of cable bacteria is the co-cultivation with plants and the reduction of methane emissions (see Fig. 3b). With methane as an important greenhouse gas and high methane emissions caused by rice fields, Scholz et al. investigated the impact of cable bacteria in the context of methane emissions from rice fields [62]. The basic idea here is that a reduction in methane release can be achieved by the competition of sulphate reducers with methanogens for the same substrates. Scholz et al. achieved a reduction of methane emissions by 93% compared to control trials by inoculating rice cultures with cable bacteria. Additionally, the presence of cable bacteria in the rhizosphere of different aquatic plants, like *Oryza sativa* (rice) amongst others, was shown [63]. The oxygen released from plant roots might offer an advantage for the occurrence of cable bacteria, which then can as well solely grow in the anoxic sediment layer. By oxidation of sulphide species by the cable bacteria, plant roots might be protected from toxic sulphide, giving this co-occurrence a positive effect for both partners. Concludingly, the application of cable bacteria in rice fields is a promising approach for the reduction of greenhouse gas with a positive effect as well for rice plants and cable bacteria. Since the co-occurrence of cable bacteria with other plants such as seagrass [41], the freshwater plant *Littorella uniflora* [63] and the marsh plant *Lobelia cardinalis* or *Salicornia europaea*, which is growing on tidal flats [63], has been described, the productive co-cultivation of cable bacteria and plants could be an emerging research field in the future. Further investigations on larger-scale inoculation methods for rice fields with cable bacteria, as well as the possibility of transferring the co-cultivation with cable bacteria on different plants, could lead to a successful application of cable bacteria in this area in the future.

As is already reviewed by Bonné et al., another possible application field of cable bacteria is in biodegradable electronics (see Fig. 3c), due to the high conductivity of cable bacteria filaments [64]. Thus, this topic is only discussed here shortly. With electronic waste being an increasing problem worldwide [65], the development of alternative, biodegradable solutions is an emerging field of research. Cable bacteria reach conductivities over 10 S cm^{-1} [21], which are comparable to polymer nanowires [66] and higher than other bacterial nanowires such as from the electroactive *G. sulfurreducens* and *S. oneidensis* [28, 29, 67]. Furthermore, with electron transport over several centimetres compared to ranges of about 10 μm in polymer nanowires and other bacterial nanowires, cable bacteria exceed with a factor of over 1,000. Additionally, Bonné et al. could show a degradation of cable bacteria in 9–30 days, which gives them further advantages as a possible material in biodegradable electronics [64]. Possible applications are hereby the use as electrical connections or as transistors [68].

Due to their high electrical conductivity, a future application in the field of electrochemical bioprocesses is also a promising possibility (see Fig. 3d). So far, no application of cable bacteria in electrochemical bioprocesses has been published. However, Reimers et al. describe the finding of cable bacteria filaments growing attached to a carbon fibre brush used as the anode of a benthic microbial fuel cell, when analysing the microbial community after 1 year of incubation in the sea floor [69]. The benthic microbial fuel cell thereby reached potentials of ~170–250 mV measured against a standard hydrogen electrode. This allowed the cable bacteria naturally occurring in the sea floor to use the anode as an alternative electron sink instead of oxygen. Based on these findings, Li et al. tried to induce the attachment of cable bacteria to an electrode [70]. Therefore, the marine cable bacterium *Candidatus* Electrothrix was cultivated in its natural sediment in a sediment bioreactor consisting of a polycarbonate tube. Into the polycarbonate tube, carbon fibre brushes were integrated as electrodes, with the anode positioned over the sediment in the overlaying seawater. By using anaerobic seawater, oxygen was excluded as electron sink. A potential of 300 mV was applied, and growth of cable bacteria filaments could be shown, proving the possibility of cable bacteria using electrodes in this case as an electron sink. Bonné et al. evaluated the effect of carbon fibre and carbon-felt electrodes on cable bacteria at a potential of 200 mV [71]. Hereby, cable bacteria growing in the sediment were attracted to the electrode as long as a potential was applied, and an increase in current could be observed. When switching off the current again, the cable bacteria retracted again from the electrode. These studies show that an induction of cable bacteria growth on electrodes is possible and can also be used for obtaining further insights into the electron transfer mechanisms of cable bacteria. Thus, this is a first step towards application of cable bacteria in bioelectrochemical systems, for which however further investigations are necessary. Another possibility instead of using cable bacteria for energy generation by using an anode as electron sink could be the application of a negative potential and the supply of electrons to the cable bacteria via the cathode. This could be used for CO_2-based electrobiochemical production processes, for which, however, the product spectrum of cable bacteria still needs to be examined in more detail with regard to industrially relevant products.

6 Conclusion

In the short years since their discovery, cable bacteria have grown into an emerging research field, revealing more and more important information about their unique growth and metabolism. Hereby, especially their ability for CO_2 fixation and very high conductivity make cable bacteria promising microorganisms for a future application in biotechnology. However, for a successful application of cable bacteria, further research concerning cultivation conditions and possible products is necessary. For once, the establishment of reproducible cultivation methods is necessary, namely the development of defined cultivation media, differing from the currently used natural sediment. Furthermore, specialized reactor systems must respond to the special requirements of the cable bacteria, such as the presence of gradients in the sediment, and have sensors adapted to the special conditions. This would thus allow the necessary cultivation under controlled conditions. Various fields of application are already emerging in the literature, with a focus on sustainable and resource-conserving applications. For once, the use in bioremediation offers the possibility for an effective and environmentally friendly decontamination of ecosystems, especially soils and sediments. Co-cultivation with plants could offer different advantages, such as reduced methane emissions from rice fields, but as well better growth of both plants and cable bacteria. High conductivity and biodegradability open up possibilities for use in bioelectronics. The ability for CO_2 fixation in combination with high conductivity gives cable bacteria furthermore the potential for the development of sustainable, biotechnological production processes, especially in the field of electrochemical bioprocesses. In conclusion, despite the great potential of cable bacteria, many more developments are needed before they can be used in biotechnology. With the great interest in cable bacteria and steadily growing field of research in this area, including a possible new, application-related focus, cable bacteria could play an increasing role in biotechnology in the future and thus achieve a further improvement in the sustainability of our society.

Acknowledgments The author thanks Prof. Dr. Lars Peter Nielsen (Center for Electromicrobiology, Aarhus University, Aarhus, Denmark) for supplying the strain *Candidatus* Electronema aureum GS.

References

1. Trojan D, Schreiber L, Bjerg JT, Bøggild A, Yang T, Kjeldsen KU, Schramm A (2016) A taxonomic framework for cable bacteria and proposal of the candidate genera Electrothrix and Electronema. Syst Appl Microbiol 39:297. https://doi.org/10.1016/j.syapm.2016.05.006
2. Risgaard-Petersen N, Kristiansen M, Frederiksen RB, Dittmer AL, Bjerg JT, Trojan D, Schreiber L, Damgaard LR, Schramm A, Nielsen LP (2015) Cable bacteria in freshwater sediments. Appl Environ Microbiol 81:6003. https://doi.org/10.1128/AEM.01064-15
3. Burdorf LDW, Tramper A, Seitaj D, Meire L, Hidalgo-Martinez S, Zetsche E-M, Boschker HTS, Meysman FJR (2017) Long-distance electron transport occurs globally in marine sediments. Biogeosciences 14:683. https://doi.org/10.5194/bg-14-683-2017
4. Nielsen LP, Risgaard-Petersen N (2015) Rethinking sediment biogeochemistry after the discovery of electric currents. Annu Rev Mar Sci 7:425. https://doi.org/10.1146/annurev-marine-010814-015708
5. Meysman FJ (2018) Cable bacteria take a new breath using long-distance electricity. Trends Microbiol 26:411. https://doi.org/10.1016/j.tim.2017.10.011
6. Pfeffer C, Larsen S, Song J, Dong M, Besenbacher F, Meyer RL, Kjeldsen KU, Schreiber L, Gorby YA, El-Naggar MY, Leung KM, Schramm A, Risgaard-Petersen N, Nielsen LP (2012) Filamentous bacteria transport electrons over centimetre distances. Nature 491:218. https://doi.org/10.1038/nature11586
7. Schauer R, Risgaard-Petersen N, Kjeldsen KU, Tataru Bjerg JJ, Jørgensen BB, Schramm A, Nielsen LP (2014) Succession of cable bacteria and electric currents in marine sediment. ISME J 8:1314. https://doi.org/10.1038/ismej.2013.239
8. Bjerg JT, Boschker HTS, Larsen S, Berry D, Schmid M, Millo D, Tataru P, Meysman FJR, Wagner M, Nielsen LP, Schramm A (2018) Long-distance electron transport in individual, living cable bacteria. Proc Natl Acad Sci USA 115:5786. https://doi.org/10.1073/pnas.1800367115
9. Scilipoti S, Koren K, Risgaard-Petersen N, Schramm A, Nielsen LP (2021) Oxygen consumption of individual cable bacteria. Sci Adv 7. https://doi.org/10.1126/sciadv.abe1870

10. Geerlings NMJ, Karman C, Trashin S, As KS, Kienhuis MVM, Hidalgo-Martinez S, Vasquez-Cardenas D, Boschker HTS, de Wael K, Middelburg JJ, Polerecky L, Meysman FJR (2020) Division of labor and growth during electrical cooperation in multicellular cable bacteria. Proc Natl Acad Sci USA 117:5478. https://doi.org/10.1073/pnas.1916244117
11. Bjerg JT, Damgaard LR, Holm SA, Schramm A, Nielsen LP (2016) Motility of electric cable bacteria. Appl Environ Microbiol 82:3816. https://doi.org/10.1128/AEM.01038-16
12. Bjerg JJ, Lustermans JJM, Marshall IPG, Mueller AJ, Brokjær S, Thorup CA, Tataru P, Schmid M, Wagner M, Nielsen LP, Schramm A (2023) Cable bacteria with electric connection to oxygen attract flocks of diverse bacteria. Nat Commun 14:1614. https://doi.org/10.1038/s41467-023-37272-8
13. Lustermans JJM, Bjerg JJ, Burdorf LDW, Nielsen LP, Schramm A, Marshall IPG (2023) Persistent flocks of diverse motile bacteria in long-term incubations of electron-conducting cable bacteria, Candidatus Electronema aureum. Front Microbiol 14. https://doi.org/10.3389/fmicb.2023.1008293
14. Dong M, Yang S, Yang X, Xu M, Hu W, Wang B, Huang Y, Xu J, Lu H, Yang Y, Chen X, Huang H, Sun G (2022) Water quality drives the distribution of freshwater cable bacteria. Sci Total Environ 841:156468. https://doi.org/10.1016/j.scitotenv.2022.156468
15. Seitaj D, Schauer R, Sulu-Gambari F, Hidalgo-Martinez S, Malkin SY, Burdorf LDW, Slomp CP, Meysman FJR (2015) Cable bacteria generate a firewall against euxinia in seasonally hypoxic basins. Proc Natl Acad Sci USA 112:13278. https://doi.org/10.1073/pnas.1510152112
16. Middelburg JJ, Levin LA (2009) Coastal hypoxia and sediment biogeochemistry. Biogeosciences 6:1273. https://doi.org/10.5194/bg-6-1273-2009
17. van de Velde S, Lesven L, Burdorf LD, Hidalgo-Martinez S, Geelhoed JS, van Rijswijk P, Gao Y, Meysman FJ (2016) The impact of electrogenic sulfur oxidation on the biogeochemistry of coastal sediments: a field study. Geochim Cosmochim Acta. https://doi.org/10.1016/j.gca.2016.08.038
18. Hermans M, Lenstra WK, Hidalgo-Martinez S, van Helmond NAGM, Witbaard R, Meysman FJR, Gonzalez S, Slomp CP (2019) Abundance and biogeochemical impact of cable bacteria in Baltic Sea sediments. Environ Sci Technol 53:7494. https://doi.org/10.1021/acs.est.9b01665
19. Hermans M, Astudillo Pascual M, Behrends T, Lenstra WK, Conley DJ, Slomp CP (2021) Coupled dynamics of iron, manganese, and phosphorus in brackish coastal sediments populated by cable bacteria. Limnol Oceanogr 66:2611. https://doi.org/10.1002/lno.11776
20. Kjeldsen KU, Schreiber L, Thorup CA, Boesen T, Bjerg JT, Yang T, Dueholm MS, Larsen S, Risgaard-Petersen N, Nierychlo M, Schmid M, Bøggild A, van de Vossenberg J, Geelhoed JS, Meysman FJR, Wagner M, Nielsen PH, Nielsen LP, Schramm A (2019) On the evolution and physiology of cable bacteria. Proc Natl Acad Sci USA 116:19116. https://doi.org/10.1073/pnas.1903514116
21. Meysman FJR, Cornelissen R, Trashin S, Bonné R, Martinez SH, van der Veen J, Blom CJ, Karman C, Hou J-L, Eachambadi RT, Geelhoed JS, de Wael K, Beaumont HJE, Cleuren B, Valcke R, van der Zant HSJ, Boschker HTS, Manca JV (2019) A highly conductive fibre network enables centimetre-scale electron transport in multicellular cable bacteria. Nat Commun 10:4120. https://doi.org/10.1038/s41467-019-12115-7
22. Cornelissen R, Bøggild A, Thiruvallur Eachambadi R, Koning RI, Kremer A, Hidalgo-Martinez S, Zetsche E-M, Damgaard LR, Bonné R, Drijkoningen J, Geelhoed JS, Boesen T, Boschker HTS, Valcke R, Nielsen LP, D'Haen J, Manca JV, Meysman FJR (2018) The cell envelope structure of cable bacteria. Front Microbiol 9. https://doi.org/10.3389/fmicb.2018.03044
23. Malkin SY, Rao AMF, Seitaj D, Vasquez-Cardenas D, Zetsche E-M, Hidalgo-Martinez S, Boschker HTS, Meysman FJR (2014) Natural occurrence of microbial sulphur oxidation by long-range electron transport in the seafloor. ISME J 8:1843. https://doi.org/10.1038/ismej.2014.41
24. Jiang Z, Zhang S, Klausen LH, Song J, Li Q, Wang Z, Stokke BT, Huang Y, Besenbacher F, Nielsen LP, Dong M (2018) In vitro single-cell dissection revealing the interior structure of cable bacteria. Proc Natl Acad Sci USA 115:8517. https://doi.org/10.1073/pnas.1807562115

25. Thiruvallur Eachambadi R, Bonné R, Cornelissen R, Hidalgo-Martinez S, Vangronsveld J, Meysman FJR, Valcke R, Cleuren B, Manca JV (2020) An ordered and fail-safe electrical network in cable bacteria. Adv Biosyst 4:e2000006. https://doi.org/10.1002/adbi.202000006
26. Boschker HTS, Cook PLM, Polerecky L, Eachambadi RT, Lozano H, Hidalgo-Martinez S, Khalenkow D, Spampinato V, Claes N, Kundu P, Wang D, Bals S, Sand KK, Cavezza F, Hauffman T, Bjerg JT, Skirtach AG, Kochan K, McKee M, Wood B, Bedolla D, Gianoncelli A, Geerlings NMJ, van Gerven N, Remaut H, Geelhoed JS, Millan-Solsona R, Fumagalli L, Nielsen LP, Franquet A, Manca JV, Gomila G, Meysman FJR (2021) Efficient long-range conduction in cable bacteria through nickel protein wires. Nat Commun 12. https://doi.org/10.1038/s41467-021-24312-4
27. Digel L, Justesen ML, Bonné R, Fransaert N, Wouters K, Jensen PB, Plum-Jensen LE, Marshall IPG, Nicolas-Asselineau L, Drace T, Bøggild A, Hansen JL, Schramm A, Bøjesen ED, Nielsen LP, Manca JV, Boesen T (2023) Comparative electric and ultrastructural studies of cable bacteria reveal new components of conduction machinery. bioRxiv. https://doi.org/10.1101/2023.05.24.541955
28. Polizzi NF, Skourtis SS, Beratan DN (2012) Physical constraints on charge transport through bacterial nanowires. Faraday Discuss 155:43. https://doi.org/10.1039/c1fd00098e
29. Lovley DR, Yao J (2021) Intrinsically conductive microbial nanowires for 'Green' electronics with novel functions. Trends Biotechnol 39:940. https://doi.org/10.1016/j.tibtech.2020.12.005
30. Le T-H, Kim Y, Yoon H (2017) Electrical and electrochemical properties of conducting polymers. Polymers 9. https://doi.org/10.3390/polym9040150
31. Vasquez-Cardenas D, van de Vossenberg J, Polerecky L, Malkin SY, Schauer R, Hidalgo-Martinez S, Confurius V, Middelburg JJ, Meysman FJR, Boschker HTS (2015) Microbial carbon metabolism associated with electrogenic sulphur oxidation in coastal sediments. ISME J 9:1966. https://doi.org/10.1038/ismej.2015.10
32. Ragsdale SW, Pierce E (2008) Acetogenesis and the Wood-Ljungdahl pathway of CO(2) fixation. Biochim Biophys Acta 1784:1873. https://doi.org/10.1016/j.bbapap.2008.08.012
33. Huang Y, Hu W, Dong M, Yang Y, Yang X, Huang H, Yang S, Jia W, Wang B, Xu M (2023) Cable bacteria accelerate the anaerobic removal of pyrene in black odorous river sediments. J Hazard Mater 443:130305. https://doi.org/10.1016/j.jhazmat.2022.130305
34. Vasquez-Cardenas D, Meysman FJR, Boschker HTS (2020) A cross-system comparison of dark carbon fixation in coastal sediments. Glob Biogeochem Cycles 34:e2019GB006298. https://doi.org/10.1029/2019GB006298
35. Malkin S, Cardini U (2021) Facilitative interactions on the rise: cable bacteria associate with diverse aquatic plants. New Phytol 232:1897. https://doi.org/10.1111/nph.17664
36. Geerlings NMJ, Kienhuis MVM, Hidalgo-Martinez S, Hageman R, Vasquez-Cardenas D, Middelburg JJ, Meysman FJR, Polerecky L (2022) Polyphosphate dynamics in cable bacteria. Front Microbiol 13. https://doi.org/10.3389/fmicb.2022.883807
37. Sulu-Gambari F, Seitaj D, Meysman FJR, Schauer R, Polerecky L, Slomp CP (2016) Cable bacteria control iron-phosphorus dynamics in sediments of a coastal hypoxic basin. Environ Sci Technol 50:1227. https://doi.org/10.1021/acs.est.5b04369
38. Geerlings NMJ, Zetsche E-M, Hidalgo-Martinez S, Middelburg JJ, Meysman FJR (2019) Mineral formation induced by cable bacteria performing long-distance electron transport in marine sediments. Biogeosciences 16:811. https://doi.org/10.5194/bg-16-811-2019
39. Sandfeld T, Marzocchi U, Petro C, Schramm A, Risgaard-Petersen N (2020) Electrogenic sulfide oxidation mediated by cable bacteria stimulates sulfate reduction in freshwater sediments. ISME J 14:1233. https://doi.org/10.1038/s41396-020-0607-5
40. Thiruvallur Eachambadi R, Boschker HTS, Franquet A, Spampinato V, Hidalgo-Martinez S, Valcke R, Meysman FJR, Manca JV (2021) Enhanced laterally resolved ToF-SIMS and AFM imaging of the electrically conductive structures in cable bacteria. Anal Chem 93:7226. https://doi.org/10.1021/acs.analchem.1c00298
41. Martin BC, Bougoure J, Ryan MH, Bennett WW, Colmer TD, Joyce NK, Olsen YS, Kendrick GA (2019) Oxygen loss from seagrass roots coincides with colonisation of sulphide-oxidising

cable bacteria and reduces sulphide stress. ISME J 13:707. https://doi.org/10.1038/s41396-018-0308-5
42. Thorup C, Petro C, Bøggild A, Ebsen TS, Brokjær S, Nielsen LP, Schramm A, Bjerg JJ (2021) How to grow your cable bacteria: establishment of a stable single-strain culture in sediment and proposal of Candidatus Electronema aureum GS. Syst Appl Microbiol 44:126236. https://doi.org/10.1016/j.syapm.2021.126236
43. Li C, Reimers CE, Chace PJ (2022) Protocol for using autoclaved intertidal sediment as a medium to enrich marine cable bacteria. STAR Protoc 3:101604. https://doi.org/10.1016/j.xpro.2022.101604
44. Dam A-S, Marshall IPG, Risgaard-Petersen N, Burdorf LDW, Marzocchi U (2021) Effect of salinity on cable bacteria species composition and diversity. Environ Microbiol 23:2605. https://doi.org/10.1111/1462-2920.15484
45. Sachs C, Kanaparthi D, Kublik S, Szalay AR, Schloter M, Damgaard LR, Schramm A, Lueders T (2022) Tracing long-distance electron transfer and cable bacteria in freshwater sediments by agar pillar gradient columns. FEMS Microbiol Ecol 98. https://doi.org/10.1093/femsec/fiac042
46. Müller H, Bosch J, Griebler C, Damgaard LR, Nielsen LP, Lueders T, Meckenstock RU (2016) Long-distance electron transfer by cable bacteria in aquifer sediments. ISME J 10:2010. https://doi.org/10.1038/ismej.2015.250
47. Geerlings NMJ, Geelhoed JS, Vasquez-Cardenas D, Kienhuis MVM, Hidalgo-Martinez S, Boschker HTS, Middelburg JJ, Meysman FJR, Polerecky L (2021) Cell cycle, filament growth and synchronized cell division in multicellular cable bacteria. Front Microbiol 12. https://doi.org/10.3389/fmicb.2021.620807
48. Liau P, Kim C, Saxton MA, Malkin SY (2022) Microbial succession in a marine sediment: inferring interspecific microbial interactions with marine cable bacteria. Environ Microbiol 24: 6348. https://doi.org/10.1111/1462-2920.16230
49. Wang Z, Liu F, Li E, Yuan Y, Yang Y, Xu M, Qiu R (2022) Network analysis reveals microbe-mediated impacts of aeration on deep sediment layer microbial communities. Front Microbiol 13. https://doi.org/10.3389/fmicb.2022.931585
50. Yin H, Aller RC, Zhu Q, Aller JY (2021) The dynamics of cable bacteria colonization in surface sediments: a 2D view. Sci Rep 11:7167. https://doi.org/10.1038/s41598-021-86365-1
51. Müller H, Marozava S, Probst AJ, Meckenstock RU (2020) Groundwater cable bacteria conserve energy by sulfur disproportionation. ISME J 14:623. https://doi.org/10.1038/s41396-019-0554-1
52. Marzocchi U, Palma E, Rossetti S, Aulenta F, Scoma A (2020) Parallel artificial and biological electric circuits power petroleum decontamination: the case of snorkel and cable bacteria. Water Res 173:115520. https://doi.org/10.1016/j.watres.2020.115520
53. Marzocchi U, Thorup C, Dam A-S, Schramm A, Risgaard-Petersen N (2022) Dissimilatory nitrate reduction by a freshwater cable bacterium. ISME J 16:50. https://doi.org/10.1038/s41396-021-01048-z
54. Geelhoed JS, van de Velde SJ, Meysman FJR (2020) Quantification of cable bacteria in marine sediments via qPCR. Front Microbiol 11. https://doi.org/10.3389/fmicb.2020.01506
55. Marzocchi U, Trojan D, Larsen S, Meyer RL, Revsbech NP, Schramm A, Nielsen LP, Risgaard-Petersen N (2014) Electric coupling between distant nitrate reduction and sulfide oxidation in marine sediment. ISME J 8:1682. https://doi.org/10.1038/ismej.2014.19
56. Nielsen LP, Risgaard-Petersen N, Fossing H, Christensen PB, Sayama M (2010) Electric currents couple spatially separated biogeochemical processes in marine sediment. Nature 463: 1071. https://doi.org/10.1038/nature08790
57. Risgaard-Petersen N, Revil A, Meister P, Nielsen LP (2012) Sulfur, iron-, and calcium cycling associated with natural electric currents running through marine sediment. Geochim Cosmochim Acta 92:1. https://doi.org/10.1016/j.gca.2012.05.036
58. Meysman FJ, Risgaard-Petersen N, Malkin SY, Nielsen LP (2015) The geochemical fingerprint of microbial long-distance electron transport in the seafloor. Geochim Cosmochim Acta 152: 122. https://doi.org/10.1016/j.gca.2014.12.014

59. Yuan Y, Zhou L, Hou R, Wang Y, Zhou S (2020) Centimeter-long microbial electron transport for bioremediation applications. Trends Biotechnol 39:181. https://doi.org/10.1016/j.tibtech.2020.06.011
60. Matturro B, Cruz Viggi C, Aulenta F, Rossetti S (2017) Cable bacteria and the bioelectrochemical snorkel: the natural and engineered facets playing a role in hydrocarbons degradation in marine sediments. Front Microbiol 8. https://doi.org/10.3389/fmicb.2017.00952
61. Liu F, Wang Z, Wu B, Bjerg JT, Hu W, Guo X, Guo J, Nielsen LP, Qiu R, Xu M (2021) Cable bacteria extend the impacts of elevated dissolved oxygen into anoxic sediments. ISME J 15: 1551. https://doi.org/10.1038/s41396-020-00869-8
62. Scholz VV, Meckenstock RU, Nielsen LP, Risgaard-Petersen N (2020) Cable bacteria reduce methane emissions from rice-vegetated soils. Nat Commun 11:1878. https://doi.org/10.1038/s41467-020-15812-w
63. Scholz VV, Müller H, Koren K, Nielsen LP, Meckenstock RU (2019) The rhizosphere of aquatic plants is a habitat for cable bacteria. FEMS Microbiol Ecol 95. https://doi.org/10.1093/femsec/fiz062
64. Bonné R, Wouters K, Lustermans JJM, Manca JV (2022) Biomaterials and electroactive bacteria for biodegradable electronics. Front Microbiol 13. https://doi.org/10.3389/fmicb.2022.906363
65. Perkins DN, Brune Drisse M-N, Nxele T, Sly PD (2014) E-waste: a global hazard. Ann Glob Health. https://doi.org/10.1016/j.aogh.2014.10.001
66. Groenendaal L, Jonas F, Freitag D, Pielartzik H, Reynolds JR (2000) Poly (3,4-ethylenedioxythiophene) and its derivatives: past, present, and future. Adv Mater 12:481. https://doi.org/10.1002/(SICI)1521-4095(200004)12:7<481:AID-ADMA481>3.0.CO;2-C
67. Lampa-Pastirk S, Veazey JP, Walsh KA, Feliciano GT, Steidl RJ, Tessmer SH, Reguera G (2016) Thermally activated charge transport in microbial protein nanowires. Sci Rep 6. https://doi.org/10.1038/srep23517
68. Bonné R, Hou J-L, Hustings J, Wouters K, Meert M, Hidalgo-Martinez S, Cornelissen R, Morini F, Thijs S, Vangronsveld J, Valcke R, Cleuren B, Meysman FJR, Manca JV (2020) Intrinsic electrical properties of cable bacteria reveal an Arrhenius temperature dependence. Sci Rep 10:19798. https://doi.org/10.1038/s41598-020-76671-5
69. Reimers CE, Li C, Graw MF, Schrader PS, Wolf M (2017) The identification of cable bacteria attached to the anode of a benthic microbial fuel cell: evidence of long distance extracellular electron transport to electrodes. Front Microbiol 8. https://doi.org/10.3389/fmicb.2017.02055
70. Li C, Reimers CE, Alleau Y (2020) Inducing the attachment of cable bacteria on oxidizing electrodes. Biogeosciences 17:597. https://doi.org/10.5194/bg-17-597-2020
71. Bonné R, Marshall IP, Bjerg J, Marzocchi U, Manca J, Nielsen LP, Aiyer K (2023) Electrically controlled interaction between cable bacteria and carbon electrodes. bioRxiv. https://doi.org/10.1101/2023.08.14.553267

Adv Biochem Eng Biotechnol (2025) 192: 215–228
https://doi.org/10.1007/10_2025_283

Published online: 25 March 2025

The Amoeba *Dictyostelium discoideum* as Novel Production Host for Complex Substances

Johann E. Kufs, Christin Reimer, and Lars Regestein

Contents

Abstract In this chapter, we discuss the necessity of novel chassis organisms for the production of natural products to steer away from petrochemical approaches and the usage of common model organisms. We present the social amoeba *Dictyostelium discoideum* as a novel host for the production of complex organic substances and exploration of cryptic biosynthetic routes of secondary metabolites. We shed light on the genetic repertoire of the amoeba in terms of natural product biosyntheses and give an overview of growth characteristics, genetic engineering tools, and cultivation methodologies from shake flasks to stirred-tank bioreactors. Finally, an outlook is

J. E. Kufs (✉) and C. Reimer
Research Group Genome Engineering and Editing, Faculty of Technology, Bielefeld University, Bielefeld, Germany
e-mail: johann.kufs@uni-bielefeld.de

L. Regestein (✉)
Bio Pilot Plant, Leibniz Institute for Natural Product Research and Infection Biology – Hans Knöll Institute (Leibniz-HKI), Jena, Germany
e-mail: lars.regestein@leibniz-hki.de

made on the perspective of *D. discoideum* as the chassis for biotechnological production and discovery of novel active substances.

Keywords Active substances, Amoebae, Biosynthetic gene clusters, Metabolic engineering, Microbial cell factories, Natural products, Phytochemicals, Synthetic biology

1 Natural Products: A Treasure Trove of Bioactive Substances

Natural products are low-molecular-weight organic compounds derived from the secondary metabolism and are produced by a plethora of organisms ranging from simple prokaryotes to complex eukaryotes. Their synthesis starts from simple building blocks of the primary metabolism, such as acyl-CoAs or amino acids (Fig. 1). According to the biosynthetic origin and involved enzymes, secondary metabolites can be classified into four different groups, namely terpenes, polyketides, non-ribosomal peptides (NRPs), and ribosomally synthesized and

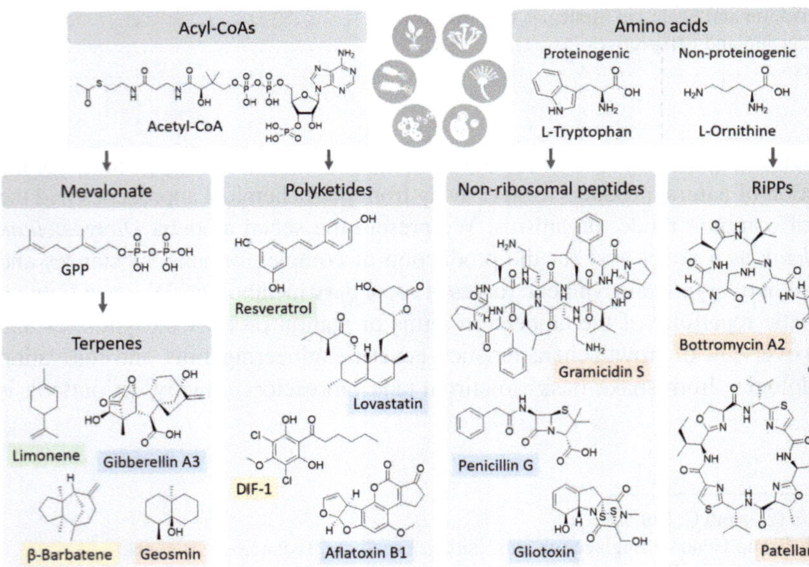

Fig. 1 Main classes of natural products and their typical building blocks. The illustration shows selected metabolites from bacteria (red), fungi (blue), plants (green), and amoebae (orange) as representative examples for the main classes of secondary metabolites, namely polyketides, terpenes, non-ribosomal peptides, and ribosomally synthesized and posttranslationally modified peptides (RiPPs). *CoA* coenzyme A, *GPP* geranyl pyrophosphate, *DIF-1* differentiation-inducing factor 1. (Reproduced from Reimer, 2022 [6])

posttranslationally modified peptides (RiPPs) [1]. However, substances are also known that are classified as mixed forms, such as meroterpenoids derived from polyketides and terpenes, or polyketide-NRP hybrids [2]. The high structural diversity of natural products (NPs) also results in high degree of functionalities, ranging from beneficial effects for human health, e.g., antioxidant (resveratrol), antimicrobial (penicillin), or cholesterol-lowering (lovastatin) activities, over flavorings and fragrances (e.g., limonene, geosmin), to harmful characteristics like the toxicity of aflatoxins [3–5].

Due to these features, secondary metabolites have proved to be a valuable source of pharmaceuticals, as most of their active substances are derived from natural products [7]. However, access to novel compounds is becoming increasingly difficult because the most abundant core structures have already been discovered. One reason for this difficulty is that many microbes do not produce their complete chemical bouquet under laboratory conditions. Natural products are derived from secondary metabolism and are not crucial for the normal growth and survival of organisms, which is why a continuous production is not required. External triggers are needed in order to induce the expression of corresponding biosynthetic genes. Two approaches exist to address this issue. On the one side, research is going away from monocultures into more realistic habitat-simulating co-cultivations with two or even more microbial partners. This enables interspecies communications by small molecules or physical contact [8] leading to activation of biosynthetic gene expression and synthesis of secondary metabolites. On the other side, tools of synthetic biology can be exploited to extract the genetic information of NP synthesis and transfer it into microbial hosts. However, these organisms need to be equipped with the accessory machinery for secondary metabolism. This includes a sufficient supply with precursors, such as malonyl-CoA or geranyl pyrophosphate, as well as activating enzymes for polyketide synthases (PKSs) or non-ribosomal peptide synthetases (NRPSs) like phosphopantetheinyl transferases (PPTases) [9].

2 Synthetic Biotechnology for NP Discovery

The transition from a petrochemical-based to a circular and renewable economy is one of the major challenges of modern society. White biotechnology in particular is already providing a substantial input in the form of sustainable manufacturing processes of formerly synthetically produced natural products [10]. However, so far, the percentage of biotechnologically produced compounds especially in the chemical and pharmaceutical industry sector is still very low. The conventional approach for developing sustainable processes is based on the use of established microbial model organisms such as the baker's yeast *Saccharomyces cerevisiae*, the lab workhorse *Escherichia coli*, or the filamentous fungus *Aspergillus niger*. Although the feasibility of the processes can be demonstrated, minor product yields possess obstacles for commercialization. Also, the approach of generating minimal organisms that only contain minimalistic metabolic and regulatory elements in the genome did not bring the expected revenues, yet [11]. Therefore, it is becoming more and more important to investigate abilities of other organisms proving better suitability for certain product classes. In the field of natural product discovery and production, mainly the naturally producing organisms are used to synthesize precursors that are further modified by chemical synthesis to obtain the final active ingredient. Synthetic biotechnology has revolutionized both basic and applied research and enables the simultaneous expression of multiple genes in an organism of choice to provide the molecular building machinery required for final product formation. With these technologies in hand, it is nowadays possible to engineer species-independent biosynthetic pathways which are implemented in an appropriate host system [12]. And especially for this reason, there is an urgent need for versatile host organisms that are, on the one hand, easily cultivable, scalable, and genetically tractable and, on the other hand, tailored for the supply of necessary precursors for the desired secondary metabolic pathway.

3 Amoebae as a Novel Biotechnological Host System

The social amoeba *Dictyostelium discoideum* belongs to the taxonomic clade of Amoebozoa that evolved after plants, but before the transition to fungi and animals (Fig. 2). As a free-living unicellular organism with a size of 10–20 μm, the amoeba replicates by binary fission during vegetative growth and feeds on bacteria as a natural nutrient source by phagocytosis [13, 14]. Due to this ability, *D. discoideum* has been widely used to study host-pathogen interactions and microbial infections [13]. Under starvation conditions, amoeba cells stream together, aggregate, and develop a multicellular fruiting body consisting of a stalk and a head of spores that start germinating, when a food source becomes available [15, 16]. This remarkable life cycle makes the amoeba a unique model for studying fundamental biological processes, especially the evolutionary transition from unicellular cells to multicellular organisms [15].

In addition to a good cultivability and the availability of genome information [17], the genetic tractability of an organism is of particular importance to serve as a novel host system. There are multiple tools available for the genetic modification of *D. discoideum*, starting from simple ectopic expression of individual genes using vector systems based on constitutive actin promotors [18] to established TetON/OFF systems for inducible gene expression using doxycycline [19]. These inducible vectors have been further developed to allow the polycistronic expression of multiple genes based on viral 2A peptides [20–22]. As the amoeba is used as a model system for fundamental and developmental studies, there are also multiple genome editing tools available, such as knockout strategies based on selection markers and the Cre-loxP system [23], as well as CRISPR/Cas9-mediated targeting of multiple genes [24, 25]. For comfortable access to genome and transcriptome data, the public community database, dictyBase, provides a comprehensive repository and also harbors the Dicty Stock Center for plasmid and strain procurement [26, 27].

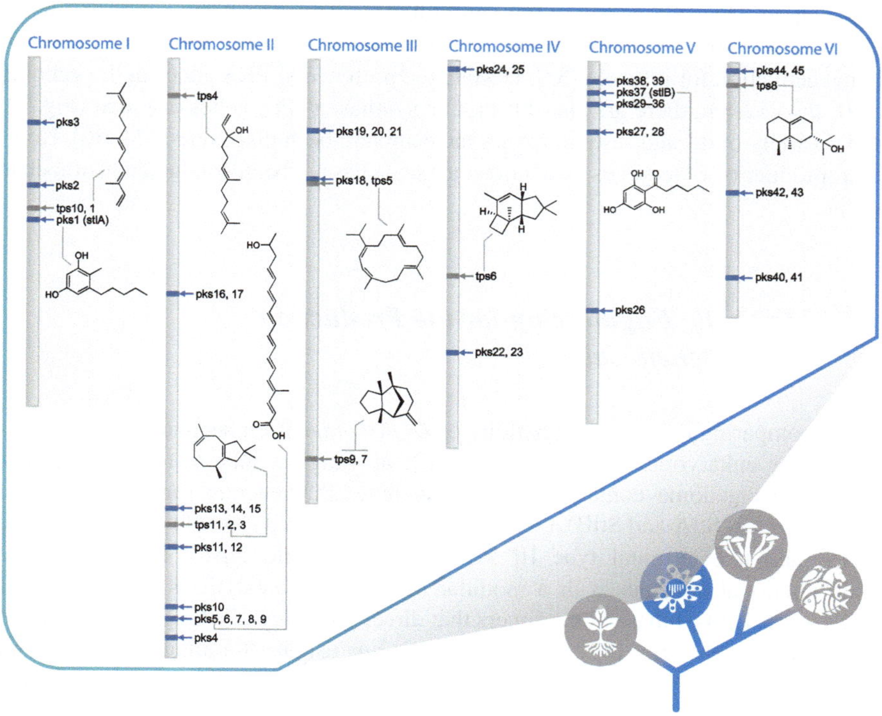

Fig. 2 Natural product potential of the amoeba *Dictyostelium discoideum*. The social amoeba *D. discoideum* belongs to the taxonomic clade of Amoebozoa that evolved after plants, but before the transition to fungi and animals, as shown in the phylogenetic tree of life. The amoeba harbors several polyketide synthase (PKS, blue) and terpene synthase (TPS, grey) genes with a high potential for secondary metabolites, distributed over all six chromosomes as indicated

3.1 The Natural Product Potential of D. discoideum

The amoeba *D. discoideum* exhibits the genetic prerequisite for the production of numerous secondary metabolites and novel biosynthetic routes. For instance, the amoeba encodes up to 45 potential polyketide synthase (PKS) genes (Fig. 2) in its genome [17, 28]. Despite this major genetic predisposition, less than ten polyketides have been isolated from amoebae to date, most of which were found to be involved in differentiation during the developmental cycle [29–31]. Recently, extrachromosomal expression of selected PKS genes resulted in production of the corresponding polyketides methylolivetol and phlorocaprophenone (PCP) and also led to the discovery of novel, deep-yellow polyketide pigments in *D. discoideum*, termed dictyodenes [21, 32]. The latter ones are involved in the maintenance of spore dormancy within mature fruiting bodies and hence play a crucial role during the life cycle of amoebae [32].

As *D. discoideum* is a native producer of polyketides, it also possesses the required accessory machineries to synthesize and modify the corresponding metabolites. This also opens up the opportunity to investigate polyketide synthases originating from other amoebae species, without the need for additional transcriptional activation for instance [33]. Despite the presence of PKS genes in the genome of *D. discoideum*, there are also 11 terpene synthase (TPS) genes encoded (Fig. 2), and a variety of di- and sesquiterpenes have already been discovered [34–36]. However, production of terpenes on a bioreactor scale has not been conducted in amoebae so far.

3.2 *Metabolic Engineering for the Production of Phytochemicals*

When comparing the genetic repertoire of *D. discoideum* for secondary metabolites with other eukaryotic microorganisms such as fungi, a unique feature becomes obvious. Its genome contains two special genes that encode for unique hybrid PKS enzymes (StlA and StlB), comprised of an *N*-terminal type I fatty acid synthase (FAS) and a *C*-terminal type III PKS (Fig. 3). While type I PKSs are large multifunctional enzymes with a modular assembly of catalytic domains, type III PKSs act as ketosynthase homodimers that directly use acyl-CoA esters for iterative condensation reactions [37]. The latter ones are usually found in the plant and bacterial kingdom [38]. Prominent examples of plant enzymes are stilbene and chalcone synthases responsible for the synthesis of resveratrol and naringenin chalcone from coumaric acid, respectively [39]. Interestingly, although no isolated structures would suggest it, there are three coumaroyl-CoA ligases encoded in the genome of *D. discoideum*. Their functional expression was verified by the formation of resveratrol upon expression of a stilbene synthase from the grapevine *Vitis*

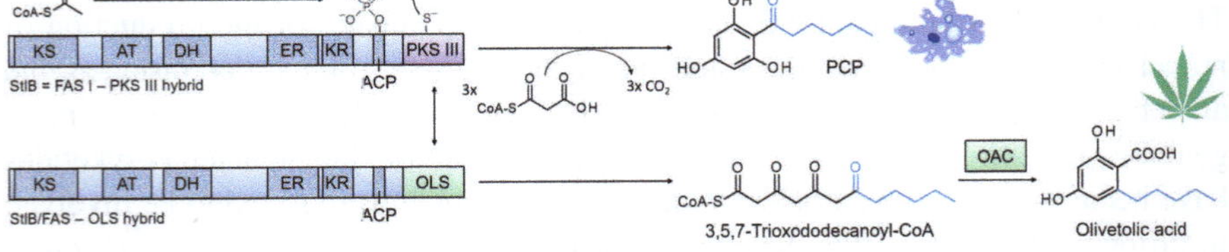

Fig. 3 Olivetolic acid production in amoeba using an inter-kingdom PKS hybrid enzyme. Illustration of the native StlB enzyme of *D. discoideum* consisting of the *N*-terminal type I FAS (blue) and the *C*-terminal type III PKS (purple) (top). The StlB/FAS domain produces a hexanoyl intermediate (blue), which is used as a starter unit by the PKS III domain for further elongation and PCP biosynthesis (top). By swapping the StlB/PKS III domain with the olivetol synthase (OLS, green) of *C. sativa*, the amoeba FAS domain can provide the hexanoyl moiety for OA production by OLS and olivetolic acid cyclase (OAC, bottom). *KS* ketosynthase, *AT* acyltransferase, *DH* dehydratase, *ER* enoyl reductase, *KR* ketoreductase, *ACP* acyl-carrier protein. (Reproduced with permission from Reimer et al., *Nature Biotechnology*, 2022 [21])

vinifera and supplementation with coumaric acid [21]. Olivetolic acid (OA) is the key intermediate in the cannabinoid biosynthetic pathway [40] and a structural isomer of PCP. Hence, a synthetic pathway was constructed to produce OA in amoeba starting from hexanoic acid and utilizing a bacterial medium-chain fatty acid CoA-ligase (FadK) [21]. To allow OA production directly from primary metabolites, it has been shown that the amoeba FAS domain can be used to provide the hexanoyl moiety for OA synthesis by fusion with the olivetol synthase (OLS) and simultaneous expression of the olivetolic acid cyclase (OAC) from *Cannabis sativa*. This amoeba/plant inter-kingdom hybrid enzyme enabled OA production without precursor supplementation and represents a shortcut in the cannabinoid biosynthetic route (Fig. 3). All these examples underline the enormous potential of *D. discoideum* as the hidden source of novel secondary metabolites and also its suitability to serve as an alternative chassis for the production of plant-derived natural products.

3.3 Co-culture Processes for NP Discovery

Due to natural predator-prey interactions between bacteria and amoebae, the presence of *D. discoideum* may trigger the formation of secondary metabolites as a defense mechanism of bacteria. However, the establishment of co-cultivations with amoebae and their natural nutrient sources under defined conditions is challenging and most rewarding at the same time. It has been shown that there are species at the bacterial site that can counteract predation by amoebae due to the formation of amoebacidal substances [41]. Therefore, it is conceivable that if the suitable interaction partner is present, also gene expression for novel secondary metabolites is induced in amoebae [42].

In contrast to co-cultures of bacteria and fungi, amoeba-based co-cultivations are more complex since both partners are not only concurring on the same nutrients but growth is directly linked due to the interaction between predator and prey. Hence, the best approach for a controllable setup would be to select a carbon source only usable by bacteria, so that the amoebae are directly dependent on the growth rate of their prey. Such an experiment has already been carried out in minimal medium using *E. coli* as food source for wild-type amoebae [43]. Interval cultivation using repeated pulses with glucose demonstrated the dependency of wild-type amoebae on bacteria rather than the utilization of glucose. However, this approach is not feasible for bioprocess development, as wild-type cells are not able to grow in media on classical soluble carbon sources without their prey [44, 45]. To avoid the need of co-cultivations during precultures and strain maintenance, axenically growing strains AX2 and AX4 have been generated that are able to proliferate in complex media by taking up nutrients via macropinocytosis [39]. Establishing co-cultures with those axenic strains would require the usage of different carbon sources that are not metabolized by amoebae.

4 Cultivation and Scale-Up Strategies for Amoebae

In general, *D. discoideum* can be grown in three different ways: on solid substrate, in liquid medium attached to the culture vessel, or planktonically in liquid medium [46]. For the latter approach, a physical movement is necessary to avoid attachment of the cells. Axenically growing strains can be cultivated in complex but also in chemically defined media upon an adaptation phase of approximately 3 weeks [47]. A general preculture regime for a liquid culture with axenic strains can be started by inoculating medium-filled petri dish with spores from a glycerol stock solution. After hatching of amoebae, the cells grow adherently to the bottom of the culture dish. Important to note, selective markers such as G418 or hygromycin should not be used directly, but rather added by exchanging with fresh medium on the next day. Depending on the strain-specific growth rates, the cells are grown at 22°C in the dark for 2–4 days, until they reach a confluency of ~80% and get subcultured. It is recommended to grow the cells not longer than 4–6 weeks to circumvent undesired mutations [46]. To set up a liquid culture, first nonviable detached cells are removed by washing before the cells are scraped off and transferred to a shake flask containing fresh medium. After overnight cultivation, cell concentration and viability are determined, usually using a cell counter device. For inoculating a new amoeba culture, cell concentrations below 1×10^4 cells mL^{-1} are not recommended, as the growth rate will be drastically reduced. For further preculture steps, the cells should remain in an exponential growth phase (~4×10^6 cells mL^{-1}); otherwise, they may change their gene expression to enter the developmental cycle. Usually, cell concentrations of $2–3 \times 10^7$ cells mL^{-1} can be reached in a shake flask and $5 \times 10^7–1 \times 10^8$ cells mL^{-1} in a stirred-tank bioreactor setting [48].

Regardless of the chosen bioreactor system, the following characteristics must be considered to successfully cultivate amoebae planktonically (Fig. 4). Amoebae have a maximum growth rate (μ_{max}), which depends largely on the chosen carbon source. For cultivations based on macropinocytosis, e.g., liquid media containing glucose, the growth rate can be up to 0.065 h^{-1} [48]. For (co-)cultivations in which bacteria

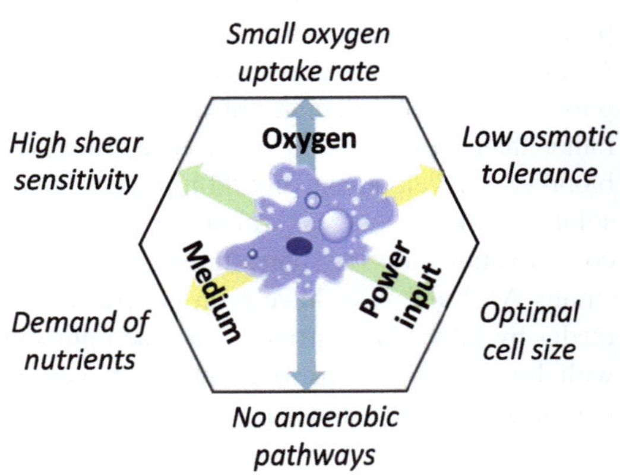

Fig. 4 Operational window for amoeba cultivations in terms of the selected essential parameters: medium composition, power input, and oxygen availability

are available as a nutrient source (phagocytosis), μ_{max} can also be larger than 0.1 h^{-1} [43]. In addition, the amoeba reaches a maximum cell density of approx. 10^7 cells mL^{-1}, which corresponds to a cell dry weight of 1–3 g L^{-1}, depending on the size of each amoeba [48].

The low growth rate and cell density result in an oxygen uptake rate of less than 10 mmol L^{-1} h^{-1}, which is comparable to plant cell cultivations [49]. To ensure that the selected conditions cover this oxygen demand, a calculation according to Eq. (1) from Meier et al. (2016) can be performed for shake flask cultivations [50]. Depending on the filling volume (V_L), shaking frequency (n), shaking diameter (d_0), diameter of the shake flask (d), pressure of reactor (ρ_R), oxygen mole fraction in the gas phase (Y_{*O2}) and osmolarity of the medium (Osmol), cultivation conditions can be easily determined that prevent a limitation of the oxygen supply. This is crucial as the amoeba does not have anaerobic metabolic pathways to compensate for a deficient supply of oxygen [51]:

$$OTR_{max} = 3.72 \times 10^{-7} \cdot Osmol^{0.05} \cdot n^{(1.18 - Osmol/10.1)} \cdot V_L^{-0.74} \cdot d_0^{0.33} \cdot d^{1.88} \cdot p_R \cdot y^*_{O2} \quad (1)$$

For cultivations in microtiter plates, a calculation according to Lattermann et al. (2014) is recommended [52]. Besides filling volume and shaking frequency, the perimeter (Peri) of the well must be known:

$$OTR_{max} = 2.5 \times 10^{-7} \cdot Peri^6 \cdot n^{2.37} \cdot V_L^{-0.64} + 3.3 \times 10^{-4} \quad (2)$$

Despite having a direct effect on the oxygen transfer rate (OTR), the osmolarity of the medium is also a critical cultivation parameter as amoebae have a low osmotic tolerance. In previous literature, no quantitative statements have been reported about this parameter, but the calculated conductivities of the media used are all in the range of small 10 μS cm^{-2}, also comparable to cultivations of plant cells. Hence, higher concentrations of salt, buffer, or carbon sources should be avoided to ensure optimal growth conditions. Literature reported that cultivations with excess of glucose are limited to an upper value of 10 g L^{-1} [47], suggesting the use of fed-batch approaches to circumvent this hurdle [53]. On the scale of a microtiter plate and shake flask, it is possible to use slow-release systems for the buffer and carbon source [54, 55]. Moreover, for submerged aerated cultivations, it is also important to avoid the use of antifoaming agents, if possible, which often applies to osmo-intolerant microbial systems.

Another important parameter for amoeba-based cultivations is the shear sensitivity of these organisms due to a lacking cell wall. The pressure difference between a moving part of the reactor (wall of a shake flask, stirrer blade, etc.) and the amoeba can therefore lead to cell damages. This is essential in two respects: (1) shear sensitivity is the most critical parameter in scaling the bioprocess, and (2) it affects the cell size and density in the reaction space, which has a direct impact on possible product formation [48].

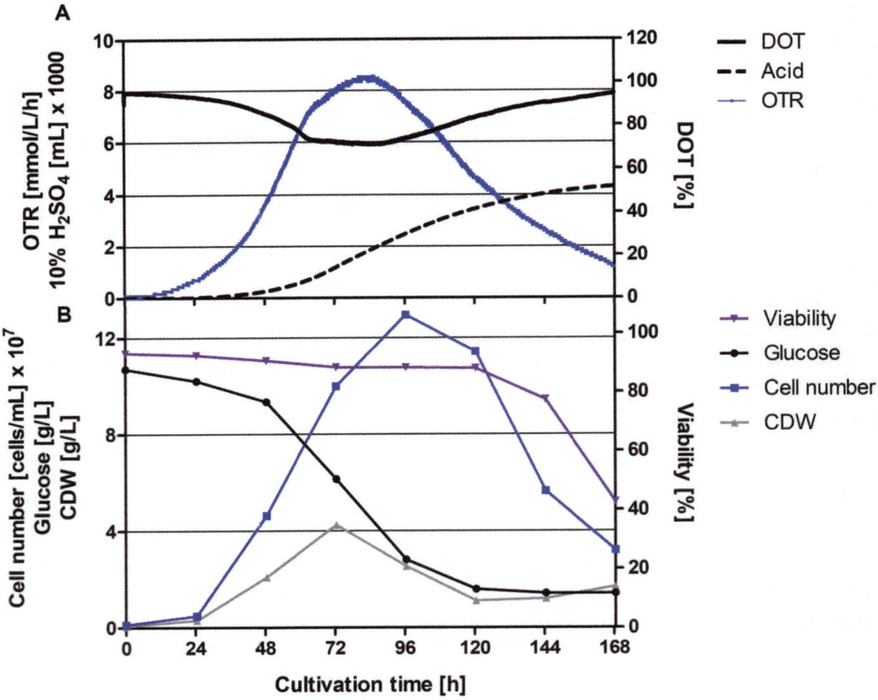

Fig. 5 Batch culture of OA-producing amoebae in a 300 L stirred-tank bioreactor. After expanding the preculture regime from 1.6 L over 7–75 L STR, cells were transferred into a 300 L STR at a cell density of 1×10^7 cells mL^{-1}. (**a**) Online parameters of batch fermentation including oxygen transfer rate (OTR), dissolved oxygen tension (DOT), and sulfuric acid consumption (acid) were monitored. (**b**) Offline parameters such as cell dry weight (CDW), cell number, glucose consumption, and viability are depicted over the course of the fermentation. Culture conditions: Filling volume = 200 L; HL5 medium with 10 g L^{-1} glucose, 20 µg mL^{-1} G418, and 20 µg mL^{-1} doxycycline; pH = 6.5 ± 0.05 was maintained by the addition of 10% H_2SO_4; tip speed = 2.62 m s^{-1}; stirring rate = 255 rpm; aeration rate = 0.3 vvm; temperature = 22°C. (Reproduced from Kufs et al., *Microbial Cell Factories*, 2022 [48])

To keep the shear stress constant in all cultivation systems and thus enable the process scale-up, the power input (P), maximum local energy dissipation (ε_{max}), and circulation time play a particularly important role. These points usually apply to all shear-sensitive microbial systems with, e.g., mycelium-like growth. For shake flasks, the ratio of maximum power input to average power input is recommended as a scale-up parameter, whereby for stirred systems, ε_{max} or the quotient of energy dissipation and circulation time is recommended [48, 56, 57]. The latter reflects not only the maximum shear stress that the amoeba is subjected to, but also how often this occurs during cultivation. Interestingly, amoeba cells can be adapted to a certain hydromechanical stress level prior to scale-up experiments. Thereby, a stirrer profile is performed to the desired power input within 15 min and held for at least 8 h. In this way, the lag phase for individual reactor stages can be minimized, and amoeba-based processes can be scaled up to industrially relevant scales such as 300 L bioreactors (Fig. 5).

5 Conclusion and Perspectives

Is the amoeba really a future workhorse of biotechnology or rather an exciting organism to gain new insights that can be transferred to other classical microbial hosts? Regarding the discovery of new active substances, the amoeba *D. discoideum* represents definitively a valuable source for novel biosynthetic pathways and a suitable chassis for the production of native and plant-derived secondary metabolites. Especially for the production of complex molecules like phytochemicals, the amoeba has the same *raison d'être* as other sophisticated hosts such as plant and animal cells. In addition, there are currently very few patents in the field of amoeba research, so there is a lot of *freedom to operate* in order to develop new products and markets.

References

1. Skellam E (2019) Strategies for engineering natural product biosynthesis in fungi. Trends Biotechnol 37:416–427
2. Han J, Jiang L, Zhang L, Quinn RJ, Liu X, Feng Y (2021) Peculiarities of meroterpenoids and their bioproduction. Appl Microbiol Biotechnol 105:3987–4003
3. Brakhage AA (2013) Regulation of fungal secondary metabolism. Nat Rev Microbiol 11:21–32
4. De La Lastra CA, Villegas I (2007) Resveratrol as an antioxidant and pro-oxidant agent: mechanisms and clinical implications. Biochem Soc Trans 35:1156–1160
5. Watson S, Jüttner F (2019) Biological production of taste and odour compounds. In: Taste and odour in source and drinking water: causes, controls, and consequences. IWA Publishing, London, pp 63–112
6. Reimer C (2022) Engineering of polyketide production routes in the amoeba *Dictyostelium discoideum*. Friedrich-Schiller-Universität Jena
7. Harvey AL (2008) Natural products in drug discovery. Drug Discov Today 13:894–901
8. Krespach MKC, Stroe MC, Netzker T, Rosin M, Zehner LM, Komor AJ, Beilmann JM, Krüger T, Scherlach K, Kniemeyer O et al (2023) Streptomyces polyketides mediate bacteria–fungi interactions across soil environments. Nat Microbiol 8:1348–1361
9. Walsh CT, Gehring AM, Weinreb PH, Quadri LEN, Flugel RS (1997) Post-translational modification of polyketide and nonribosomal peptide synthases. Curr Opin Chem Biol 1:309–315
10. Jullesson D, David F, Pfleger B, Nielsen J (2015) Impact of synthetic biology and metabolic engineering on industrial production of fine chemicals. Biotechnol Adv 33:1395–1402
11. Gil R, Peretó J (2015) Small genomes and the difficulty to define minimal translation and metabolic machineries. Front Ecol Evol 3. https://doi.org/10.3389/fevo.2015.00123
12. Kufs JE, Hoefgen S, Rautschek J, Bissell AU, Graf C, Fiedler J, Braga D, Regestein L, Rosenbaum MA, Thiele J et al (2020) Rational design of flavonoid production routes using combinatorial and precursor-directed biosynthesis. ACS Synth Biol 9:1823–1832

13. Annesley SJ, Fisher PR (2009) *Dictyostelium discoideum* – a model for many reasons. Mol Cell Biochem 329:73–91
14. Schilde C, Schaap P (2013) The Amoebozoa. Methods Mol Biol 983:1–15
15. Bozzaro S (2019) The past, present and future of *Dictyostelium* as a model system. Int J Dev Biol 63:321–331
16. Chen ZH, Schaap P (2016) Secreted cyclic di-GMP induces stalk cell differentiation in the eukaryote *Dictyostelium discoideum*. J Bacteriol 198:27–31
17. Eichinger L, Pachebat JA, Glockner G, Rajandream MA, Sucgang R, Berriman M, Song J, Olsen R, Szafranski K, Xu Q et al (2005) The genome of the social amoeba *Dictyostelium discoideum*. Nature 435:43–57
18. Veltman DM, Akar G, Bosgraaf L, Van Haastert PJ (2009) A new set of small, extrachromosomal expression vectors for *Dictyostelium discoideum*. Plasmid 61:110–118
19. Veltman DM, Keizer-Gunnink I, Haastert PJ (2009) An extrachromosomal, inducible expression system for *Dictyostelium discoideum*. Plasmid 61:119–125
20. Zhu X, Ricci-Tam C, Hager ER, Sgro AE (2023) Self-cleaving peptides for expression of multiple genes in *Dictyostelium discoideum*. PLoS One 18:e0281211
21. Reimer C, Kufs JE, Rautschek J, Regestein L, Valiante V, Hillmann F (2022) Engineering the amoeba *Dictyostelium discoideum* for biosynthesis of a cannabinoid precursor and other polyketides. Nat Biotechnol 40:751–758
22. Hoefgen S, Lin J, Fricke J, Stroe MC, Mattern DJ, Kufs JE, Hortschansky P, Brakhage AA, Hoffmeister D, Valiante V (2018) Facile assembly and fluorescence-based screening method for heterologous expression of biosynthetic pathways in fungi. Metab Eng 48:44–51
23. Faix J, Kreppel L, Shaulsky G, Schleicher M, Kimmel AR (2004) A rapid and efficient method to generate multiple gene disruptions in *Dictyostelium discoideum* using a single selectable marker and the Cre-loxP system. Nucleic Acids Res 32:e143
24. Muramoto T, Iriki H, Watanabe J, Kawata T (2019) Recent advances in CRISPR/Cas9-mediated genome editing in *Dictyostelium*. Cells 8:46
25. Sekine R, Kawata T, Muramoto T (2018) CRISPR/Cas9 mediated targeting of multiple genes in *Dictyostelium*. Sci Rep 8:8471
26. Fey P, Dodson RJ, Basu S, Hartline EC, Chisholm RL (2019) dictyBase and the Dicty Stock Center (version 2.0) – a progress report. Int J Dev Biol 63:563–572
27. Fey P, Dodson RJ, Basu S, Chisholm RL (2013) One stop shop for everything *Dictyostelium*: DictyBase and the Dicty Stock Center in 2012. Methods Mol Biol 983:59–92
28. Zucko J, Skunca N, Curk T, Zupan B, Long PF, Cullum J, Kessin RH, Hranueli D (2007) Polyketide synthase genes and the natural products potential of *Dictyostelium discoideum*. Bioinformatics 23:2543–2549
29. Morris HR, Taylor GW, Masento MS, Jermyn KA, Kay RR (1987) Chemical structure of the morphogen differentiation inducing factor from *Dictyostelium discoideum*. Nature 328:811–814
30. Ghosh R, Chhabra A, Phatale PA, Samrat SK, Sharma J, Gosain A, Mohanty D, Saran S, Gokhale RS (2008) Dissecting the functional role of polyketide synthases in *Dictyostelium discoideum*: biosynthesis of the differentiation regulating factor 4-methyl-5-pentylbenzene-1,3-diol. J Biol Chem 283:11348–11354
31. Saito T, Taylor GW, Yang JC, Neuhaus D, Stetsenko D, Kato A, Kay RR (2006) Identification of new differentiation inducing factors from *Dictyostelium discoideum*. Biochim Biophys Acta 1760:754–761
32. Günther M, Reimer C, Herbst R, Kufs JE, Rautschek J, Ueberschaar N, Zhang S, Peschel G, Reimer L, Regestein L et al (2022) Yellow polyketide pigment suppresses premature hatching in social amoeba. Proc Natl Acad Sci USA 119:e2116122119
33. Kubohara Y, Kikuchi H (2019) Dictyostelium: an important source of structural and functional diversity in drug discovery. Cells 8:6
34. Chen X, Köllner TG, Jia Q, Norris A, Santhanam B, Rabe P, Dickschat JS, Shaulsky G, Gershenzon J, Chen F (2016) Terpene synthase genes in eukaryotes beyond plants and fungi: occurrence in social amoebae. Proc Natl Acad Sci USA 113:12132

35. Rabe P, Rinkel J, Nubbemeyer B, Köllner TG, Chen F, Dickschat JS (2016) Terpene cyclases from social amoebae. Angew Chem Int Ed Engl 55:15420–15423
36. Kufs JE, Reimer C, Stallforth P, Hillmann F, Regestein L (2022) The potential of amoeba-based processes for natural product syntheses. Curr Opin Biotechnol 77:102766
37. Abe I, Morita H (2010) Structure and function of the chalcone synthase superfamily of plant type III polyketide synthases. Nat Prod Rep 27:809–838
38. Austin MB, Noel JP (2003) The chalcone synthase superfamily of type III polyketide synthases. Nat Prod Rep 20:79–110
39. Austin MB, Saito T, Bowman ME, Haydock S, Kato A, Moore BS, Kay RR, Noel JP (2006) Biosynthesis of *Dictyostelium discoideum* differentiation-inducing factor by a hybrid type I fatty acid-type III polyketide synthase. Nat Chem Biol 2:494–502
40. Carvalho Â, Hansen EH, Kayser O, Carlsen S, Stehle F (2017) Designing microorganisms for heterologous biosynthesis of cannabinoids. FEMS Yeast Res 17:fox037
41. Klapper M, Götze S, Barnett R, Willing K, Stallforth P (2016) Bacterial alkaloids prevent amoebal predation. Angew Chem Int Ed Engl 55:8944–8947
42. Barnett R, Stallforth P (2018) Natural products from social amoebae. Eur J Chem 24:4202–4214
43. Tsuchiya HM, Drake JF, Jost JL, Fredrickson AG (1972) Predator-prey interactions of *Dictyostelium discoideum* and *Escherichia coli* in continuous culture. J Bacteriol 110:1147–1153
44. Dunn JD, Bosmani C, Barisch C, Raykov L, Lefrancois LH, Cardenal-Munoz E, Lopez-Jimenez AT, Soldati T (1906) Eat Prey, Live: *Dictyostelium discoideum* as a model for cell-autonomous defenses. Front Immunol 2017:8
45. Van Haastert PJ (2010) Chemotaxis: insights from the extending pseudopod. J Cell Sci 123: 3031–3037
46. Fey P, Kowal AS, Gaudet P, Pilcher KE, Chisholm RL (2007) Protocols for growth and development of *Dictyostelium discoideum*. Nat Protoc 2:1307–1316
47. Han S-I, Friehs K, Flaschel E (2004) Improvement of a synthetic medium for *Dictyostelium discoideum*. Process Biochem 39:925–930
48. Kufs JE, Reimer C, Steyer E, Valiante V, Hillmann F, Regestein L (2022) Scale-up of an amoeba-based process for the production of the cannabinoid precursor olivetolic acid. Microb Cell Factories 21:217
49. Ullisch DA, Müller CA, Maibaum S, Kirchhoff J, Schiermeyer A, Schillberg S, Roberts JL, Treffenfeldt W, Büchs J (2012) Comprehensive characterization of two different *Nicotiana tabacum* cell lines leads to doubled GFP and HA protein production by media optimization. J Biosci Bioeng 113:242–248
50. Meier K, Klöckner W, Bonhage B, Antonov E, Regestein L, Büchs J (2016) Correlation for the maximum oxygen transfer capacity in shake flasks for a wide range of operating conditions and for different culture media. Biochem Eng J 109:228–235
51. Srinivas UK, Katz ER (1980) Oxygen utilization in *Dictyostelium discoideum*. FEMS Microbiol Lett 9:53–55
52. Lattermann C, Funke M, Hansen S, Diederichs S, Büchs J (2014) Cross-section perimeter is a suitable parameter to describe the effects of different baffle geometries in shaken microtiter plates. J Biol Eng 8:18
53. Han S-I, Friehs K, Flaschel E (2004) Cultivation of *Dictyostelium discoideum* on an improved synthetic medium in a conventional bioreactor. Process Biochem 39:585–589
54. Keil T, Dittrich B, Lattermann C, Habicher T, Büchs J (2019) Polymer-based controlled-release fed-batch microtiter plate – diminishing the gap between early process development and production conditions. J Biol Eng 13:18
55. Habicher T, Czotscher V, Klein T, Daub A, Keil T, Büchs J (2019) Glucose-containing polymer rings enable fed-batch operation in microtiter plates with parallel online measurement of scattered light, fluorescence, dissolved oxygen tension, and pH. Biotechnol Bioeng 116: 2250–2262

56. Hardy N, Augier F, Nienow AW, Béal C, Ben Chaabane F (2017) Scale-up agitation criteria for Trichoderma reesei fermentation. Chem Eng Sci 172:158–168
57. Daub A, Böhm M, Delueg S, Mühlmann M, Schneider G, Büchs J (2014) Maximum stable drop size measurements indicate turbulence attenuation by aeration in a $3m^3$ aerated stirred tank. Biochem Eng J 86:24–32

Adv Biochem Eng Biotechnol (2025) 192: 229–250
https://doi.org/10.1007/10_2025_287

Published online: 25 September 2025

Variable Bioproduction with *Euglena gracilis*: A Function of Light Flux or Carbon Source, Supplements, and Time

Dominik Cholewa and Tulsi Wormuth

Contents

Abstract *Euglena gracilis* is neither a plant nor an animal. It generates its energy from light and CO_2 purely photoautotrophically or it assimilates a carbon source chemoheterotrophically and transforms its chloroplasts into proplastids resulting in an animal cell structure. *E. gracilis* is a unicellular protist with a length of about 50 μm and developed by secondary endosymbiosis. For this reason, the chloroplasts have three membranes instead of a double membrane with a positive effect on the lipid content. It has no cell wall and is therefore easily bioavailable to humans. *Euglena* produces large amounts of vitamin E α-tocopherol and the β-1,3-glucan paramylon in granule form and has a good amount of lipids. Thanks to its contractile vacuole, *Euglena* is able to grow in a wide pH range from around pH 1–11.

D. Cholewa (✉)
Working Group Multiscale Bioengineering, Technical Faculty, Bielefeld University, Bielefeld, Germany
e-mail: dominik.cholewa@uni-bielefeld.de

T. Wormuth
Algenium GmbH & Co. KG, Bielefeld, Germany
e-mail: wormuth@algenium.de

Cultivation in the acidic range thus simplifies cultivation on a technical scale under axenic conditions and enhances the solubility of solids and trace elements.

Graphical Abstract

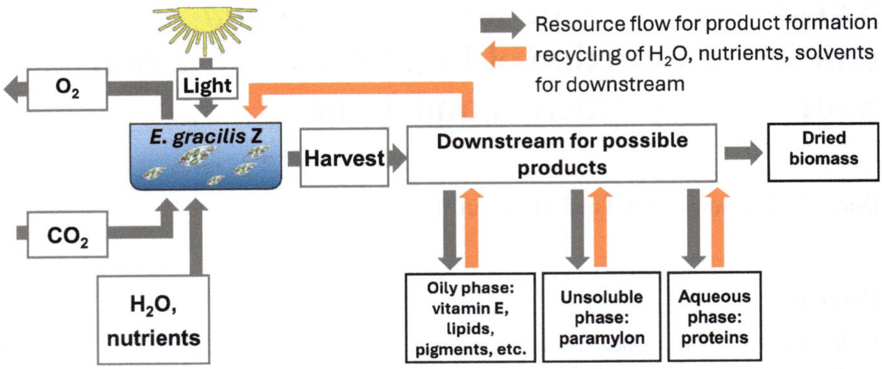

Keywords α-Tocopherol, β-1,3-Glucan, Acidophilic, Bioproduction, Biorefinery, Extremophilic, Lipids, Microalgae, Paramylon, Photoautotrophic growth

Abbreviations

CM	Cramer-Myers
DCW	Dry cell weight
kDa	kilo Dalton
LED	Light-emitting diode
PAR	Photosynthetically active radiation

1 Introduction to Bioproduction with Microalgae

In today's world, awareness of the need for a sustainable lifestyle is growing more than ever. This awareness is fostered by, among other things, climate change, negative effects of pollution on people's health, and foreseeable depletion of the world's oil reserves [1, 2]. However, humanity faces the fundamental challenge of transforming its consumerist lifestyle into a more sustainable way based on more realistic needs and natural and renewable resources [3]. To achieve this goal and make efficient use of natural resources, technological innovation must be further developed. One technology that offers the possibility of harnessing organic material from renewable resources is the concept of a biorefinery. In general, a biorefinery describes a facility for converting renewable organic materials, such as biomass, into low-, medium-, and high-value products by using (bio)chemical processes or whole microorganisms in the most environmentally friendly way possible [4, 5]. Therefore, many countries, including Germany, France, the United States, Brazil, and South Korea, are engaged in the economic development of various biomass chains with the aim of optimizing the use of biomass [4]. Potential products that can be produced by a biorefinery concept are energy, biofuels, and chemicals [6, 7]. In addition, fertilizers, animal feed, or even food supplements for human consumption can be obtained [8]. To receive these products, a biorefinery concept can be driven in different ways [5]. Examples of biorefinery concepts used around the world rely on sugarcane or corn as feedstock [4, 9]. A major bottleneck in using plants as renewable substrate is the need for farmland, as space is becoming a valuable commodity due to the fast-growing world population. This also leads to the commonly known "plate or tank" debate. Furthermore, plants require a large amount of water to grow; e.g., to produce 9 tons of corn per hectare requires about 7 million litres of water per hectare [10]. However, it has been clear for a decade that water will become a precious commodity in the future and that the way we use it will have to change drastically.

To circumvent this challenge, algae represent a promising alternative as a renewable biomass substrate [7, 8, 11].

These photosynthetically active organisms can be roughly distinguished into microalgae and macroalgae [12]. In comparison, microalgae can represent the more sustainable source, as macroalgae may be subject to country-specific conditions and seasonal fluctuations [13]. A huge advantage of algae is their ability to convert CO_2 into O_2 and macromolecular organic materials, e.g., carbohydrates, polysaccharides, lipids, pigments, proteins, and other compounds [14]. This production goes hand in hand with growth and thus biomass production. Figures show the great potential of microalgae in this respect, as, depending on the type of microalgae, 183 tons of CO_2 can be converted into 100 tons of biomass [15]. The fixation of CO_2 allows us to create a sustainable production cycle in which atmospheric CO_2 can be bound and no additional CO_2 is produced. Thus, production with microalgae can make a significant contribution to mitigating the effects of climate change. In this way, biorefineries can contribute to achieving the goal agreed at the Copenhagen climate summit in 2009 of limiting global warming by less than 2°C by reducing CO_2 emissions by 80–90% by 2050 [1]. The actual fixation rate is highly dependent on the environmental conditions and the microalgae species and may have higher rates than terrestrial plants [16]. Therefore, microalgae use sunlight five times more efficiently than terrestrial plants, whose photoconversion efficiency is between 4% and 6%. This leads to faster and more efficient growth and production [8, 15, 17]. The faster growth and associated higher productivity are also due to the asexual reproduction of the microalgae by cell division [8]. In addition, the need for fertilizers, pesticides, herbicides, water, and large cultivated areas decreases compared to agricultural crops and forest residues [18]. The reduction of water consumption can be further advanced through the use of wastewater or seawater in which microalgae thrive just as well. In some bioreactors, additional water can be saved by suitable water recirculation [8, 15]. These options of water saving permit

Table 1 Different known microalga species for commercial use

Species of interest	Commercial use case	References
Limnospira platensis	High protein content, used in food and nutraceutical industries	[26, 27]
Haematococcus pluvialis	Astaxanthin, a carotenoid with antioxidant properties	[28, 29]
Dunaliella salina	High beta-carotene content, used in food and cosmetic applications	[29]
Chlorella sp.	Proteins used in dietary supplements and food products	[30]
Nannochloropsis sp.	High lipid content, suitable for biofuel production	[31, 32]

production in areas where food crops cannot thrive, bypassing the aforementioned "plate or tank" debate [5, 8]. Microalgae also do not possess roots, leaves, or stems, making the entire biomass usable in a biorefinery concept. By recovering and separating all biomass components, waste production can be minimized and the generation of a sustainable material cycle is possible [8, 19]. Depending on the species and cultivation conditions, microalgae can reach biomass productivities of up to 7 g L^{-1} h^{-1} and have a carbon content of about 50% by weight. Additionally, up to 70% of the biomass can consist of lipids of which in turn 10% up to 20% can be polyunsaturated fatty acids [5, 8, 20, 21]. These fatty acids cover a wide spectrum from C12 to C24 and are used in the food or biofuel industry [22, 23]. Under consideration of the forecasts for these sectors, the high demand and potential become clear. As mentioned by Eppink et al., the Food and Agriculture Organization of the United Nations predicts a necessary increase in food production by 70% until 2050 to satisfy the growing world population [24]. Seventy-five percent of algae biomass and derived products are already sold in the food industry [8]. Furthermore, the demand for bioenergy is expected to serve a market of up to 100 tons of microalgae biodiesel per hectare and year [5].

The choice of desired microalgae for the biorefinery concept seems to be unlimited. In 2025, over 58,000 algae species were identified in the largest algae database, algaebase.org [25]. Only a small proportion of the known species are used commercially in industry (Table 1). This could be related to the fact that the usability depends on many parameters such as the accumulation and storage of potential products in dependence to nutrient availability or light intensity [8, 33]. Nevertheless, this biodiversity represents a great opportunity for new possibilities and applications, even if the quantity of microalgae biomass produced is only around 15,000 tons per year with costs of up to 100,000 US$ per ton [34]. These high production installation and operating costs as well as fluctuating culture conditions due to inconsistent light supply or varying weather conditions and the risk of contamination with other algae or bacteria are some of the challenges in commercializing microalgae cultivation [15, 33].

In particular, the reduction in availability of light with an increase in biomass can lead to increasing costs as a function of culture thickness [19]. For establishment, it is assumed that the cost of microalgal biomass needs to be reduced from 2–20 € kg^{-1} to less than 1 € kg^{-1} [8]. To ensure profitability, the microalgae should produce

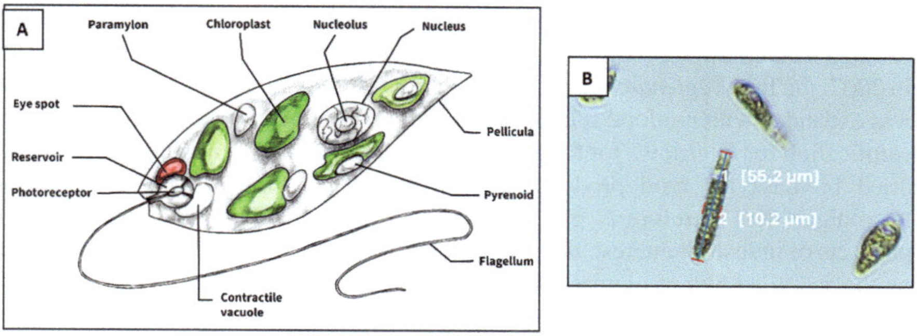

Fig. 1 Different representations of a photoautotrophically grown *Euglena gracilis Z* cell. (**a**) Schematic cell illustration with labeling of all important cell components and organelles (adapted from [23]). (**b**) Light microscopic cell image and measurement of the length and width of a freshly divided cell

many other byproducts with high market value such as lipids, carbohydrates, proteins, vitamins, and pigments, in addition to high biomass concentrations, and should be able to readily assert themselves or withstand extreme conditions. In addition, a low-cost nutrient source as well as easy harvesting of the products is crucial [8, 15, 33]. One microalga with the ability to meet these requirements and whose components can be fully valorized is *Euglena gracilis*.

1.1 The protist Euglena gracilis

E. gracilis is a unicellular protist, belonging to the green algae, and one of the most comprehensively studied organisms within the class of *Euglenida*, a group of various flagellates characterized by a striated cell surface or pellicle [2, 14, 35]. Typical photoautotrophic grown cells of *E. gracilis* can be seen in Fig. 1 with a characteristic length of 50 μm, while the schematic illustration shows the cellular components.

The genome comprises more than 500 Mb with an estimation of around 36,000 protein-encoding genes [36]. Thus, the unicellular phototrophic protist can grow in different ways. The wide growth spectrum includes photoautotrophic cultivations based on sunlight, heterotrophic cultivations with an external organic carbon source, and combination of both known as mixotrophic cultivation [37]. In addition, *E. gracilis* tolerates several external stresses like pH values between 0.9 and 11 or temperatures from 1 to 38°C with an optimum between 28 and 30°C in terms of specific growth rate [23, 38, 39]. The resulting ease and fast cultivation as well as simple harvest of products have helped *Euglena* to become one of the best studied model organisms in the context of microalgae [40].

Due to the general high nutrient content of microalgae, initial studies to investigate and determine the potential and value of *E. gracilis* as a supplier of biomass for food and other products started in the 1970s. The commercialization of *E. gracilis*-

related products for the Japanese market got under way in 2005 with the foundation of Euglena Co., Ltd., a company with the aim of producing and supplying *Euglena*. In 2007, the first *Euglena*-containing food products were launched. Later, the market was expanded with nonfood-related products. In 2021, the company acquired GRAS certification for *E. gracilis* for food safety in the United States, enabling the protist as dried biomass for food products to be sold in the United States [41]. Since microalgae, and therefore *E. gracilis*, produce, accumulate, and store a variety of products of industrial interest, the commercialization of *Euglena* production is based on the so-called 5Fs of biomass strategy. The 5Fs refer to the products food, fiber, feed, fertilizer, and fuel from biomass [42]. Four important products that were examined in more detail as part of this chapter are biomass itself, paramylon, α-tocopherol, and lipids.

Biomass itself can be seen as a double-edged product. On the one hand, biomass acts as the end product. On the other hand, biomass serves as an intermediate product from which the abovementioned biologically active substances can be isolated [22]. Under photoautotrophic conditions, *E. gracilis* achieves biomass concentrations of around 3 g L^{-1}, which is approximately 20-fold less than under mixotrophic or heterotrophic conditions [11, 43]. Thereby, under phototrophic conditions, specific growth rates of 0.9 up to 1.1 day^{-1} are achieved, whereas cultivations under heterotrophic conditions may well exceed these values [39, 44]. One product that can be obtained from *E. gracilis* biomass by anaerobic digestion with a consortium of microorganisms is biogas. Previous studies have shown that *E. gracilis* cell mass is suitable for a yield of around 650 or 800 mL biogas per g biomass under phototrophic or heterotrophic cultivation conditions, which is up to ten times higher compared to other microalgae [11, 45]. Because of the large number of biologically active compounds in *Euglena*'s biomass, the algae can be seen as a new functional food. For this reason, the EFSA Panel on Nutrition, Novel Foods and Food Allergens approved the dried whole cells as a novel food in May 2020 [46]. In developing countries, this can help to combat malnutrition and maintain the health of people around the world, as some studies have suggested a positive effect on human health. For example, anti-inflammatory effects by *E. gracilis* supplements can be a promising approach for treating overweight and obesity and related diseases such as atherosclerosis or diabetes [47–49].

1.2 Bioproduction with E. gracilis

E. gracilis grows under very flexible conditions, which is advantageous for generally robust biomass production on a technical scale [42, 50]. A good overview of possible bioproducts reported over the last 20 years is given by Gissibl and colleagues [51]. Depending on the need or target product, you can choose between photoautotrophic, α-tocopherol mixotrophic, and heterotrophic cultivation methods. The organism achieves high specific growth rates and high cell titers using relatively modest culture media [52, 53]. To promote axenic grow conditions for *E. gracilis* on a technical scale, growth at low pH around pH 2.0 is suitable for this organism

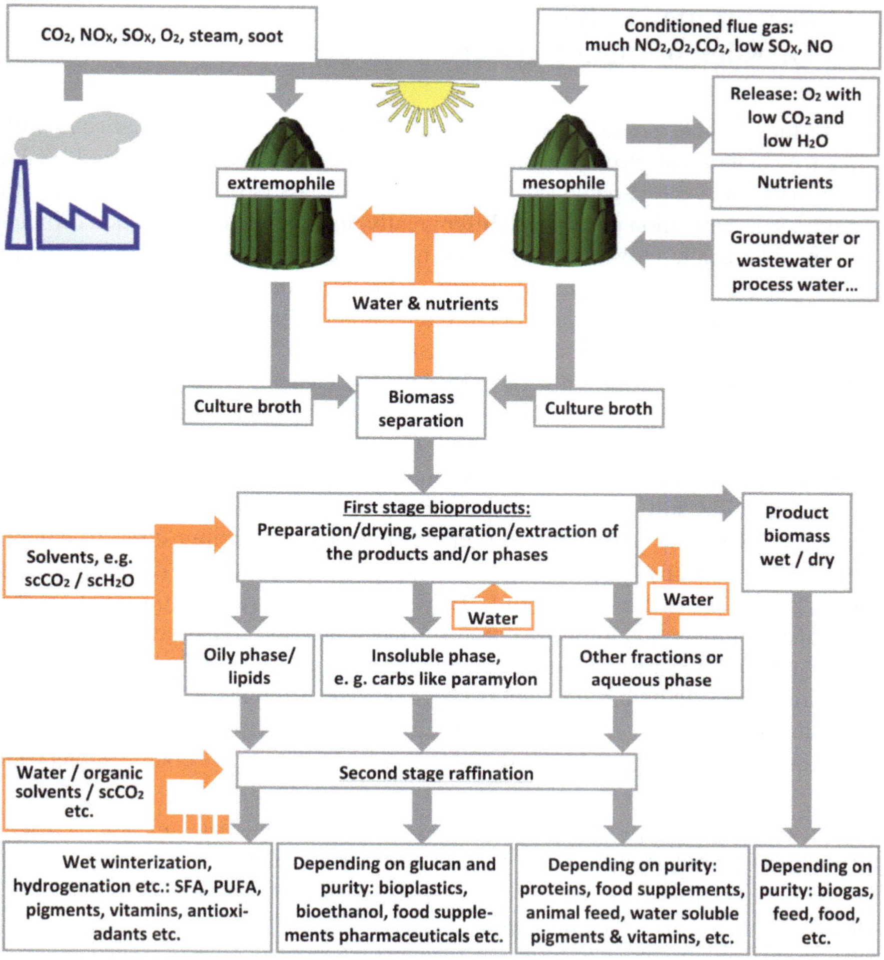

Fig. 2 Bioproduction with microalgae. An extremophilic microalga can be cocultured separately in the concept, for example in the case of flue gas with a high sulfur oxide load. Material flows to the end products: arrows in gray, recycling circuits: arrows in orange. (Scheme adapted from [23])

[23, 42, 54]. Under various conditions, further acidification of the nutrient medium by *E. gracilis* can be observed, which increasingly limits cultivation to acidophilic organisms and promotes axenic cultivation [55]. Biomass harvested by scalable low-energy cell settler is ultimately determined by the size of the microorganism [23]. The biomass can be used in its entirety or separated into fractions. Potential high-value products would be paramylon, proteins, fatty acids, and α-tocopherol [11, 56]. Combined with another extremophilic microalga, such as *Galdieria sulphuraria*, flue gases can also be used as a carbon dioxide source for production on a technical scale, which are contaminated with sulfur. If necessary, the sulfur-tolerant *G. sulphuraria* takes over the conditioning of the flue gas, as can be seen in Fig. 2.

2 Concentration of Biomass and Products: Is It All a Question of Light Intensity?

Due to the variable possibilities of cultivating *Euglena*, direct comparisons with references in terms of growth behavior are only possible to a limited extent [38, 39, 57]. In particular, the variable cell volume and dry mass depending on the growth conditions or cultivation phases can differ greatly, making direct correlation difficult [56, 58]. The availability of light and the nutrient composition of the medium are decisive factors for a direct manipulation of the cell composition; for example, a variation of the nitrogen source led to different protein contents [59, 60]. Nutritional sources like molasses, corn steep, or yeast extract can be used as the basis for complex media as carbon and nitrogen sources. Generally, complex nutritional sources may influence growth and production due to fluctuating batch qualities [61]. In contrast, already in 1952, Marian Cramer and Jack Myers created a base synthetic medium that is still used today known as CM medium [62]. In the last few decades, there is still a small amount of relevant publications dealing with the increase of biomass content due to optimization of media composition or culture conditions. Generally, the main focus for possible products is on proteins, paramylon, lipids, and α-tocopherol – Euglena also shows a high potential for the simultaneous production for further products like β-carotene, tyrosine, ascorbic acid, or provitamin A [51]. Nevertheless, the publication output over all scientific topics with *Euglena* is still high because of the high relevance as a model organism. Please note that sometimes not *E. gracilis* Z but *E. gracilis* var. *saccharophila*, *E. gracilis* Klebs, or other species is mentioned. Table 2 shows a selection of publications with

Table 2 Overview of achieved high dry mass concentrations by cultivations of *E. gracilis* Z with different trophic conditions presented by different authors in the last decade

Media	Cultivation mode	Duration (day)	X_{Max} (g L^{-1})	Par (g g^{-1})	References
Mod. CM + 25 g L^{-1} Glc, feed[a]	mix/het (STR, x)	12	86.2	0.78	[63]
37.5% PL+ 15 g L^{-1} Glc, feed[b]	het (STR, fed-batch)	6	17.5	0.77	[64]
30% CSS + 20 g L^{-1} Fru	het (shake flask)	2,8	16.6	0.62	[61]
75% PL + 20 g L^{-1} Glc	het (STR, batch)	3,7	16.1	0.70	[65]
CM + 20 g L^{-1} Glc	het (shake flask)	7	12.3	0.83	[11]
PP + 20 g L^{-1} Glc	het (shake flask)	10	12.3	0.80	[56]
Mod. Hutner	phot (STR, batch)	14	5	0.53	[59]
Mod. CM	phot (shake flask)	14	4.1	0.58	[60]

X_{Max} dry mass concentration, *Par* amount of paramylon in gram paramylon per gram biomass, *Mod.* modified, *CM* Cramer-Myers, *het* heterotrophically grown, *mix* mixotrophically grown, *x* two-stage cultivation mode, where the second stage was a fed batch with seven additions of ~25 g L^{-1} Glc, *PL* potato liquor, *Glc* glucose, *CSS* corn steep solid, *Fru* fructose, *PP* proteose peptone, *phot* photo-autotrophically grown, *STR* stirred tank reactor

[a] 20× mod. CM with 500 g L^{-1} glucose

[b] Pure potato liquor supplemented with 66.7 g L^{-1} glucose

the aim of production of high dry mass concentrations. The overview in Table 2 indicates that high dry biomass concentrations were accomplished more successfully in heterotrophically grown cultures. The selected data demonstrates the ability of *E. gracilis* to utilize a broad range of carbon sources. But for high cell densities, there is a need for sophisticated process and feed strategies [63, 66].

Under photoautotrophic conditions, the absorption coefficient (ε) of the *E. gracilis* cells was estimated to be 139 $m^2 kg^{-1}$ using the Lambert-Beer's equation. That is why photoautotrophically grown cultures became fast light limiting, especially with increasing culture thickness. A large-scale experiment demonstrated by Suzuki with an illuminated culture thickness of 4 cm and photoautotrophic conditions achieved 0.55 g L^{-1} [42]. For higher dry mass concentrations, a light limitation should be avoided, e.g., by reducing the culture thickness or by increasing the light flux. Experimental designs in shake flasks with CM medium and a culture layer thickness of only 2.3 cm (in the resting state) at a high light flux of 1,000 μmol m^{-2} s^{-1} delivered 1.4, 3.1, and 3.4 g L^{-1} dry cell weight after 4, 7, and 10 days, respectively [11]. At a very low light flux of 20 μmol m^{-2} s^{-1} in a 200 mL shake flask and a working volume of 100 mL, only 0.3 g L^{-1} dry cell weight was observed after about 5 days' cultivation [55]. On a 1,000 L scale with a high culture layer thicknesses, a dry cell weight of 0.9 g L^{-1} was measured using CM medium and a light flux of 960 μmol m^{-2} s^{-1} after 10 days' course of cultivation and about 11% (v/v) CO_2 (from flue gases) in the supply air [67]. Takeyama et al. [66] described a dry cell weight of 1 g L^{-1} after 10 days of cultivation for photoautotrophically grown cultures in CM medium and a light intensity of 18 W m^{-2} in a 500 mL hexagonal flask with 400 mL volume. The limited growth of *Euglena gracilis* based on CM medium led to the development of the EG medium for photoautotrophic growth conditions with high biomass yields and high concentrations of potential bioproducts. To elucidate the whole spectrum of this medium, cultivations were done with only 1 cm thickness and different light fluxes. The performance of the defined EG 5.1 medium is shown in Fig. 3.

Figure 3 shows a significant increase in the maximum biomass concentration reached on day 8 in ascending order of the light flux from 130 to 1,000 μmol m^{-2} s^{-1}. Cultures exposed to 1,000 μmol m^{-2} s^{-1} light flux showed an increasing yellow coloration from day 8, whereas this phenomenon was not observed in the less illuminated cultures. Microscopic examination showed an increasing number of dead cells in the 1,000 μmol m^{-2} s^{-1} cultures from day 8, whereas cyst formation increased in the lower-illuminated cultures [23]. Wang and colleagues achieved comparable dry biomasses to Fig. 3 of around 4 g L^{-1} after 10 days using a well-illuminated stirred tank reactor with 400 μmol m^{-2} s^{-1} at 30°C and a synthetic medium [59]. At all, a comparison of the data of the maximum dry mass concentrations in Fig. 3 and the discussed studies above indicate that photoautotrophically grown cultures are hard depending on the light regime. Using high light fluxes and low culture layer thickness generates dry mass weights above 6 g L^{-1} after 8 days' cultivation time.

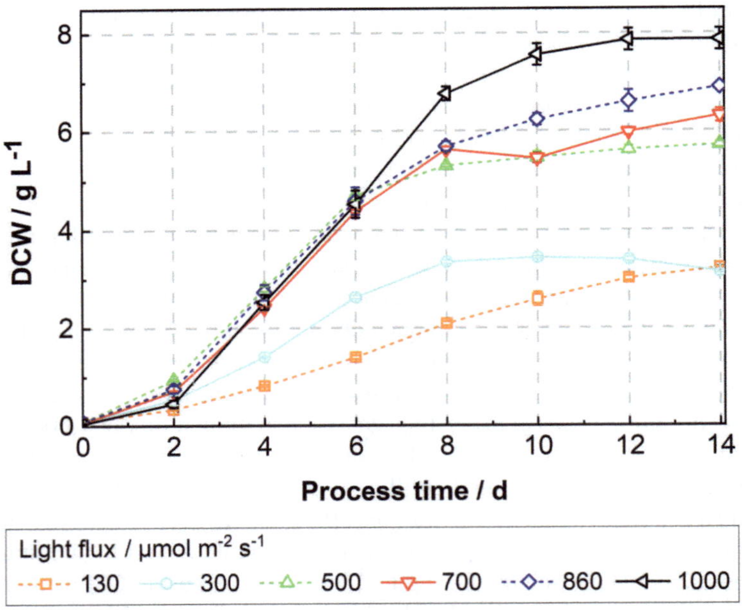

Fig. 3 Dry mass weights of 50 mL *E. gracilis* Z cultures in 300 mL shake flasks with resulting 1 cm culture layer thickness and different light fluxes (referred to as PAR) in EG 5.1 medium with an initial pH of 3.0. Culture layer thicknesses were corrected by weighing every 2 days through the addition of sterile water. The entire contents of the shake flasks were used for sampling. Cultures were grown at 29°C on bottom-illuminated LED shakers at a rotation frequency of 120 min^{-1} in a gas atmosphere of 5% (v/v) CO_2 and 95% (v/v) air. $n = 9$. (Diagram adapted from [23])

2.1 The Commercial Use and Production of α-Tocopherol

α-Tocopherol is the most bioactive form of vitamin E. *Euglena gracilis* is able to produce α-tocopherol in both mitochondria and chloroplasts [68–70]. Compared to chemically synthesized α-tocopherol, the natural one is produced by photosynthetically active organisms and exhibits a higher bioactivity [71, 72]. Other advantages of vitamin E production with *E. gracilis* are high productivity and product homogeneity. Thus, a comparison of over 200 organisms elects *E. gracilis* to be the best α-tocopherol producer with a high content of 97% of the total tocopherol [73, 74]. Depending on the type of cultivation, α-tocopherol concentrations can reach around 2.5 mg g^{-1} biomass with peak values of over 7 mg g^{-1} [11, 73]. These results are several times higher than concentrations observed in crop oils [37]. Naturally, α-tocopherol acts as an antioxidant to minimize oxidative stress by reacting with free radicals [74]. Based on its potential in the treatment of Alzheimer's and other disorders, as well as the general positive effects on human health and the body, the market for α-tocopherol as food supplement and product in pharmacy and cosmetics is increasingly large [75–78]. The anticipated global market value of tocopherol is estimated to grow exponentially to USD 3.8 billion by 2032 from USD 1.8 billion

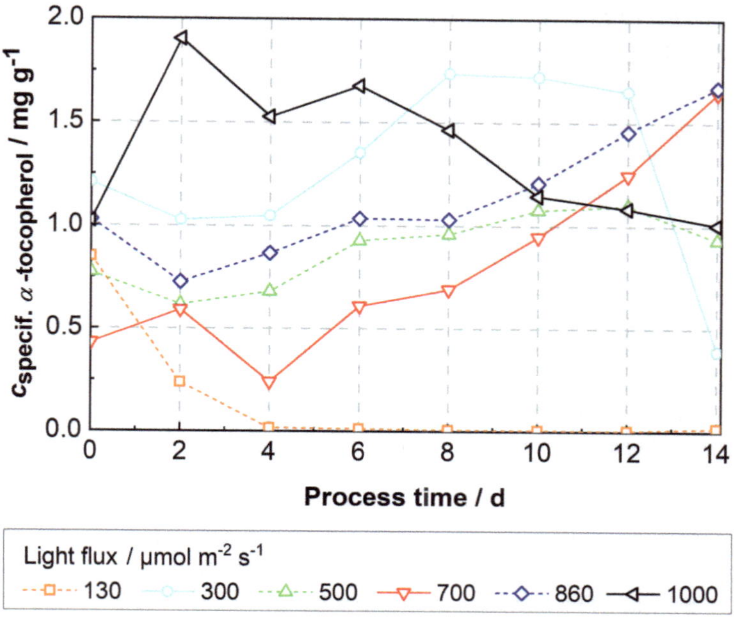

Fig. 4 Bioproduction of α-tocopherol by *E. gracilis* Z cultures; the corresponding biomass is shown in Fig. 3. The α-tocopherol was extracted from the dry mass with methanol and then measured via RP-HPLC. $n = 6$. (Diagram adapted from [23])

in 2023 [79]. This forecast highlights the future potential and need in α-tocopherol production. Regarding the light flux-dependent biomass development in Fig. 3, a direct dependence was also found for tocopherol [23], as can be seen in Fig. 4.

Thus, low photon fluxes only result in low tocopherol concentrations. Takeyama and colleagues determined an α-tocopherol-specific concentration of 1.5 mg g^{-1} under rather weak light conditions with 18 W m^{-2} with photoautotrophic Cramer-Myers medium after 10 days of cultivation. Assuming that the photosynthetically active spectrum was meant by the value for light power, the 18 W m^{-2} would be converted to about 82 µmol m^{-2} s^{-1} according to the calculation of McCree [80]. Using the Cramer-Myers medium, 6.8 and 8.6 mg L^{-1} α-tocopherol were determined with a light flux of 1,000 µmol m^{-2} s^{-1} after 7 and 10 days, respectively. If, in addition to a strong light flux, a very nutrient-rich medium was used for photoheterotrophic cultivation, a photon flux of 740 µmol m^{-2} s^{-1} in the complex medium "*Euglena* broth" allowed up to 7.2 mg g^{-1} after 100 h of cultivation [81]. The highest reported α-tocopherol concentrations of 44.2 mg L^{-1} in culture or 1.1 mg g^{-1} dry mass were reached with *E. gracilis* in a heterotrophic fed-batch cultivation with a medium containing ethanol as an additional carbon source after 455 h [68]. Comparable concentrations of 40 mg L^{-1} culture but much higher specific concentrations of 3.7 mg g^{-1} dry mass were achieved with a bleached strain in a batch cultivation with a medium for heterotrophic cultivation with ethanol, glutamate, and malate [37].

2.2 *The diverse uses of paramylon and its bioproduction with* E. gracilis

Paramylon is a water-insoluble storage polysaccharide consisting of linear glucan chains with β-1,3-linkages synthesized from uridine diphosphate glucose. Its molecular weight is estimated between 100 and 500 kDa [82]. The crystalline glucan is present as discoidal paramylon granules in the cytoplasm of *E. gracilis* cells [52]. Under photoautotrophic conditions, up to 50% paramylon content was achieved by many authors after 8 days of cultivation [11, 23, 56, 64, 83]. Heterotrophically cultivated cells accumulate up to 70–90% paramylon [61, 63]. Thus, even under photoautotrophic conditions, *E. gracilis* produces larger quantities of β-1,3-glucan than plants, fungi, or yeasts [84]. These properties show a high potential for the commercial use of this beta-glucan. Please note that the market size for beta-glucans is expected to reach $894.3 million by 2030, rising at a market growth of 6.3% compound annual growth rate (CAGR) during the forecast period from 2024 to 2030 [85]. The areas of application are not only nutritional – many health-promoting properties are attributed to paramylon. Besides its anti-inflammatory effects, paramylon is also expected to have anti-tumor and anti-HIV effects as well as other immunomodulatory functions [82, 86, 87]. Actually, *Saccharomyces cerevisiae* is well known for production on an industrial scale, whereby the glucan is extracted with an optimum yield of only 5–7% of the dry mass [61]. Furthermore, paramylon is a valuable product for the food industry and for medical or pharmaceutical applications. A usage as bioplastic is possible, too. Even first applications with nanofibers from *E. gracilis* paramylon are described [88, 89]. This versatility of paramylon and the above-average production by *E. gracilis* result in great potential, especially since the global market for β-1,3-glucan is projected to reach 868 million US$ by 2027 with an annual growth rate of 7% [85].

It is known that the light source can have further effects on paramylon production [90–92]. Table 2 shows the range of paramylon production by *E. gracilis* at different trophic conditions. In Fig. 5, higher intensities of light flux have an increasingly positive effect on paramylon accumulation, which was postulated already in the early 1960s by Cook as an effect of excess energy [93]. At all, photon fluxes above 700 μmol m^{-2} s^{-1} generate a paramylon content of around 50%, and low photon fluxes generate contents below or around 10%. In volumetric concentrations, the cultivation series with 500 μmol m^{-2} s^{-1} generates up to 1.7 g L^{-1} paramylon. The highest concentrations were achieved with 3.9 and 4.0 g L^{-1} in the cultivation series with light fluxes of 860 and 1,000 μmol m^{-2} s^{-1} [23].

From the fourth day of cultivation, the different light fluxes strongly affected the rate of paramylon formation. Compared with Fig. 5 at 1,000 μmol m^{-2} s^{-1} with 4.0 g L^{-1} (0.51 g g^{-1}), similar cultivation series were achieved by Grimm *et al.* with the Cramer-Myers medium after 10 days with paramylon concentrations up to 3.3 g L^{-1} [11].

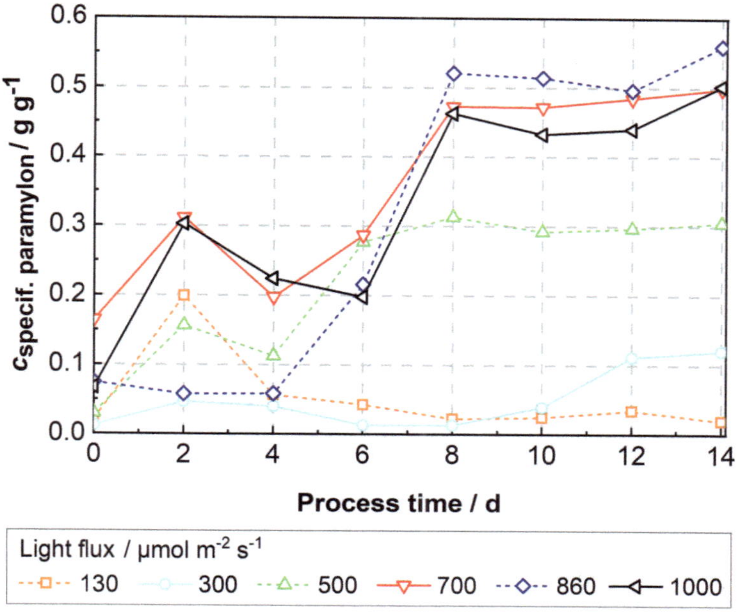

Fig. 5 Bioproduction of paramylon by *E. gracilis* Z cultures; the corresponding biomass is shown in Fig. 3. Paramylon was extracted and washed several times with different solvents and digested to remove the amino acid-pellicle of the granules. After drying, it was measured gravimetrically. $n = 6$. (Diagram adapted from [23])

In addition to the medium, the cultivation mode has a major influence on the paramylon dry mass concentration and the paramylon selectivity. When there is an excess of adequate carbon source in the culture medium, synthesized *E. gracilis* has high amounts of the storage polysaccharide paramylon [40]. A comparison with identical culture conditions and with Cramer-Myers medium: Grimm described heterotrophic cultivations of *E. gracilis* with 20 g L^{-1} glucose and a maximum paramylon dry mass concentration of 9.3 g L^{-1}. For photoheterotrophic conditions (light flux 1,000 µmol m^{-2} s^{-1}) with 20 g L^{-1} glucose, maximum paramylon dry matter concentration is 3.1 g L^{-1}, and for photoautotrophically grown cells, maximum paramylon dry mass concentration is 0.73 g L^{-1} at 1,000 µmol m^{-2} s^{-1} after a cultivation time of 7 days [94].

Regarding industrial paramylon production, Wu et al. investigated the possibility of cultivation with combined growth conditions at pilot scale under mixotrophic and heterotrophic conditions. The upstream mixotrophic cultivation of *E. gracilis* enables an initial accumulation of biomass with high viability and important metabolites that are beneficial for the growth and productivity of the algae. The specific paramylon concentration of 0.23 g g^{-1} DW could be increased to a maximum of 0.78 g g^{-1} by switching to heterotrophic conditions. Metabolomic analysis showed that under heterotrophic conditions, carbon flux from various metabolic pathways, including glycerophospholipid biosynthesis and amino acids, was diverted to the paramylon synthesis pathway [63].

2.3 *The bioproduction of lipids with* Euglena *and their general use*

Another product that will be discussed in this chapter is lipids. The lipid content of *E. gracilis* can account for up to 25% of the biomass [95]. These lipids are manly localized in the membranes and chloroplasts of *E. gracilis* cells. In this context, lipid quantity may increase under illumination and photoautotrophic conditions, as *E. gracilis* membrane components increase [42, 90]. For example, *E. gracilis* develops functional thylakoid membranes containing lipids and proteins for photosynthesis [96]. Fatty acids and wax esters have the largest shares in the total lipid fraction [97]. Especially fatty acids can accumulate within the total lipids [90, 98]. The overall lipid content in relation to biomass is almost independent of the growth phase, but the composition of lipids, especially fatty acids, could change depending on the cultivation conditions [56, 59]. This property can be used to control desired yields in lipid composition for specific applications, which is a major advantage of production with *E. gracilis*. Possible production areas are the bioenergy sector, e.g., for biodiesel, the food sector, e.g., as functional food or supplements, and the pharmaceutical industry [56]. Under consideration of the future market development, the need and potential for sustainable production with *E. gracilis* become clear once again. One forecast expects an increase in global energy demand of 1.2% per year with fossil fuels still accounting for 78% by 2040 [99]. An exemplary course of lipid production with *E. gracilis* is shown in Fig. 6.

The maxima of the lipid concentration for all cultures in Fig. 6 are in a range between 0.8 and 1.0 g L^{-1} at day 8 corresponding to the specific concentrations of 0.17–0.4 g g^{-1}. Here, we observed that the weaker the light flux, the more intense the green coloring of the cells under the microscope and of the cultures in the shake flasks. Depending on the energy requirement and amount of light, as well as the age of the culture, *Euglena gracilis* is able to regulate the number of chloroplasts between 6 and up to 25 [101]. The more chloroplasts are produced for photoacclimation, the higher the lipid content per cell through the lipid membranes of the chloroplasts. As the cultivation progressed in Fig. 6, the lipid concentrations which correspond to specific lipid concentrations decreased and reached a proportion of 15–24% during the transition to the stationary phase [23]. These lipid contents correspond to Becker's data of 14–20% for photoautotrophically grown *E. gracilis* cultures [102]. Wang and colleagues used comparable conditions for experiments on lipid production in *Euglena* and observed contents of 19% in moderate light regime after 14 days of course of cultivation [59]. The values for the lipid content are also in those of Tucci et al.'s results stated with a range for the lipid content of wild-type strains of 10–30% [97]. Recently, He et al. presented a

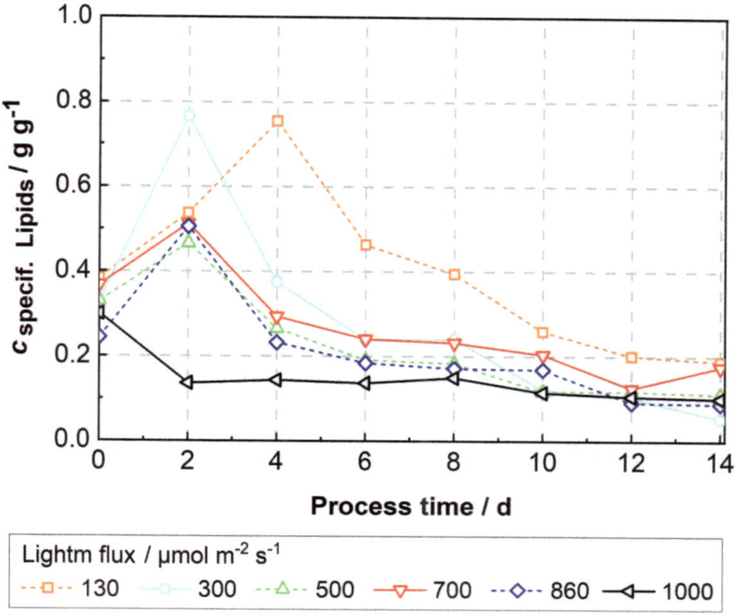

Fig. 6 Bioproduction of lipids by *E. gracilis* Z cultures; the corresponding biomass is shown in Fig. 3. The lipid fraction was extracted from the dry biomass. An environmentally friendly lipid extraction according to Smedes [100] was carried out, which extracts about 20% less lipids compared to lipid extractions with chloroform according to Bligh and Dyer. $n = 6$. (Diagram adapted from [23])

content of 42% with a modified CM medium after 14 days of cultivation using a low photon flux [60].

3 Conclusion

Despite its great potential and versatile applications, products with and from *E. gracilis* are sold primarily in Japan. One major bottleneck is higher prices for products derived from the protist compared to its commonly used and established alternatives [56]. One intention of the results discussed in this chapter should be the help to expand the global commercialization of *E. gracilis*-derived products. Here, we saw some examples that are able for the production of other substances like beta-carotene, tyrosine, ascorbic acid, and further more [51]. The first steps for *Euglena* as a food ingredient are already running in Europe and the United States.

The aim was to show an optimal culture duration and simultaneously harvest several products. This examination may, for example, help to reduce production costs and lead to faster development of a sustainable and economically profitable biorefinery concept.

References

1. Stephens E, Ross IL, Mussgnug JH, Wagner LD, Borowitzka MA, Posten C, Kruse O, Hankamer B (2010) Future prospects of microalgal biofuel production systems. Trends Plant Sci 15(10):554–564. https://doi.org/10.1016/j.tplants.2010.06.003
2. Khatiwada B, Sunna A, Nevalainen H (2020) Molecular tools and applications of Euglena gracilis: from biorefineries to bioremediation. Biotechnol Bioeng 117(12):3952–3967. https://doi.org/10.1002/bit.27516
3. Clark JH, Luque R, Matharu AS (2012) Green chemistry, biofuels, and biorefinery. Annu Rev Chem Biomol Eng 3:183–207. https://doi.org/10.1146/annurev-chembioeng-062011-081014
4. Vaz Jr S (2019) Sugarcane-biorefinery. Adv Biochem Eng Biotechnol 166:125–136. https://doi.org/10.1007/10_2016_70
5. Hariskos I, Posten C (2014) Biorefinery of microalgae – opportunities and constraints for different production scenarios. Biotechnol J 9(6):739–752. https://doi.org/10.1002/biot.201300142
6. Espinoza Pérez AT, Camargo M, Narváez Rincón P, Alfaro Marchant M (2017) Key challenges and requirements for sustainable and industrialized biorefinery supply chain design and management: a bibliographic analysis. Renew Sustain Energy Rev 69:350–359. https://doi.org/10.1016/j.rser.2016.11.084
7. Chew KW, Yap JY, Show PL, Suan NH, Juan JC, Ling TC, Lee DJ, Chang JS (2017) Microalgae biorefinery: high value products perspectives. Bioresour Technol 229:53–62. https://doi.org/10.1016/j.biortech.2017.01.006
8. Rosello Sastre R, Posten C (2010) Die vielfältige Anwendung von Mikroalgen als nachwachsende Rohstoffe. Chem Ingenieur Tech 82(11):1925–1939. https://doi.org/10.1002/cite.201000124
9. Ohara H (2003) Biorefinery. Appl Microbiol Biotechnol 62(5–6):474–477. https://doi.org/10.1007/s00253-003-1383-7
10. Pimentel D, Marklein A, Toth MA et al (2009) Food versus biofuels: environmental and economic costs. Hum Ecol 37:1. https://doi.org/10.1007/s10745-009-9215-8
11. Grimm P, Risse JM, Cholewa D, Müller JM, Beshay U, Friehs K, Flaschel E (2015) Applicability of Euglena gracilis for biorefineries demonstrated by the production of α-tocopherol and paramylon followed by anaerobic digestion. J Biotechnol 215:72–79. https://doi.org/10.1016/j.jbiotec.2015.04.004
12. Vieira V, Cadoret J-P, Acien FG, Benemann J (2022) Clarification of most relevant concepts related to the microalgae production sector. Processes 10(1):175. https://doi.org/10.3390/pr10010175
13. Behera B, Venkata Supraja K, Paramasivan B (2021) Integrated microalgal biorefinery for the production and application of biostimulants in circular bioeconomy. Bioresour Technol 339:125588. https://doi.org/10.1016/j.biortech.2021.125588
14. Salami R, Kordi M, Bolouri P, Delangiz N, Asgari Lajayer B (2021) Algae-based biorefinery as a sustainable renewable resource. Circ Econ Sustain 1(4):1349–1365. https://doi.org/10.1007/s43615-021-00088-z
15. Sarma S, Sharma S, Rudakiya D, Upadhyay J, Rathod V, Patel A, Narra M (2021) Valorization of microalgae biomass into bioproducts promoting circular bioeconomy: a holistic approach of bioremediation and biorefinery. 3 Biotech 11(8):378. https://doi.org/10.1007/s13205-021-02911-8
16. Almomani F (2020) Kinetic modeling of microalgae growth and CO2 bio-fixation using central composite design statistical approach. Sci Total Environ 720:137594. https://doi.org/10.1016/j.scitotenv.2020.137594
17. Yen HW, Hu IC, Chen CY, Ho SH, Lee DJ, Chang JS (2013) Microalgae-based biorefinery-from biofuels to natural products. Bioresour Technol 135:166–174. https://doi.org/10.1016/j.biortech.2012.10.099

18. Ullah K, Sharma VK, Ahmad M, Lv P, Krahl J, Wang Z, Sofia (2017) The insight views of advanced technologies and its application in bio-origin fuel synthesis from lignocellulose biomasses waste, a review. Renew Sustain Energy Rev 82(3):3992–4008. https://doi.org/10.1016/j.rser.2017.10.074
19. Gifuni I, Pollio A, Safi C, Marzocchella A, Olivieri G (2018) Current bottlenecks and challenges of the microalgal biorefinery. Trends Biotechnol 37(3):242–252. https://doi.org/10.1016/j.tibtech.2018.09.006
20. Chen CY, Yeh KL, Aisyah R, Lee DJ, Chang JS (2011) Cultivation, photobioreactor design and harvesting of microalgae for biodiesel production: a critical review. Bioresour Technol 102(1):71–81. https://doi.org/10.1016/j.biortech.2010.06.159
21. Liang Y, Sarkany N, Cui Y (2009) Biomass and lipid productivities of Chlorella vulgaris under autotrophic, heterotrophic and mixotrophic growth conditions. Biotechnol Lett 31(7): 1043–1049. https://doi.org/10.1007/s10529-009-9975-7
22. Dolganyuk V, Belova D, Babich O et al (2020) Microalgae: a promising source of valuable bioproducts. Biomolecules 10(8):1153. https://doi.org/10.3390/biom10081153
23. Cholewa D (2017) Ein Bioraffineriekonzept mit Euglena gracilis. Universität Bielefeld, Bielefeld
24. Eppink MHM, Olivieri G, Reith H, van den Berg C, Barbosa MJ, Wijffels RH (2019) From current algae products to future biorefinery practices: a review. Adv Biochem Eng Biotechnol 166:99–123. https://doi.org/10.1007/10_2016_64
25. Guiry MD, Guiry GM (2025) AlgaeBase. World-Wide Electronic Publication, University of Galway. https://www.algaebase.org. Accessed 18 Apr 2025
26. Maddiboyina B, Vanamamalai H, Roy HR, Gandhi S, Kavisri M, Moovendhan M (2023) Food and drug industry applications of microalgae Spirulina platensis: a review. J Basic Microbiol 63:573–583. https://doi.org/10.1002/jobm.202200704
27. Lupatini A, Colla L, Canan C, Colla E (2017) Potential application of microalga Spirulina platensis as a protein source. J Sci Food Agric 97(3):724–732. https://doi.org/10.1002/jsfa.7987
28. Shah M, Liang Y, Cheng J, Daroch M (2016) Astaxanthin-producing green microalga Haematococcus pluvialis: from single cell to high value commercial products. Front Plant Sci 7:531. https://doi.org/10.3389/fpls.2016.00531
29. Rammuni M, Ariyadasa T, Nimarshana P, Attalage R (2019) Comparative assessment on the extraction of carotenoids from microalgal sources: astaxanthin from H. pluvialis and β-carotene from D. salina. Food Chem 277:128–134. https://doi.org/10.1016/j.foodchem.2018.10.066
30. Görs M, Schumann R, Hepperle D, Karsten U (2010) Quality analysis of commercial Chlorella products used as dietary supplement in human nutrition. J Appl Phycol 22:265–276. https://doi.org/10.1007/s10811-009-9455-4
31. Sukenik A, Zmora O, Carmeli Y (1993) Biochemical quality of marine unicellular algae with special emphasis on lipid composition. II: Nannochloropsis sp. Aquaculture 117:313–326. https://doi.org/10.1016/0044-8486(93)90328-V
32. Chua E, Schenk P (2017) A biorefinery for Nannochloropsis: induction, harvesting, and extraction of EPA-rich oil and high-value protein. Bioresour Technol 244(Pt 2):1416–1424. https://doi.org/10.1016/j.biortech.2017.05.124
33. Barkia I, Saari N, Manning SR (2019) Microalgae for high-value products towards human health and nutrition. Mar Drugs 17(5):304. https://doi.org/10.3390/md17050304
34. Benemann J (2013) Microalgae for biofuels and animal feeds. Energies 6(11):5869–5886. https://doi.org/10.3390/en6115869
35. Adl SM, Bass D, Lane CE, Lukeš J, Schoch CL, Smirnov A, Agatha S, Berney C, Brown MW, Burki F, Cárdenas P, Čepička I, Chistyakova L, Del Campo J, Dunthorn M, Edvardsen B, Eglit Y, Guillou L, Hampl V, Heiss AA, Hoppenrath M, James TY, Karnkowska A, Karpov S, Kim E, Kolisko M, Kudryavtsev A, Lahr DJG, Lara E, Le Gall L, Lynn DH, Mann DG, Massana R, Mitchell EAD, Morrow C, Park JS, Pawlowski JW, Powell MJ, Richter DJ,

Rueckert S, Shadwick L, Shimano S, Spiegel FW, Torruella G, Youssef N, Zlatogursky V, Zhang Q (2019) Revisions to the classification, nomenclature, and diversity of eukaryotes. J Eukaryot Microbiol 66(1):4–119. https://doi.org/10.1111/jeu.12691
36. Ebenezer TE, Zoltner M, Burrell A, Nenarokova A, Novák Vanclová AMG, Prasad B, Soukal P, Santana-Molina C, O'Neill E, Nankissoor NN, Vadakedath N, Daiker V, Obado S, Silva-Pereira S, Jackson AP, Devos DP, Lukeš J, Lebert M, Vaughan S, Hampl V, Carrington M, Ginger ML, Dacks JB, Kelly S, Field MC (2019) Transcriptome, proteome and draft genome of Euglena gracilis. BMC Biol 17:11. https://doi.org/10.1186/s12915-019-0626-8
37. Rodríguez-Zavala JS, Ortiz-Cruz MA, Mendoza-Hernández G, Moreno-Sánchez R (2010) Increased synthesis of α-tocopherol, paramylon and tyrosine by Euglena gracilis under conditions of high biomass production. J Appl Microbiol 109(6):2160–2172. https://doi.org/10.1111/j.1365-2672.2010.04848.x
38. Jones CR, Cook JR (1978) Culture pH, CO2 tension, and cell division in Euglena gracilis Z. J Cell Physiol 96(2):253–259. https://doi.org/10.1002/jcp.1040960214
39. Kitaya Y, Azuma H, Kiyota M (2005) Effects of temperature, CO2/O2 concentrations and light intensity on cellular multiplication of microalgae, Euglena gracilis. Adv Space Res 35(9): 1584–1588. https://doi.org/10.1016/j.asr.2005.03.039
40. Kitaoka S, Nakano Y, Miyatake K, Yokota A (1989) Enzymes and their functional location. In: Buetow DE (ed) The biology of Euglena, vol 4. Academic Press Limited, London, pp 2–119
41. GRAS notice for dried Euglena gracilis (ATCC PTA-123017). GRAS notice (GRN) No. 697. https://www.fda.gov/media/106482/download
42. Suzuki K (2017) Large-scale cultivation of Euglena. In: Schwartzbach SD, Shigeoka S (eds) Euglena: biochemistry, cell and molecular biology. Advances in experimental medicine and biology, vol 979. Springer, pp 285–293. https://doi.org/10.1007/978-3-319-54910-1_14
43. Hasan MT, Sun A, Mirzaei M, Te'o J, Hobba G, Sunna A et al (2017) A comprehensive assessment of the biosynthetic pathways of ascorbate, α-tocopherol and free amino acids in Euglena gracilis var. saccharophila. Algal Res 27:140–151. https://doi.org/10.1016/j.algal.2017.08.029
44. Ogbonna JC, Ichige E, Tanaka H (2002) Interactions between photoautotrophic and heterotrophic metabolism in photoheterotrophic cultures of Euglena gracilis. Appl Microbiol Biotechnol 58(4):532–538. https://doi.org/10.1007/s00253-001-0901-8
45. Behera S, Singh R, Arora R, Sharma NK, Shukla M, Kumar S (2014) Scope of algae as third generation biofuels. Front Bioeng Biotechnol 2:1–13. https://doi.org/10.3389/fbioe.2014.00090
46. EFSA NDA Panel (EFSA Panel on Nutrition, Novel Foods and Food Allergens) (2020) Scientific opinion on the safety of dried whole cell Euglena gracilis as a novel food pursuant to Regulation (EU) 2015/2283. EFSA J 18:e06100. https://doi.org/10.2903/j.efsa.2020.6100
47. Shimada R, Fujita M, Yuasa M, Sawamura H, Watanabe T, Nakashima A, Suzuki K (2016) Oral administration of green algae, Euglena gracilis, inhibits hyperglycemia in OLETF rats, a model of spontaneous type 2 diabetes. Food Funct 7:4655–4659. https://doi.org/10.1039/C6FO00606J
48. Sugimoto R, Ishibashi-ohgo N, Atsuji K, Miwa Y, Iwata O (2018) Euglena extract suppresses adipocyte-differentiation in human adipose-derived stem cells. PLoS One 13:e0192404. https://doi.org/10.1371/journal.pone.0192404
49. Stienstra R, Duval C, Michael M, Kersten S (2007) PPARs, obesity, and inflammation. PPAR Res 2007:95974. https://doi.org/10.1155/2007/95974
50. Bakku RK, Yamamoto Y, Inaba Y, Hiranuma T, Gianino E, Amarianto L, Mahrous W, Suzuki H, Suzuki K (2023) New insights into raceway cultivation of Euglena gracilis under long-term semi-continuous nitrogen starvation. Sci Rep 13:7123. https://doi.org/10.1038/s41598-023-34164-1

51. Gissibl A, Sun A, Care A, Nevalainen H, Sunna A (2019) Bioproducts from Euglena gracilis: synthesis and applications. Front Bioeng Biotechnol 7. https://doi.org/10.3389/fbioe.2019.00108
52. Tanaka Y, Goto K, Luo J, Nishino K, Ogawa T, Maruta T, Ishikawa T (2022) Identification of glucanases and phosphorylases involved in hypoxic paramylon degradation in Euglena gracilis. Algal Res 67:102829. https://doi.org/10.1016/j.algal.2022.102829
53. Buetow DE (2011) Euglena. In: Encyclopedia of life sciences. John Wiley & Sons Ltd., Chichester. https://doi.org/10.1002/9780470015902.a0001964.pub3
54. Ogbonna JC (2009) Microbiological production of tocopherols: current state and prospects. Appl Microbiol Biotechnol 84:217–225. https://doi.org/10.1007/s00253-009-2104-7
55. Yamane Y, Utsunomiya T, Watanabe M, Sasaki K (2001) Biomass production in mixotrophic culture of Euglena gracilis under acidic condition and its growth energetics. Biotechnol Lett 23:1223–1228. https://doi.org/10.1023/A:1010573218863
56. Schwarzhans J-P, Cholewa D, Grimm P, Beshay U, Risse J-M, Friehs K, Flaschel E (2015) Dependency of the fatty acid composition of Euglena gracilis on growth phase and culture conditions. J Appl Phycol 27:1389–1399. https://doi.org/10.1007/s10811-014-0458-4
57. Barsanti L, Vismara R, Passarelli V, Gualtieri P (2001) Paramylon (1,3-β-glucan) content in wild type and WZSL mutant of Euglena gracilis: effects of growth conditions. J Appl Phycol 13:59–65. https://doi.org/10.1023/A:1008105416065
58. Buetow DE (1962) Differential effects of temperature on the growth of Euglena gracilis. Exp Cell Res 27:137–142. https://doi.org/10.1016/0014-4827(62)90051-4
59. Wang Y, Seppänen-Laakso T, Rischer H, Wiebe MG (2018) Euglena gracilis growth and cell composition under different temperature, light and trophic conditions. PLoS One 13(4): e0195329. https://doi.org/10.1371/journal.pone.0195329
60. He Q, Zhang X, Xu H, Wang X, Zhang X, Hao Y, Song X, Cao X (2025) Cultivation of high-protein Euglena gracilis for enhanced protein production under inorganic nitrogen sources: mechanisms revealed by proteomics. Bioresour Technol 430:132560. https://doi.org/10.1016/j.biortech.2025.132560
61. Ivušić F, Šantek B (2015) Optimization of complex medium composition for heterotrophic cultivation of Euglena gracilis and paramylon production. Bioprocess Biosyst Eng 38:1103–1112. https://doi.org/10.1007/s00449-015-1353-3
62. Cramer M, Myers J (1952) Growth and photosynthetic characteristics of Euglena gracilis. Arch Mikrobiol 17:384–402. https://doi.org/10.1007/BF00410835
63. Wu M, Qin H, Deng J, Liu Y, Lei A, Zhu H, Hu Z, Wang J (2021) A new pilot-scale fermentation mode enhances Euglena gracilis biomass and paramylon (β-1,3-glucan) production. J Clean Prod 321:128996. https://doi.org/10.1016/j.jclepro.2021.128996
64. Šantek B, Felski M, Friehs K, Lotz M, Flaschel E (2012) Production of paramylon, a β-1,3-glucan, by heterotrophic growth of Euglena gracilis on potato liquor in fed-batch and repeated-batch mode of cultivation. Eng Life Sci 12:89–94. https://doi.org/10.1002/elsc.201100025
65. Šantek B, Felski M, Friehs K, Lotz M, Flaschel E (2010) Production of paramylon, a β-1,3-glucan, by heterotrophic cultivation of Euglena gracilis on potato liquor. Eng Life Sci 10:165–170. https://doi.org/10.1002/elsc.200900077
66. Takeyama H, Kanarmaru A, Yoshino Y, Kakuta H, Kawamura Y, Matsunaga T (1997) Production of antioxidant vitamins, β-carotene, vitamin C, and vitamin E, by two-step culture of Euglena gracilis Z. Biotechnol Bioeng 53:185–190. https://doi.org/10.1002/(SICI)1097-0290(19970120)53:2<185::AID-BIT8>3.0.CO;2-K
67. Chae SR, Hwang EJ, Shin HS (2006) Single cell protein production of Euglena gracilis and carbon dioxide fixation in an innovative photo-bioreactor. Bioresour Technol 97:322–329. https://doi.org/10.1016/j.biortech.2005.02.037
68. Ogbonna JC, Tanaka H, Tomiyamal S (1998) Heterotrophic cultivation of Euglena gracilis Z for efficient production of α-tocopherol. J Appl Phycol 10:67–74. https://doi.org/10.1023/A:1008011201437

69. Kusmic C, Barsacchi R, Barsanti L, Gualtieri P, Passarelli V (1999) Euglena gracilis as source of the antioxidant vitamin E. Effects of culture conditions in the wild strain and in the natural mutant WZSL. J Appl Phycol 10:555–559. https://doi.org/10.1023/A:1008022305865
70. Shigeoka S, Onishi T, Nakano Y, Kitaoka S (1986) The contents and subcellular distribution of tocopherols in Euglena gracilis. Agric Biol Chem 50:1063–1065. https://doi.org/10.1080/00021369.1986.10867518
71. Weiss WP, Hogan JS, Wyatt DJ (2009) Relative bioavailability of all-rac and RRR vitamin E based on neutrophil function and total alpha-tocopherol and isomer concentrations in periparturient dairy cows and their calves. J Dairy Sci 92(2):720–731. https://doi.org/10.3168/jds.2008-1635
72. Wilburn EE, Mahan DC, Hill DA, Shipp TE, Yang H (2008) An evaluation of natural (RRR-alpha-tocopheryl acetate) and synthetic (all-rac-alpha-tocopheryl acetate) vitamin E fortification in the diet or drinking water of weanling pigs. J Anim Sci 86(3):584–591. https://doi.org/10.2527/jas.2007-0377
73. Tani Y, Tsumura H (1989) Screening for tocopherol-producing microorganisms and α-tocopherol production by Euglena gracilis Z. Agric Biol Chem 53:305–312. https://doi.org/10.1080/00021369.1989.10869324
74. Fryer MJ (1992) The antioxidant effects of thylakoid vitamin E (alpha-tocopherol). Plant Cell Environ 15:381–392. https://doi.org/10.1111/j.1365-3040.1992.tb00988.x
75. Berman K, Brodaty H (2004) Tocopherol (vitamin E) in Alzheimer's disease and other neurodegenerative disorders. CNS Drugs 18(12):807–825. https://doi.org/10.2165/00023210-200418120-00005
76. Abraham A, Kattoor AJ, Saldeen T, Mehta JL (2019) Vitamin E and its anticancer effects. Crit Rev Food Sci Nutr 59(17):2831–2838. https://doi.org/10.1080/10408398.2018.1474169
77. Sen CK, Khanna S, Roy S (2007) Tocotrienols in health and disease: the other half of the natural vitamin E family. Mol Asp Med 28(5–6):692–728. https://doi.org/10.1016/j.mam.2007.03.001
78. Khadangi F, Azzi A (2019) Vitamin E – the next 100 years. IUBMB Life 71(4):411–415. https://doi.org/10.1002/iub.1990
79. Anuradha B (2024) Tocopherol market research report 2033. Dataintelo. FB-52227. https://dataintelo.com/report/tocopherol-market
80. McCree KJ (1981) Photosynthetically active radiation. In: Lange OL, Nobel PS, Osmond CB, Ziegler H (eds) Physiological plant ecology. Encyclopedia of plant physiology, new series, vol 12A. Springer Verlag, Berlin, pp 41–55. https://doi.org/10.1007/978-3-642-68090-8_3
81. Ruggeri BA, Gray RJH, Watkins TR, Tomlins RI (1985) Effects of low-temperature acclimation and oxygen stress on Tocopheron production in Euglena gracilis Z. Appl Environ Microbiol 50:1404–1408. https://doi.org/10.1128/aem.50.6.1404-1408.1985
82. Barsanti L, Passarelli V, Evangelista V, Frassanito AM, Gualtieri P (2011) Chemistry, physico-chemistry and applications linked to biological activities of β-glucans. Nat Prod Rep 28:457–466. https://doi.org/10.1039/C0NP00018C
83. Šantek B, Felski M, Friehs K, Lotz M, Flaschel E (2009) Production of α-tocopherol, a β-1,3-glucan, by heterotrophic cultivation of Euglena gracilis on a synthetic medium. Eng Life Sci 9:23–28. https://doi.org/10.1002/elsc.200700032
84. Zhu F, Du B, Xu B (2016) A critical review on production and industrial applications of beta-glucans. Food Hydrocoll 52:275–288. https://doi.org/10.1016/j.foodhyd.2015.07.003
85. GlobeNewswire (2020) Global beta-glucan industry. https://www.globenewswire.com/news-release/2020/08/14/2078387/0/en/Global-Beta-Glucan-Industry.html
86. Watanabe T, Shimada R, Matsuyama A, Yuasa M, Sawamura H, Yoshida E, Suzuki K (2013) Antitumor activity of the β-glucan paramylon from Euglena against preneoplastic colonic aberrant crypt foci in mice. Food Funct 4(11):1685–1690. https://doi.org/10.1039/c3fo60256g
87. Koizumi N, Sakagami H, Utsumi A, Fujinaga S, Takeda M, Asano K, Sugawara I, Ichikawa S, Kondo H, Mori S et al (1993) Anti-HIV (human immunodeficiency virus) activity of sulfated paramylon. Antivir Res 21(1):1–14. https://doi.org/10.1016/0166-3542(93)90063-o

88. Kawahara Y, Koganemaru A (2006) Development of novel film using paramylon prepared from Euglena gracilis. J Appl Polym Sci 102:3495–3497. https://doi.org/10.1002/app.24618
89. Shibakami M, Tsubouchi G, Hayashi M (2014) Thermoplasticization of euglenoid β-1,3-glucans by mixed esterification. Carbohydr Polym 105:90–96. https://doi.org/10.1016/j.carbpol.2014.01.053
90. Rosenberg A (1963) A comparison of lipid patterns in photosynthesizing and nonphotosynthesizing cells of Euglena gracilis. Biochemistry 2:1148–1154. https://doi.org/10.1021/bi00905a042
91. Barras DR, Stone BA (1968) Carbohydrate composition. In: Buetow DE (ed) The biology of Euglena, vol 2. Academic Press, New York, pp 149–191
92. Pohl P (1973) Light-induced changes of radioactivities in the ^{14}C-labeled lipids and fatty acids of dark grown Euglena gracilis. Z Naturforsch 28C:264–269. https://doi.org/10.1515/znc-1973-5-607
93. Cook JR (1963) Adaptations in growth and division in *Euglena* effected by energy supply. J Protozool 10:436–444. https://doi.org/10.1111/j.1550-7408.1963.tb01703.x
94. Grimm P (2016) Bioraffineriekonzepte mit dem Euglenoid Euglena grazilis Z: Gewinnung von alpha-tocopherol, paramylon und biogas. Universität Bielefeld, Bielefeld
95. Mahapatra DM, Chanakya HN, Ramachandra TV (2013) Euglena sp. as a suitable source of lipids for potential use as biofuel and sustainable wastewater treatment. J Appl Phycol 25:855–865. https://doi.org/10.1007/s10811-013-9979-5
96. Shibata S, Arimura SI, Ishikawa T, Awai K (2018) Alterations of membrane lipid content correlated with chloroplast and mitochondria development in Euglena gracilis. Front Plant Sci 9:370. https://doi.org/10.3389/fpls.2018.00370
97. Tucci S, Vacula R, Krajcovic J, Proksch P, Martin W (2010) Variability of wax ester fermentation in natural and bleached *Euglena gracilis* strains in response to oxygen and the elongase inhibitor Flufenacet. J Eukaryot Microbiol 57:63–69. https://doi.org/10.1111/j.1550-7408.2009.00452.x
98. Korn ED (1964) The fatty acids of Euglena gracilis. J Lipid Res 5(3):352–362. https://doi.org/10.1016/S0022-2275(20)40204-4
99. Koyama K (2017) The role and future of fossil fuel. IEEJ Energy J Special Issue 2017:80–83
100. Smedes F (1999) Determination of total lipid using non-chlorinated solvents. Analyst 124:1711–1718. https://doi.org/10.1039/A905904K
101. Cook J (1973) Unbalanced growth and replication of chloroplast populations in *Euglena gracilis*. J Gen Microbiol 75:51–60. https://doi.org/10.1099/00221287-75-1-51
102. Becker EW (1994) Microalgae: biotechnology and microbiology. Reprint 1995. Cambridge University Press, Cambridge, pp 177–185

MIX
Papier aus verantwortungsvollen Quellen
Paper from responsible sources
FSC® C105338

If you have any concerns about our products,
you can contact us on
ProductSafety@springernature.com

In case Publisher is established outside the EU,
the EU authorized representative is:
Springer Nature Customer Service Center GmbH
Europaplatz 3, 69115 Heidelberg, Germany

Printed by Libri Plureos GmbH
in Hamburg, Germany